DATE DUE

MAY 01 2009			
GAYLORD			PRINTED IN U.S.A.

Container Terminals and Automated Transport Systems

Hans-Otto Günther
Kap Hwan Kim
Editors

Container Terminals and Automated Transport Systems

Logistics Control Issues
and Quantitative Decision Support

Professor Dr. Hans-Otto Günther
TU Berlin
Department of Production Management
Wilmersdorfer Straße 148
10585 Berlin
Germany
Email: ho.guenther@pm-berlin.net

Professor Kap Hwan Kim
Pusan National University
Department of Industrial Engineering
Jangjeon-dong, Kumjeong-ku
Pusan 609-735
Korea
Email: kapkim@pusan.ac.kr

Parts of the papers of this volume have been published in the journal *OR Spectrum*.

Cataloging-in-Publication Data applied for

A catalog record for this book is available from the Library of Congress.

Bibliographic information published by Die Deutsche Bibliothek
Die Deutsche Bibliothek lists this publication in the Deutsche Nationalbibliografie; detailed bibliographic data available in the internet at *http://dnb.ddb.de*

ISBN 3-540-22328-2 Springer Berlin Heidelberg New York

Springer is a part of Springer Science+Business Media
springeronline.com

Cover design: Erich Kirchner, Heidelberg
Production: Helmut Petri
Typesetting: Steingraeber
Printing: betz-druck
Printed on acid-free paper – 42/3130 – 5 4 3 2 1 0

Preface

Logistics control issues of container terminals and automated transportation systems

Hans-Otto Günther[1] and Kap Hwan Kim[2]

[1] Department of Production Management, Technical University Berlin, Wilmersdorfer Str. 148, 10585 Berlin, Germany (e-mail: ho.guenther@pm-berlin.net)
[2] Deptartment of Industrial Engineering, Pusan National University, Jangjeon-dong, Kumjeong-ku, Busan, 609-735, Korea (e-mail: kapkim@pusan.ac.kr)

Marine container industry has grown dramatically in the last 30 years. As a result, container transportation has become a predominant mode of inter-continental cargo traffic. Container terminals have been playing an important role as multi-modal interfaces between sea and land transport. In order to benefit from the economy of scale, the size of container ships has significantly increased during the last decade. Frequently, a large container ship requires thousands of container lifts in a port terminal during one call. Since a container ship involves a major capital investment and significant daily operating costs, customer service has become an important issue for container port terminals and many container terminals are attempting to improve their throughput and to reduce the turnaround times of vessels and customers' trucks. With growing containerization, the number of container terminals worldwide increased considerably and the competition among them got stronger and stronger. In the academic world, issues related to container terminal operations have been neglected for quite a long time. Only recently, due to the ever-increasing importance of inter-continental cargo traffic and higher competition among container terminals, these issues have attracted the attention of the academic community.

In most existing container terminals, computers are employed to schedule and control different kinds of handling operations. A container terminal is a complex system with various interrelated components. Hence, there are many difficult decisions that operators and planners have to make. Because computer systems have capabilities to maintain a large volume of data and analyze them in a short time, they have been utilized to assist human experts in making decisions on the design of a container terminal, in developing operational plans and supporting real time decisions, for instance.

Furthermore, there is an ongoing trend in the development of seaport container terminal configurations to use automated container handling and transportation technology, particularly, in countries with high labour costs. This in turn requires a much more sophisticated control strategy in order to meet the desired performance measures. As a result, quantitative methodology has received considerable attention to analyze and support the design, operations, and control issues arising in automated container terminals. Clearly, container terminal logistics is a challenging field for many research disciplines, e.g. industrial engineering, automation, operations research, and management. Yet, given the recent developments in information technology, automated handling and transportation equipment, optimization algorithms and modelling tools, the overall productivity of container terminals can be significantly increased.

Handling operations and equipment in seaport container terminals

Seaport container terminals usually employ four different types of yard-side equipment: the on-chassis system, the carrier-direct system, the combined system of straddle carriers and yard trucks (straddle-carrier-relay system), and the combined system of yard cranes and prime movers. For the latter one, either yard trucks or automated guided vehicles (AGVs) could be used (yard-crane-relay system). According to the different types of yard-side equipment, handling systems can be classified into two groups. One is called "*direct transfer system*", which includes the on-chassis system and the carrier-direct system, and the other one is the "*indirect transfer system*", which includes the straddle-carrier-relay and the yard-crane-relay systems. These two groups of systems are explained in more detail below.

In direct transfer systems, no yard cranes are needed. The same equipment is used to pick up (put down) a container from (into) the marshalling yard, deliver it to (from) the apron, and transfer it to (from) a quay crane. In an *on-chassis system* which is illustrated in Figure 1, every container is stacked on a chassis and a tractor pulls the chassis between the apron and the marshalling yard. In a *carrier-direct system* which is illustrated in Figure 2, containers are stacked in multiple tiers and straddle carriers pick up (put down) containers from (into) the yard and deliver them between the apron and the marshalling yard.

In indirect transfer systems, a prime mover delivers a container between the apron and the marshalling yard. Straddle carriers or yard cranes transfer containers between prime movers and yard stacks in the marshalling yard. In the *straddle-carrier-relay system*, straddle carriers are used to transfer containers, while yard

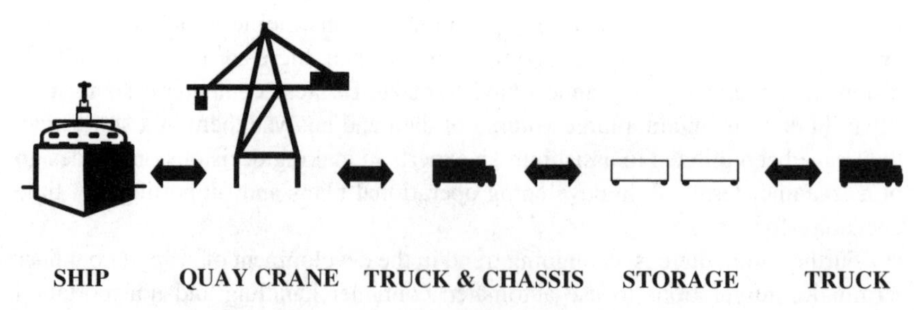

Fig. 1. Container flows in an on-chassis system

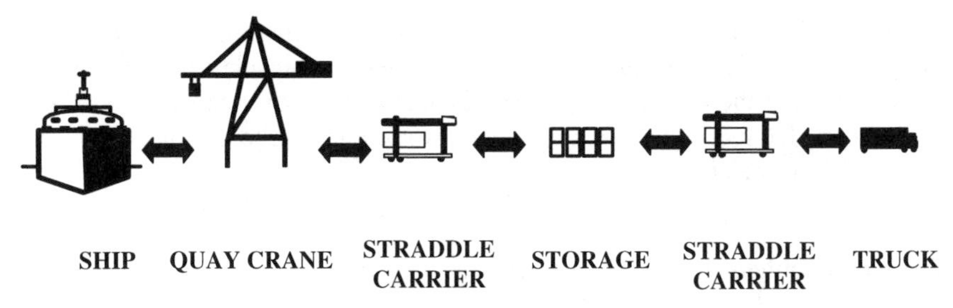

Fig. 2. Container flows in a carrier-direct system

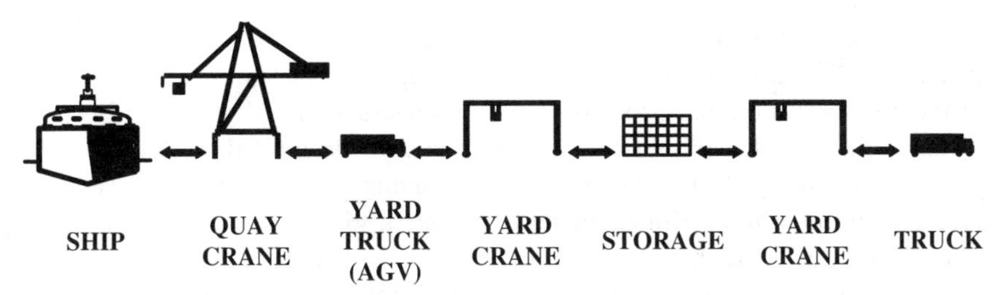

Fig. 3. Container flows in a yard-crane-relay system

cranes do so in the *yard-crane-relay system* which is illustrated in Figure 3. Throughout the world all automated container terminals adopted this type of yard-side equipment. In the ECT terminal of Rotterdam and the CTA terminal in Hamburg, automated guided vehicles (AGVs) are being used as prime movers. The type of yard cranes employed are automated stacking cranes, double rail-mounted gantry cranes (RMGC), and overhead bridge cranes in ECT, CTA, and Parsir Panjang terminal in Singapore, respectively.

From the perspective of operational decisions, the most difficult decisions have to be made in the case of the yard-crane-relay system because of its higher stacks of containers compared to the other handling systems. In the following explanation of the operation of container terminals, we will assume the yard-crane-relay system, unless another type of handling system is indicated explicitly.

The handling operations in container terminals comprise three types of operations: vessel operations associated with container ships, receiving / delivery operations for outside trucks, and container handling and storage operations in a yard. Vessel operations include the discharging operation, during which containers in a vessel are unloaded from the vessel and stacked in a marshalling yard, and the loading operation, during which containers are handled in the reverse direction of the discharging operation. During the discharging operations, quay cranes transfer containers from a ship to a prime mover which can be a yard truck or an AGV. Then, the prime mover delivers the inbound (import / discharging) container to a yard crane that picks it up and stacks it into a position in a marshalling yard. For the loading operation, the process is carried out in the opposite direction.

During receiving and delivery operations, when a container arrives at a container terminal by an outside truck, the container is inspected at a gate whether all documents are ready and the container is undamaged. Also, at the gate, information regarding where an export container is to be stored and where an import container is located, is provided to the outside truck. When the outside truck arrives at a transfer point of the yard, either yard cranes or straddle carriers receive a container from the truck ("receiving operation") or deliver a container to the truck ("delivery operation"), respectively.

Decision problems in container terminals

There are three different types of decision problems in container terminals, which call for quantitative decision support: design problems, operational planning problems, and real time control problems. Design problems have to be solved by facility planners in the initial planning stage of developing terminal configurations. Most of the problems are related to investment in construction and facilities. Because resources in container terminals are very expensive and limited, the usage of the resources and the impact of the operational planning systems have to be carefully evaluated in order to maximize the performance of the entire terminal configuration. During the actual handling operation, decisions on matching handling tasks with the required resources must be made in real time. These issues are referred to as real time control problems.

The first chapter of this book by *Dirk Steenken, Stefan Voß* and *Robert Stahlbock* provides an overview and classification of container terminal operations and related decision problems as well as a comprehensive review of the relevant literature. The authors not only describe the configuration of modern container port terminals and the different types of handling equipment employed, but analyze the corresponding logistics processes and present a survey of methods for their optimization. This is complemented by an extensive list of references.

Design problems

Design problems include the determination of the type of handling equipment in the yard, the number of berths, quay cranes, yard trucks, yard cranes, storage slots, and human operators, the yard layout, and the degree of automation of transportation and handling equipment. A related issue of considerable importance is the estimation of various performance measures of the intended terminal configuration. Queuing theory and simulation have been widely used to support design problems.

This book includes two studies which address the selection of transportation equipment. *Iris F.A. Vis* and *Ismael Harika* present a detailed comparison and sensitivity analysis of different types of automated transport vehicles. They examine the effects of using automated guided vehicles (AGVs) and automated lifting vehicles (ALVs) on unloading times of a vessel by means of a simulation study. In contrast to AGVs, ALVs are capable of lifting a container from the ground by itself. The study supports the choice for a certain type of equipment and the determination of the number of vehicles required.

Another simulation study of container terminal operations is provided by *Chang Ho Yang*, *Yong Seok Choi*, and *Tae Young Ha*. They also evaluate AGVs and ALVs as two competitive types of automated transport systems in automated container terminals. From the results of a detailed simulation analysis, they determine the comparative effect by cycle time and the required number of vehicles. The study demonstrates that the ALV is superior to the AGV in terms of productivity and efficiency.

The contribution by *Sönke Hartmann* is the development of a practical tool for generating scenarios of sea port container terminals. The scenarios can be used as input data for the development of simulation models as well as for testing the efficiency of optimization algorithms for different problems of container terminal operations. The scenario generator has been successfully applied in a major simulation project for the design of a new container terminal.

Another design issue, investigated by *Pyung Hoi Koo*, *Woon Seek Lee*, and *Dong Won Jang* refers to the problem of fleet sizing and vehicle routing for containers to be moved by trucks between container terminals and off-the-dock container yards. Their study is motivated by the situation in Busan where several small container yards are scattered in the city and relocating containers causes tremendous traffic problems. The approach suggested employs an optimization model to produce a lower bound on the required fleet size and a tabu search based heuristic to generate vehicle routes.

Operational planning problems

Before handling operations in container terminals actually happen, human planners or computerized control systems usually schedule them in advance to maximize the efficiency of the operations. Typically, target resources are in limited supply and thus priorities among handling activities that require the resources at the same time must be determined. Key resources include berths, handling equipment such as quay cranes, yard cranes, yard space, and human operators.

Ship operation plans represent one of the most important operational plans. The corresponding planning process consists of berth scheduling, quay crane scheduling (in practice, called work scheduling), and discharge and load sequencing of containers. In the initial stage of berth scheduling, the berthing time and the position of a container ship are determined.

The problem of allocating berth space for vessels in container terminals is examined in the book chapter by *Yongpei Guan* and *Raymond K. Cheung*. Motivated by such a problem arising in the port of Hong Kong, they develop models and solution methods which aim at minimizing the waiting and operating time of a vessel. Computational experiments are reported which demonstrate the efficiency of the suggested methods.

Through the quay crane scheduling process, the sequence of ship-bays that each quay crane will serve and the time schedule for the service are specified. For quay crane scheduling, relevant information, such as a stowage plan of the ship and the time interval in which each quay crane is available, are usually given. The stowage plan consists of multiple cross-sectional views, each corresponding to a ship-bay. Each cross-sectional view shows slots that a specific group of containers must be loaded into or picked up from.

Young-Man Park and *Kap Hwan Kim* discuss a method for simultaneously scheduling berth and quay cranes, which are critical resources in port container terminals. In contrast to the study by Guan and Cheung, they consider the fact that the berthing time of a vessel can be reduced by assigning more quay cranes to the vessel. An integer programming model is formulated by considering various practical constraints. A two-phase solution procedure is suggested for solving the mathematical model.

After constructing the quay crane schedule, the sequence of containers for discharging and loading operations as well as the corresponding storage locations have to be determined. In practice, heuristic rules are commonly used to construct the unloading and loading sequence. Superior solutions to the load sequencing problem, however, can often be obtained by using numerical search algorithms.

In the chapter by *Kap Hwan Kim, Jin Soo Kang,* and *Kwang Ryel Ryu*, the authors address the load sequencing of outbound containers. The solution to this complex decision problem requires the determination of the travel routes of transfer cranes and the number of containers to be picked up at each yard-bay as well as the determination of the load sequence for individual containers. A beam search based solution approach is proposed which considers many practical constraints.

In addition to berth and quay cranes, storage space may be pre-assigned for containers arriving at the marshalling yard in future so that the loading / discharge operation can be performed efficiently. In general, in order to expedite the loading operation, space for containers bound for the same ship should be assigned to locations close to each other. Storage space for inbound containers is usually determined in real time at the moment of discharge. Also, other resources such as yard cranes, manpower, and prime movers can be pre-assigned to a specific type of operation.

Real time control problems

Actual commitment of resource assignment is performed in real time and triggered by certain events or specific conditions. For resources, such as the berths, quay cranes, and storage space, the allocation of the resources has already been completed in the preceding planning stage, but the final commitment and detailed decisions on the assignment of resources are usually made in real time. For resources, such as prime movers, yard cranes, and manpower, detailed operational schedules can only be developed for a very short-term planning horizon, most often only for a few minutes, because the dynamic nature of the terminal operations and frequently occurring disturbances do not allow for comprehensive scheduling of future operations. Therefore, the assignment of resources to tasks must be made in real time.

In his contribution, *Sönke Hartmann* proposes a general model for various scheduling problems that occur in container terminals, e.g. scheduling cranes and vehicles. This model considers the assignment of jobs to resources and the temporal arrangement of the jobs subject to precedence constraints and sequence-dependent setup times. To support real-world scheduling problems, priority rule based heuristics and a genetic algorithm are discussed.

The chapter by *Martin Grunow*, *Hans-Otto Günther*, and *Matthias Lehmann* presents an efficient priority rule based dispatching algorithm for multi-load AGVs in highly automated seaport container terminals. This approach is well suited for practical application within an online logistics control system. The performance of the proposed heuristic is evaluated against an MILP model formulation with respect to total lateness of the AGVs. Numerical results also reveal the superiority of multi-load compared to single-load carriers.

Automated transportation systems

Apart from seaport container terminals, automated technology is being applied to other types of transportation systems and terminals, such as freight terminals in air ports or rail stations. This book comprises additional chapters which present case studies and applications of quantitative methodology for automated transportation systems.

In the chapter by *Matthieu van der Heijden*, *Mark Ebben*, *Noud Gademann*, and *Aart van Harten*, motivated by a Dutch pilot project on an underground cargo transportation system in Amsterdam Airport Schiphol using AGVs, the authors address empty vehicle management in large-scale automated transportation systems with the objective of minimizing cargo waiting times. They propose several heuristic rules and algorithms for empty vehicle management, varying from trivial First-Come, First-Served (FCFS) via look-ahead rules to integral planning.

Mark Ebben, *Matthieu van der Heijden*, *Johann Hurink*, and *Marco Schutten* address a scheduling problem for an underground cargo transportation system in which the finite capacity of resources (such as vehicles, docks, parking places) are considered as critical resources. They propose a flexible modeling methodology which allows to construct, evaluate, and improve feasible solutions.

Knut Alicke proposes a method for scheduling transshipment tasks in an inter-modal transport terminal. The approach handles sequence-dependent duration of empty moves, alternative assignments of containers to cranes and a sequence-dependent number of operations in a rail terminal. An optimization model based on constraint satisfaction technique is formulated and heuristics for the search procedure, especially value and variable ordering, are developed.

Jae Kook Lim, *Kap Hwan Kim*, *Kazuho Yoshimoto*, *Jun Ho Lee*, and *Teruo Takahashi* suggest a dispatching method for automated guided vehicles by using an auction algorithm. The dispatching method in this study is distributed in the sense that the dispatching decisions are made through communication among related vehicles and equipment. The theoretical rationale behind the distributed dispatching method is based on the auction algorithm for solving the assignment problem.

In the chapter by *Peter Brucker*, *Silvia Heitmann*, and *Sigrid Knust*, the problem of rescheduling trains is addressed in the case where one track of a railway section consisting of two tracks in opposite directions is closed due to construction activities. A polynomial algorithm is suggested for finding an optimal schedule with minimal lateness. Based on this algorithm, a local search procedure is proposed for the general problem of finding good schedules.

Container loading

Loading individual items into a container represents a planning problem of considerable mathematical complexity. Specific decision issues refer to the location of packaging units within the three-dimensional container space, the generation of packaging layers or stacking columns, and the consideration of several side constraints, such as orientation of individual items or load stability and safety considerations.

The final book chapter by *Michael Eley* addresses the container loading problem for stacking several three dimensional, rectangular items in one or more containers in such a way that the utilization of the container space is maximized. This study considers the situation where the consignment to be loaded is accommodated in more than one container.

Final remarks

This book discusses logistics control issues of container terminals and automated transportation systems and provides approaches of quantitative decision support for design, operational planning and real time control problems. It shows how quantitative methodology, e.g. optimization and heuristic algorithms, local search procedures, and simulation can be applied successfully to both theoretical analysis and practical decision support. This is demonstrated primarily for novel problem instances that emerge from recent developments in automated transportation and handling equipment. Sixteen papers previously published in *"OR Spectrum – Quantitative Approaches in Management"*, have been selected for publication in this volume. All papers have been peer-reviewed according to the standard of the journal. The book comprises reports on the state of the art, applications of quantitative models and methods, as well as case studies and simulation results. It is divided into three parts. Part I is focussed on strategic design, operational planning, and logistics control issues arising in container port terminals. Part II considers automated transportation systems which are of increasing importance in many application environments. Finally, Part III addresses problems of container loading.

This book has greatly benefited from the cooperation among the authors, reviewers and editors. We would like to express our sincere thanks to the reviewers for their excellent and timely refereeing. Last, but not least, we thank all authors for their contributions which made this book possible.

Hans-Otto Günther
Kap Hwan Kim
Editors

Table of Contents

Preface

Günther, H.-O., Kim, K.H.
Logistics control issues of container terminals
and automated transportation systems V

Container Terminals

Steenken, D., Voß, S., Stahlbock, R.
Container terminal operation and operations research
– a classification and literature review 3

Vis, I.F.A., Harika, I.
Comparison of vehicle types
at an automated container terminal 51

Yang, C H., Choi, Y.S., and Ha, T.Y.
Simulation-based performance evaluation of transport vehicles
at automated container terminals 79

Hartmann, S.
Generating scenarios for simulation and optimization
of container terminal logistics 101

Koo, P.H., Lee, W.S., Jang, D.W.
Fleet sizing and vehicle routing for container transportation
in a static environment 123

Guan, Y., Cheung, R. K.
The berth allocation problem: models and solution methods 141

Park, Y.-M., Kim, K-H.
A scheduling method for berth and quay cranes 159

Kim, K-H., Kang, J.S., Ryu, K.R.
A beam search algorithm for the load sequencing of outbound containers
in port container terminals 183

Hartmann, S.
A general framework for scheduling equipment and manpower
at container terminals 207

Grunow, M., Günther, H.-O., Lehmann, M.
Dispatching multi-load AGVs
in highly automated sea port container terminals 231

Transportation Systems

van der Heijden, M., Ebben, M., Gademann, N., van Harten A.
Scheduling vehicles in automated transportation systems
– Algorithms and case study 259

Ebben, M., van der Heijden, M., Hurink, J., Schutten, M.
Modeling of capacitated transportation systems for integral scheduling 287

Alicke, K.
Modeling and optimization of the intermodal terminal *Mega Hub* 307

Lim, J.K., Kim, K-H., Yoshimoto, K., Lee, J.H., Takahashi, T.
A dispatching method for automated guided vehicles
by using a bidding concept 325

Brucker, P., Heitmann, S., Knust, S.
Scheduling railway traffic at a construction site 345

Container Loading

Eley, M.
A bottleneck assignment approach
to the multiple container loading problem 359

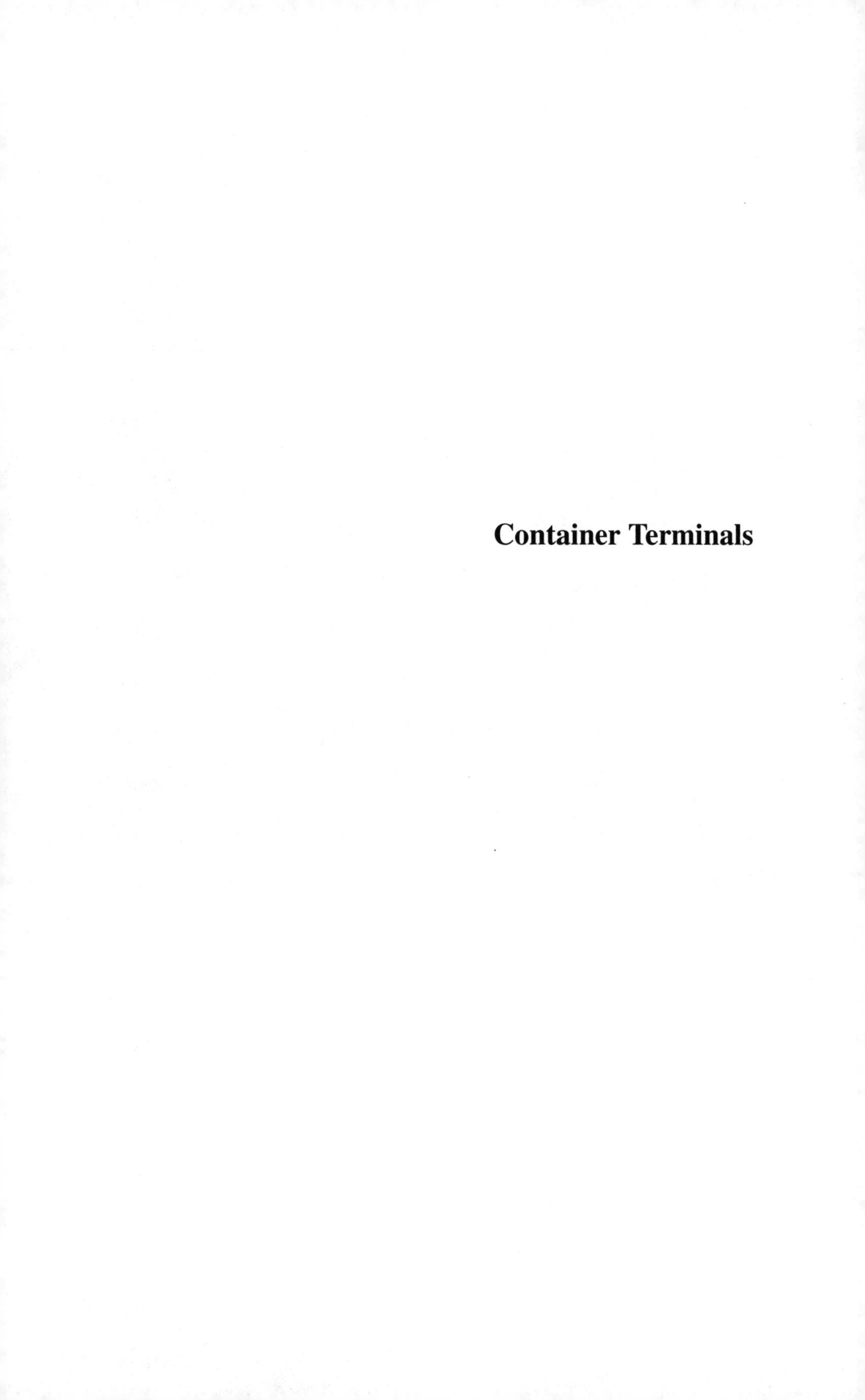

Container Terminals

Container terminal operation and operations research – a classification and literature review

Dirk Steenken[1], **Stefan Voß**[2], **and Robert Stahlbock**[2]

[1] Hamburger Hafen- und Lagerhaus AG, IS – Information Systems/Equipment Control, Bei St. Annen 1, 20457 Hamburg, Germany
(e-mail: steenken@hhla.de)

[2] Institute of Information Systems (Wirtschaftsinformatik), University of Hamburg, Von-Melle-Park 5, 20146 Hamburg, Germany
(e-mail: stefan.voss@uni-hamburg.de; stahlboc@econ.uni-hamburg.de)

Abstract. In the last four decades the container as an essential part of a unit-load-concept has achieved undoubted importance in international sea freight transportation. With ever increasing containerization the number of seaport container terminals and competition among them have become quite remarkable. Operations are nowadays unthinkable without effective and efficient use of information technology as well as appropriate optimization (operations research) methods. In this paper we describe and classify the main logistics processes and operations in container terminals and present a survey of methods for their optimization.

Keywords: Container terminal – Logistics – Planning – Optimization – Heuristics – Simulation

1 Introduction/historical overview

Containers came into the market for international conveyance of sea freight almost five decades ago. They may be regarded as well accepted and they continue to achieve even more acceptance due to the fact that containers are the foundation for a unit-load-concept. Containers are relatively uniform boxes whose contents do not have to be unpacked at each point of transfer. They have been designed for easy and fast handling of freight. Besides the advantages for the discharge and loading process, the standardization of metal boxes provides many advantages for the customers, as there are protections against weather and pilferage, and improved and simplified scheduling and controlling, resulting in a profitable physical flow of cargo. Regarding operations, we need to distinguish whether we refer just to a container (which in that sense is called a box) or we specify the type of container

Correspondence to: S. Voß

under consideration. The most common distinction refers to a so-called standard container as one which is twenty feet (20') long, describing the length of a short container. Other containers are measured by means of these containers, i.e., in twenty feet equivalent units (TEU) (e.g., 40' and 45' containers represent 2 TEU). Additional properties of containers may be specified whenever appropriate (e.g., the weight or weight class of a container, the necessity of special handling for reefer containers or oversized containers).

First regular sea container service began about 1961 with an international container service between the US East Coast and points in the Caribbean, Central and South America. The breakthrough after a slow start was achieved with large investments in specially designed ships, adapted seaport terminals with suitable equipment, and availability (purchase or leasing) of containers. A large number of container transshipments then led to economic efficiency and a rapidly growing market share. In this context, transshipment describes the transfer or change from one conveyance to another with a temporarily limited storage on the container yard.

Today over 60 % of the world's deep-sea general cargo is transported in containers, whereas some routes, especially between economically strong and stable countries, are containerized up to 100 % [140, 78]. An international containerization market analysis shows that in 1995 9.2 million TEU were in circulation. The container fleet had almost doubled in ten years from a size of 4.9 million TEU in 1985. Figure 1 shows the container turnover for the ten largest seaport terminals in the world from 1993 to 2002 [16, 17, 3, 4, 148]. Due to the positive forecast for con-

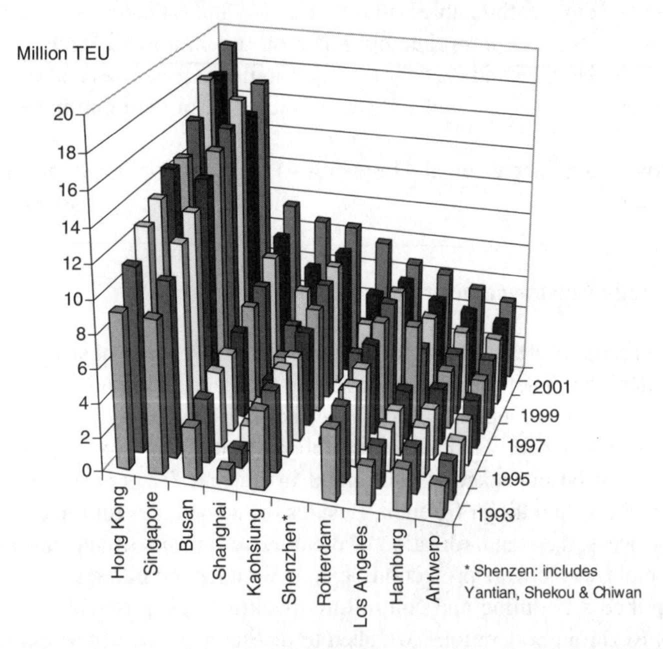

Fig. 1. Container turnover of the ten largest seaport terminals in the world from 1993 to 2002 (ranking 2002)

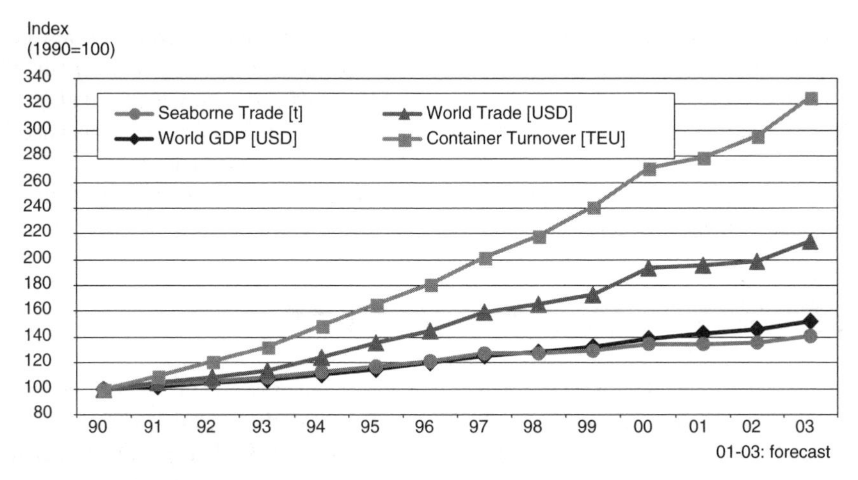

Fig. 2. Containerization trend: high growth of container turnover

tainer freight transportation, a similar development can be expected in the future. Figure 2 shows the containerization trend with high increasing rates compared with the rates of world trade, seaborne trade and the gross domestic product (GDP) of the world [198].[1]

The increasing number of container shipments causes higher demands on the seaport container terminals, container logistics, and management, as well as on technical equipment. An increased competition between seaports, especially between geographically close ones, is a result of this development. The seaports mainly compete for ocean carrier patronage and short sea operators (feeders) as well as for the land-based truck and railroad services. The competitiveness of a container seaport is marked by different success factors, particularly the time in port for ships (transshipment time) combined with low rates for loading and discharging [140, 78]. Therefore, a crucial competitive advantage is the rapid turnover of the containers, which corresponds to a reduction of the time in port of the container ships, and of the costs of the transshipment process itself. That is, as a rule of thumb one may refer to the minimization of the time a ship is at the berth as an overall objective with respect to terminal operations.

The objective of this paper is to provide an overview and a classification of container terminal operations. Moreover, based on this classification we attempt to provide a comprehensive literature review concerning operations research models and applications in this important logistics field. Usually, container terminals are characterized by means of their specific equipment and stacking facilities. Therefore, in Section 2.1 we describe possible means of handling equipment used in today's container terminals. Based on these one may classify various types of con-

[1] For detailed information about worldwide maritime transport trends see actual UNCTAD Review of Maritime Transport (via http://www.unctad.org), e. g. [189–192]. Success factors for growth in container shipping can be found in [198] or [118]. An introductory overview of intermodal freight transportation and containerization is given by [127, 140].

tainer terminals (see Section 2.2). Furthermore, we provide a general overview of the functionality of a container seaport terminal with a focus on physical container movements. In Section 3 we discuss terminal logistics and optimization methods. Here we aim at providing a considerable list of relevant references (in many cases just providing the references without going too much into detail) describing different approaches including exact methods, heuristic methods as well as simulation based approaches.[2] Finally some conclusions are given in Section 4.

2 Terminal structure and handling equipment

In general terms, container terminals can be described as open systems of material flow with two external interfaces. These interfaces are the quayside with loading and unloading of ships, and the landside where containers are loaded and unloaded on/off trucks and trains. Containers are stored in stacks thus facilitating the decoupling of quayside and landside operation.

After arrival at the port, a container vessel is assigned to a berth equipped with cranes to load and unload containers. Unloaded import containers are transported to yard positions near to the place where they will be transshipped next. Containers arriving by road or railway at the terminal are handled within the truck and train operation areas. They are picked up by the internal equipment and distributed to the respective stocks in the yard. Additional moves are performed if sheds and/or empty depots exist within a terminal; these moves encompass the transports between empty stock, packing center, and import and export container stocks (Fig. 3).

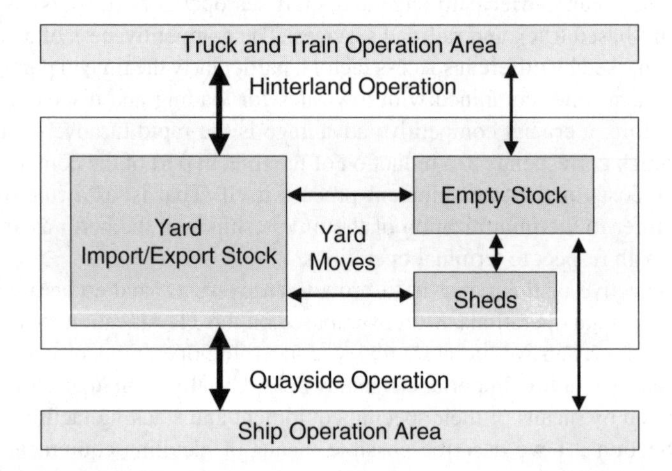

Fig. 3. Operation areas of a seaport container terminal and flow of transports

[2] All sections are moderately interleaved with references giving pointers to relevant literature. Although we try to achieve a more comprehensive list of references than other recent survey papers in this field (see, e.g., [196]) we admit that even our list is by no means complete.

It should be noted that the quayside operation or container transshipment as well as the container movement to and from the wharf is sometimes also referred to as waterside transshipment process. Correspondingly, one may find the terms hinterland transshipment processes and landside transshipment processes.

Different types of ships have to be served at the quayside. The most important ones are deep-sea vessels with a loading capacity of up to 8.000 container units (TEU) which serve the main ports of different countries and continents. Such vessels are about 320 m long with a breadth of 43 m and a draught of 13 m; on deck containers can be stowed 8 tiers high and 17 rows wide, in the hold 9 high and 15 wide. The ships' data call for respective dimensions of the cranes' height and jib length. Loading of about 2.000 boxes is common in large ports; the same is valid for unloading. Feeder vessels with a capacity of 100 to 1.200 TEU link smaller regional ports with the oversea ports delivering containers for deep-sea vessels. Inland barges are used to transport containers into the hinterland on rivers and channels. Functionally, barges are means of hinterland transportation (like trucks and trains), operationally they are ships which are served by quay cranes.

Trucks have a capacity of up to three TEU. At container terminals they are directed to transfer points where they are loaded and unloaded. To serve trains, railway stations with several tracks may be part of container terminals. The capacity of one train is about 120 TEU. Shuttle trains connecting a terminal with one specific hinterland destination obtain increased importance. The modal split of hinterland transportation is very specific for different ports which has a direct impact on the terminals' layout and type of equipment.[3]

The container storage area is usually separated into different stacks (or blocks) which are differentiated into rows, bays and tiers. Some stack areas are reserved for special containers like reefers which need electrical connection, dangerous goods, or overheight/overwidth containers which do not allow for normal stacking. Often stacks are separated into areas for export, import, special, and empty containers.

Besides in these general functions some terminals differ also in their operational units. For example, if railway stations do not exist inside the terminal, containers have to be transported by trucks or other landside transportation means between the external station and the terminal. This results in additional logistic demands.

Other differences occur if sheds exist within the terminal area. At sheds containers are stuffed and stripped, and goods are stored. Additional movements have to be performed connecting the yard stacks with the sheds. The same applies to empty depots where empty containers are stored according to the needs of shipping lines.

---

[3] The figures for Hamburg, Rotterdam, Hong Kong and Singapore illustrate this quite clearly (see, e.g., [184] for Rotterdam): Hamburg: about 47 % truck, 35 % feeder and 18 % rail; Rotterdam: about 50 % truck, 40 % feeder, 10 % rail; Hong Kong: more than 90 % truck, less than 10 % feeder, no rail; Singapore: 20 % truck, 80 % feeder, no rail.

2.1 Handling equipment

Usually, container terminals are described very specifically with respect to their equipment and stacking facilities. From a logistic point of view, however, terminals only consist of two components: stocks and transport vehicles.

The yard stacks, ships, trains, and trucks belong to the category 'stock'. Stocks are statically defined by their ability to store containers while from a dynamic point of view a stowage (or loading) instruction is necessary defining the rules how and where containers have to be stored. There is no principal difference between these different types of stocks but only a difference in capacity and complexity. Routing and scheduling of ships, trains and trucks do not belong to container terminal operation. Therefore, they can be considered statically as storage entities whereas a stowage instruction exists in any case even for trucks where at least the position of the containers to be loaded has to be defined. For specific stowage, ships and trains need instructions defining the position for every container. Transport means either transport containers in two or three dimensions. Cranes and vehicles for horizontal transport belong to this category. Their logistical specifics are that transport jobs have to be allocated to the means of transport and sequences of jobs have to be performed. The calculation of sequences is typical for the transportation means and defines a principal difference to the stocks categorized above. Not looking for these identities but being fixed on the specifics of each component and equipment applied at container terminals results in a variety of operations research approaches and solutions.

2.1.1 Types of cranes. Concerning cranes, different types are used at container terminals. The quay (or gantry) cranes (Fig. 4a) for loading and unloading ships play a major role. Two types of quay cranes can be distinguished: single-trolley cranes and dual-trolley cranes. The trolleys travel along the arm of a crane and are equipped with spreaders, which are specific devices to pick up containers. Modern spreaders allow to move two 20' containers simultaneously (twin-lift mode). Conventionally single-trolley cranes are engaged at container terminals. They move the containers from the ship to the shore either putting them on the quay or on a vehicle (and vice versa for the loading cycle). Single-trolley cranes are man-driven. Dual-trolley cranes represent a new development only applied at very few terminals. The main trolley moves the container from the ship to a platform while a second trolley picks up the container from the platform and moves it to the shore. The main trolley is man-driven while the second trolley is automatic. At modern cranes, the crane driver is supported by a semi-automatic steering system; this is both the case for one and two-trolley cranes.

The maximum performance of quay cranes depends on the crane type. The technical performance of cranes is in the range of 50–60 boxes/h, while in operation the performance is in the range of 22–30 boxes/h.

A second category of cranes is applied to stacks. There are three types of cranes, either rail mounted gantry cranes (RMG) or rubber tired gantries (RTG) and overhead bridge cranes (OBC). Rubber tired gantries are more flexible in operation while rail mounted gantries are more stable and overhead bridge cranes are mounted on

a

b

Fig. 4a,b. Quay cranes and stacking crane. **a** Quay crane (here: dual-trolley cranes). **b** Stacking crane (here: Double-RMG)

concrete or steel pillars. Commonly gantry cranes span up 8–12 rows and allow for stacking containers 4–10 high. To avoid operational interruption in case of technical failures and to increase productivity and reliability, two RMGs are often employed at one stack area (block). Containers which have to be transported from one side of the block to the other then have to be buffered in a transition area of the block. Double-RMG systems represent a new development. They consist of two RMGs of different height and width able to pass each other thus avoiding a handshake area (Fig. 4b). This results in a slightly higher productivity of the system. Although most of the gantry cranes are man-driven, the tendency is for automatic driverless gantry cranes which are in use at some terminals (e.g. Thamesport, Rotterdam, Hamburg). The technical performance of gantry cranes is approximately 20 moves/h.

Similar cranes are used for loading and unloading trains. They span several rail tracks (about six). Containers to be transferred from/to trains are pre-stowed in a buffer area alongside the tracks.

Forklifts and reachstackers are used to move and stack light containers – especially empty ones.

2.1.2 Horizontal transport means. A variety of vehicles is employed for the horizontal transport both for the ship-to-shore transportation and the landside operation. The transport vehicles can be classified into two different types. Vehicles of the first class are 'passive' vehicles in a sense that they are not able to lift containers by themselves. Loading and unloading of these vehicles is done by cranes, either quay cranes or gantry cranes. Trucks with trailers, multi-trailers and automatic guided vehicles (AGV, Fig. 5a) belong to this class. AGVs are robotics able to drive on a road network which consists of electric wires or transponders in the ground to control the position of the AGVs. AGVs can either load one 40'/45' container or two 20' containers – in the latter case multiple load operation is possible. As AGV systems demand for high investment, they are only operated where labour costs are high; they are now in operation at ECT/Rotterdam and at the HHLA/Hamburg – in combination with automatic gantry cranes.

a

b

Fig. 5a,b. Horizontal transport means: Automated guided vehicle and straddle carrier. **a** AGVs (in front of quay cranes). **b** Straddle carrier

Transport vehicles of the second class are able to lift containers by themselves. Straddle carriers (Fig. 5b), forklifts, and reachstackers belong to this class. Straddle carriers (SC) are the most important ones of it. Straddle carriers not only transport containers, but are also able to stack containers in the yard. Therefore, they can be regarded as 'cranes' not locally bound, with free access to containers independent of their position in the yard. The straddle carriers' spreader allows to transport either 20' or 40' containers; twin mode to transport/stack two 20' containers simultaneously is becoming available. Because of their properties, straddle carrier systems are very flexible and dynamic. Straddle carriers exist in numerous variants. Usually straddle carriers are man-driven and able to stack 3 or 4 containers high, i.e., they are able to move one container over 2 or 3 other containers, respectively.

During the last years progress was made to develop automatic straddle carriers. Recently, an automatic straddle carrier system has gone into production at Patrick Terminal/Brisbane, Australia. The straddle carriers are 4 high, an integrated system of differential GPS (see Section 2.1.3) and dead-reckoning serves for accurate positioning and routing. Beside this type of normal height, automatic straddle carriers of less height (height one or two) are under development. Because of the restricted height they are not provided for stacking but transport purposes only. Their ability for lifting containers allows for decoupling the work flow of transport and crane activities by using buffers at the respective interfaces. Because of the ability to lift containers, automated straddle carriers are often named Automated Lifting Vehicles (ALV).

2.1.3 Assisting systems. Besides cranes and transport vehicles, assisting systems play an eminent role for the organization and optimization of the work flow at container terminals. This is valid especially for communication and positioning systems.

Container terminal operators support a very intense communication with external parties like shipping lines, agents, forwarders, truck and rail companies, governmental authorities like customs, waterway police and others. The electronic communication is based on international standards (EDIFACT; Electronic Data Interchange For Administration, Commerce and Transport). Every change of container status is communicated between the respective parties. From the point of view of the terminal operator the most important messages are: the container loading and discharging lists which specify every container to be loaded or unloaded to/from a ship with specific data; the 'bayplan' which contains all containers of a ship with their precise data and position within the ship (it is communicated before arrival in the port); the 'stowage instruction' which describes the positions where export containers have to be located in a ship and which is the base for the stowage plan of the terminal; container pre-advices for delivery by train and truck, and the schedule and loading instruction for trains – only to name a few. Although only some of these messages – especially the stowage instruction for ships and trains – interfere directly with the operational activities of the terminal, they are very important because they serve for completeness and correctness of container data which is necessary to optimize the work flow.

Besides the communication with external partners, the internal communication systems play a major role in optimizing the terminal operation. The radio data communication, which was installed at container terminals since the middle of the 1980s, plays a key role because it is the main medium to transmit job data from the computer to cranes and transport vehicles. The radio data communication was the technical base for the implementation of operations research methods to optimize job sequences.

Since the middle of the 1990s Global Positioning Systems (GPS) were installed at container terminals. Initially they were used to automatically identify the position of the containers in the yard guaranteeing that the container yard position in the terminals' computer system is accurate. Because of the size of containers and the yard layout, differential GPS (DGPS) is necessary. DGPS components are not installed at containers but on top of the transport and stacking equipment. The position is measured, translated into yard coordinates and transmitted to the computer whenever a container is lifted or dropped. Alternatives to DGPS are optical based systems, especially Laser Radar. Sometimes both systems are integrated to assure a higher reliability. Container positioning systems like DGPS, dead-reckoning or Laser Radar constitute the technical base for the improvement of yard and stacking logistics.

Transponder and electrical circuits are used to route gantry cranes and automatic vehicles like AGVs whereas DGPS is used for the steering of automatic straddle carriers and other equipment.

Literature review

General information about technical equipment for container terminals can be found in engineering oriented journals as well as specialized outlets (see, e.g., http://www.porttechnology.org/). For different types of cranes and their use see, e.g., [147, 180]. Mobile vehicles or crane installations are also described in [127, 140, 147]. A more general insight into transport vehicles or gantry cranes is provided by, e.g., [127, 140, 106]. For a detailed overview of current state of the art handling technologies for terminal operations – like Automated Storage and Retrieval Systems (AS/RS) or AGVs – see, e.g., [85, 84, 83]. The use of DGPS at a container terminal is reported in [179]. Embedding handling equipment into more general aspects of innovation management at container terminals is considered in [199].

An interesting comparison between different types of container terminals based on specific types of equipment is provided in [168]. The authors compare the waterside productivity in different scenarios for manually operated SCs, AGVs and ALVs in a system set with yard stacking cranes. In addition they provide cost estimates based on simulation studies.

An overview of research on the potential of an integrated approach with usage of AS/RS and an AGV system is given in, e.g., [87, 5, 6]. Variations with different technical equipment – new in the field of container terminals – are shortly discussed. Effectiveness of such systems is compared with performance of current conventional systems by simulation experiments. For example, a 'Grid on Rail' concept is proposed: conventional container blocks are served by an overhead grid network of rails and a fleet of shuttle cranes moving on it. Effects are better space utilization by a more compact yard without necessity of roads between blocks and faster storage/retrieval operations than in conventional approaches with gantry cranes or straddle carriers. A pilot design is located in Hong Kong.

Details about assisting systems (without any planning functionality) can be found, e.g., on the web pages of service companies. This also includes detailed handbooks for electronic data interchange (EDI and EDIFACT) or hints for contractual agreements (see, e.g., http://www.dakosy.de and [79]).

2.2 Container terminal systems

A great variety of container terminals exists mainly depending on which type of handling equipment is combined to form a terminal system. All terminals use gantry cranes, either single- or dual-trolley, manual or semi-automatic. The transport between quay and stack can be performed either by trucks with trailers, multi-trailers, AGVs or straddle carriers. These vehicles can also serve the landside operation – except AGVs which nowadays are exclusively engaged at the quayside. Container stacking is either performed by gantry cranes or by straddle carriers.

Despite the variety of equipment combinations, two principal categories of terminals can be distinguished: pure straddle carrier systems and systems using gantry cranes for container storage.

Terminals with gantry cranes for container storage apply any kind of transport vehicles mentioned above. Even mixed systems of transport vehicles occur; e.g., multi-trailers for the quayside and straddle carriers for the landside operation. Up to now AGV terminals only exist in combination with automatic gantry cranes. Trains are normally loaded and unloaded by gantry cranes even in case of straddle carrier terminals, although in some cases straddle carriers are also used for this purpose (see Fig. 6).

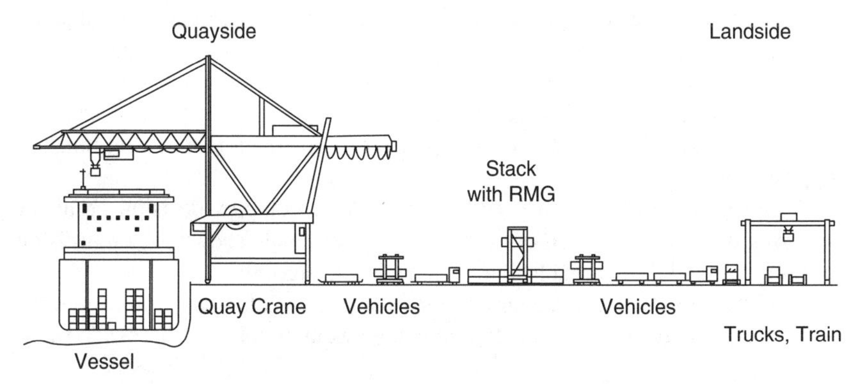

Fig. 6. Container terminal system (schematic side view, not true to size)

The decision on which equipment is used at container terminals depends on several factors. Space restrictions, economical and historical reasons play an important role. A basic factor is the dimension of the space which can be used for a terminal. If space is restricted, gantry cranes to store containers are preferred. A decision for AGVs and automated gantry cranes can be made in case of high labour costs and new terminal construction. Historical and cultural reasons have to be considered if container terminals are enhanced or modernized. Because space is becoming a scarce resource, a tendency for higher storage is to be foreseen.

Besides the mentioned two main categories, common in Europe and Asia, a third type, quite often in North America, is an on-chassis system, in which containers are stored on chassis instead of being stacked on top of each other. This system lacks of special stacking cranes, has simpler stacking logistics and is more space demanding. Its logistic aspects are covered by the other two systems.

Literature review

Container terminal operations are becoming more and more important. Therefore, an ever increasing number of publications on container terminals have appeared in the literature. While we refer to most of them in the subsequent chapters, some deserve special mention due to some of their general perspectives.

Decision problems at container terminals are comprehensively described by Vis and de Koster [196] (with some 55 references up to 2001). An overview of relevant literature for problem classes like arrival of the ship, (un)loading of a ship, transport of containers from/to ship to/from stack, stacking of containers, inter-terminal transport and complete terminals is provided.

Kozan [112] discusses major factors for the transfer efficiency of multimodal container terminals. A network model reflecting the logistic structure of a terminal and the progress of containers is shown. Its objective is the minimization of the total throughput time as the sum of handling and travelling times of containers. Earlier work of the same author is [111].

Meersmans and Dekker [132] present an overview of the use of operations research models and methods in the field of design and operation of container terminals with its decision problems on strategic, tactical and operational level.

Fung [50] presents a three-player oligopoly error-correction model for forecasting demand for Hong Kong's container handling services. Due to increasing demand and necessity of higher throughput, early construction of new terminals is suggested.

Murty et al. [141] describe various interrelated complex decision problems occurring daily during operations at a container terminal. They work on decision support tools and discuss mathematical models and algorithms.

Steenken [180] presents a comprehensive description of logistics and optimization systems in container terminals – shown by example of 'Burchardkai' (Hamburg).

For an early work on berth assignment and berth investment decisions see [45]. A general discussion of different productivity related objectives regarding transshipment terminals can be found, e.g., in [49,62]. Additional works giving more or less general descriptions of container terminals are, e.g., [34,130]. In [34] the authors view a container terminal as a production system that is represented as a network of complex substructures or platforms. The idea of platform capacity is used to represent operational aspects of a container terminal in a mathematical model for tactical planning. The problem is to allocate resources in each platform in order to minimize the total delay on the overall network and time horizon.

Konings [108] presents a survey of the possibilities for an intermodal transport concept of high quality. Conditions for best development of centers, that integrate transshipment, storage, collection and distribution of goods, are outlined. The internal transport system is identified as key element. The topic is discussed in detail for the harbour of Rotterdam.

Nam and Ha [142] investigate aspects of adoption of advanced technologies such as intelligent planning systems, operation systems and automated handling systems for container terminals. They set criteria for evaluation of different handling systems and apply them to examples in Korea. Results show that automation does not always guarantee outperformance (e.g. higher productivity) – it depends on terminal characteristics such as labour costs.

Four different types of automated container terminals are designed, analyzed and evaluated in a simulation model with very detailed cost considerations by Liu et al. [126]. The performance criteria that are used in this study to evaluate and compare different terminal systems are summarized as follows: Throughput: number of moves/hour/quay crane; throughput per acre; ship turnaround time: time it takes for a ship to get loaded/unloaded; truck turnaround time: average time it takes for a truck to enter the gate, get served, and exit the gate, minus the actual processing time at the gate; gate utilization: percent of time the gate is serving the incoming and outgoing container traffic; container dwell time: average time a container spends in the container terminal before taken away from the terminal; idle rate of equipment: percent of time the equipment is idle. The authors conclude that performance and costs of conventional terminals can be improved substantially by automation.

Important features of a terminal are related to the location of equipment and resources over the terminal. This refers, e.g., to resource allocation problems but also to some dispatching problems. Objectives may be an intelligent assignment of technical equipment (e.g., gantry cranes and straddle carriers) to the different terminal areas as well as an efficient job assignment to the utilized resources (see, e.g., [113,54,181,32,155], or [177] presenting a method for forecasting daily demand in terms of the number of container movements in a terminal based on online data in order to improve supply side decisions like allocation of handling equipment, work scheduling, etc.). Moreover, in AGV systems the layout of the network for the vehicles (in manufacturing systems it is called guidepath network) has a major impact on system effectiveness. While optimization methods for guidepath network design have been considered for various production environments (see, e.g., [162, 105]), it may also become a thread in the layout of container terminals.

While most work is related to a single terminal, some harbours even have more than one terminal. Cheu et al. [28] discuss possible savings with respect to distances travelled for the harbour of Singapore under the assumption that different terminals are combined together into one so-called mega-terminal.

It should be noted that different types of 'terminals' may have the same or only slightly modified structure compared to container terminals. This may be easily seen from carefully investigating intermodal traffic terminals or so-called megahub terminals for rail traffic or even airports. As an example the reader may be referred to intermodal traffic terminals (see, e.g., [1,2,63]).

Many of the problems in container terminal logistics can be closely related to some general classes of transportation and network routing problems (and therefore more or less standard combinatorial optimization problems) discussed comprehensively in the literature. Examples of these problems and some basic references may be given as follows: An early and very comprehensive survey on various types of routing problems is [14]. For a recent survey on the vehicle routing problem (VRP) see [187], arc routing problems are also considered in [36]. The traveling salesman problem (TSP) asks for the shortest closed path or tour through a set of cities that visits every city exactly once. It is well explained in [116]; more recent pointers can be found in [65]. The rural postman problem (RPP), which is the problem of finding a least cost closed path in a graph that includes, at least once, each edge in a specified set of arcs, is considered, e.g., in [10]. For an application in container terminal logistics see [181]. In the pickup and delivery problem a set of routes has to be constructed in order to satisfy a given number of transportation requests by a fleet of vehicles. Each vehicle has a certain capacity, an origin and a destination (depot). Each transportation request specifies the size of the load to be transported, the location where it is to be picked up and the location where it is to be delivered.

The pickup and delivery problem is considered, e.g., in [35,64]. Finally, we mention the assignment problem, which is considered in almost any basic textbook on operations research.

3 Terminal logistics and optimization methods

The need for optimization using methods of operations research in container terminal operation has become more and more important in recent years. This is because the logistics especially of large container terminals has already reached a degree of complexity that further improvements require scientific methods. The impact of concurrent methods of logistics and optimization can no longer be judged by operations experts alone. Objective methods are necessary to support decisions. Different logistic concepts, decision rules and optimization algorithms have to be compared by simulation before they are implemented into real systems.

The characteristics of container terminal operation demands online (real-time) optimization and decision. This is because most of the processes occurring at container terminals cannot be foreseen for a longer time span – in general the planning horizon for optimization is very short. Some examples shall illustrate it: although data of containers to be delivered to terminals by trucks may be pre-advised by EDI, the exact time when the containers arrive at the terminal is not known.[4] On arrival, containers have to be checked for damages, and pre-advised data may be wrong; both data influence the target stack location. As trucks have to travel to transition points where the containers are picked up by straddle carriers or cranes, the truck sequences at the gate and at the transition points need not be the same. Thus only those container jobs can be sequenced which are already released for transportation by internal terminal equipment – in general only a few. As trucks permanently arrive, recalculation has to be performed periodically or event driven. Analogous arguments hold for train operation.

A similar situation occurs for ship loading and unloading. Although in general data of containers and their positions within the ship are precisely known in advance and the preplanning process (see below) allows the calculation of job sequences, they often have to be changed because of operational disturbances. As vessels are not static and move permanently (because of tide, weather, stability), containers which are next in the sequence cannot be accessed by the crane's spreader. Crane drivers make their own decisions and may alter the pre-calculated loading or unloading sequence by themselves.

According to the classification mentioned above, the following sections describe the most important processes at container terminals that can be optimized by means of operations research methods.

[4] This is true for the north-west European ports, while East-Asian ports commonly prescribe a time-window of only several minutes when a truck has to enter a terminal.

3.1 The ship planning process

Ship planning consists of three partial processes: the berth planning, the stowage planning and the crane split.

3.1.1 Berth allocation. Before arrival of a ship, a berth has to be allocated to the ship. The schedules of large oversea vessels are known about one year in advance. They are transferred from the shipping lines to the terminal operator by means of EDI. Berth allocation ideally begins before the arrival of the first containers dedicated to this ship – on average two to three weeks before the ship's arrival. Besides technical data of ships and quay cranes – not all quay cranes can be operated at all ships – other criteria like the ship's length and the length of the crane jib have to be considered. All ships to be moored during the respective time period have to be reflected in berth allocation systems. Several objectives of optimized berth allocation exist. From a practical point of view the total sum of shore to yard distances for all containers to be loaded and unloaded should be minimized. This corresponds to maximum productivity of ship operation. Automatic and optimized berth allocation is especially important in case of ship delays because then a new berthing place has to be allocated to the ship whereas containers are already stacked in the yard.

Literature review

Berth planning problems may be formulated as different combinatorial optimization problems depending on the specific objectives and restrictions that have to be observed. As an example we mention the possibility to model berth planning by means of the resource constrained project scheduling problem. Restrictions may refer to special equipment that is needed for certain operations, as it is the case, e.g., for unavailability due to maintenance or for RoRo-ships[5] where tractor trailers need to drive into the ship. Connections of berth planning to assignment problems and especially to the quadratic semi-assignment problem are emphasized in [75]. Due to the large interdependency, berth and yard planning are frequently considered in a common optimization model [54, 19, 183].

Li et al. [120] discuss the more general problem of 'scheduling with multiple-job-on-one-processor pattern' with the goal of minimizing the makespan of the schedule. Vessels can represent jobs, a processor can be interpreted as a berth. Computational experiments show the effectiveness of a heuristic method with near-optimal results.

Lim [122] reformulates the problem as a restricted form of the two-dimensional packing problem and explores a graph theoretical representation. For this reformulation it is shown that this specific berth planning problem is NP-complete. An effective heuristic algorithm for solving the problem – applied to historical test data – is proposed.

[5] A RoRo-ship is a Roll-On Roll-Off ship, i.e., transport vehicles can enter the ship via a stern ramp.

Legato and Mazza [117] present a queuing network model and a simulation experiment of the logistic processes (arrival, berthing and departure of vessels) at a container terminal.

Nishimura et al. [146] focus on the problem of dynamic berth assignment to ships in the public berth system (not especially container ports; it is emphasized that these systems and, therefore, the shown results 'may not be suitable for most container ports of major countries'). A heuristic procedure, based on a genetic algorithm (GA), is developed – 'adaptable to real world applications'.

Similar to [146], Imai et al. [81] study berth allocation and optimization of berth utilization using a heuristic procedure, which is based on a mixed-integer programming (MIP) formulation of static and dynamic versions of the allocation problem and its Lagrangian relaxation. The same authors develop a GA-based heuristic procedure for solving the nonlinear problem of berth allocation for vessels with different service priorities [82]. Imai et al. [80] relate berth allocation to machine scheduling problems and discuss a bi-objective nonlinear optimization problem considering ship waiting times and terminal utilization.

Based on [136], Kim and Moon [98] formulate a MIP-model for determining berthing times and positions of vessels in container ports with straight-line shaped berths. They develop a simulated annealing (SA) algorithm and show near-optimal results.

Guan and Cheung [60] propose a tree-search procedure and composite heuristics for large size problems in order to minimize total weighted flow time. Efficiency of the methods is shown by computational experiments.

Park and Kim [154] combine a berth assignment approach with consideration of quay crane capacities.

Additional references dealing with berth planning are, e.g., [115,61,153].

3.1.2 Stowage planning. Stowage planning is the core of ship planning. Planning a ship's stowage is a two-step process. The first step is executed by the shipping line. The shipping line's stowage plan has to be designed for all ports of a vessel's rotation. The positions for all containers and all ports of a rotation have to be selected within the ship. Stowage planning of a shipping line usually does not act with specific containers identified by numbers, but on categories of containers. These categories or attributes are: the length or type of a container, the discharge port and the weight or weight-class of containers. Containers of these attributes are assigned to specific positions within the ship. The objective of optimization from the shipping line's point of view is to minimize the number of shifts during port operation (ship to ship or ship to shore shifts) and to maximize the ship's utilization. Constraints to be satisfied mainly result from the stability of the ship.

The stowage plan of the shipping line is transferred to the terminal operator by EDI. The stowage instruction of the shipping line is filed into the terminal's system and serves as a working instruction or pre-plan for the terminal's ship planner. The stowage instruction of the shipping line is characterized by the assignment of containers of special attribute sets to ship slots. Based on this instruction the terminal planner then assigns dedicated containers identified by numbers to the respective slots. The attribute set of the slot and the container selected in the yard have to match. The stowage planning systems of a container terminal, therefore, display both the ship's sections to be planned and the yard situation. Some of the systems allow for automatic assignment and optimization. Different objectives of optimization are possible, e.g., maximization of crane productivity, cost minimization, or minimization of yard reshuffles. From a practical point of view the minimization of yard reshuffles plays an important role. Reshuffles occur when a container has to be accessed while others on top of it have to be removed first. Reshuffling consumes time which is an offset to the transportation time between stack and shore reducing the productivity of ship operation. Because the stowage plan is generated before the ship's loading has started, this kind of optimization is offline optimization.

Although stowage planning in real terminal operation is either a manual or an offline optimization process, the process structure of ship loading applies for online optimization. This is because the loading process and the stack-to-shore transport are more complex than yet described. To achieve a high productivity for the crane operation containers have to arrive at the quay in the right time and in the order of the loading sequence; i.e., loading sequence and sequence of horizontal transport have to correspond with each other. Otherwise crane waiting times and/or queuing of transport vehicles occur. Both reduce crane productivity and extend the ship's berthing time. As a common feature, containers are more or less spread in the yard and have different distances to the crane; special containers like overheight containers need special equipment which has to be mounted before they can be transported, reefer containers have to be disconnected from the electrical circuit, and yard reshuffles occur to a respective percentage. All this consumes additional transportation time. In manually driven systems the performance additionally depends on the driver's skill and decision which path he travels. Even technical or operational disturbances of the crane operation occur which enforce to change the loading sequence. Therefore, transportation times cannot be calculated exactly even if automated equipment is in use. All reasons together mean that the stowage plan prepared in advance can be sub-optimal. Online stowage planning is a solution to omit or at least reduce these problems. In online stowage planning a stowage plan which assigns specific containers to ship positions is no longer prepared. Instead containers are selected for transportation according to the attributes assigned to ship positions in the stowage instruction of the shipping line. Containers with the same attributes are considered as equal. They are then loaded according to their arrival time at the quay crane. Thus the specific stowage plan addressing specific container data to specific ship positions is generated simultaneously to the loading activity. Online stowage planning is not yet in use at container terminals but is a future need to enhance the performance of ship loading.

Literature review

In practice, stowage planning usually is a manual or offline optimization process using respective decision support systems (see, e.g., [176]). Most of the papers below describe research work applicable to enhance existing systems by appropriate optimization functionality. Container data are assumed to be given, i.e., we do not consider the problem of loading containers (see, e.g., [26,33,171,47]).

Sculli and Hui [173] investigate distribution effects and the number of different types of containers with respect to an efficient stowage in an experimental study. Performance of stacking policies is measured by volumetric utilization, wasteful handling ratios, shortage ratio, and rejection ratio. Results indicate that the number of different types of containers has the largest impact on these measures. Effects of stacking policy and maximum store dimensions are also significant.

Avriel et al. [9,8] focus on stowage planning in order to minimize the number of unproductive shifts (temporary unloading and reloading of containers at a port before their destination ports in order to access containers below them for unloading). Aspects of ship's stability and other real-life constraints are not considered. A binary linear programming (LP) model is formulated. Due to the proven NP-hardness of the problem a so-called 'Suspensory Heuristic' – based on earlier work by Avriel and Penn [7] – is developed in order to solve even large problem instances. The heuristic assigns slots in a bay to containers dynamically with respect to the sequence of ports in a vessel's route.

Wilson and Roach [201,202] divide the container stowage process into the two subprocesses and related subproblems of strategic and tactical planning level due to complexity of a stowage plan across a number of ports. They use branch and bound algorithms for solving the first problem of assigning generalized containers to a block in a vessel. In the second step a detailed plan which assigns specific positions or locations in a block to specific containers can be found by a tabu search algorithm. Good results (not always optimal) can be found in reasonable time. The same principles are described by Wilson et al. [203,167]. They present a computer system for generating solutions for the decomposed stowage (pre-)planning problem illustrated in a case study. The authors present a GA approach in order to generate strategic stowage plans automatically. Initial computational experiments show effective sub-optimal solutions.

Haghani and Kaisar [66] propose a MIP model for developing loading plans in order to minimize the time that a vessel spends at port, and the container handling cost which is highly influenced by the number of unproductive but necessary shifts caused by an unsatisfactory arrangement of containers. Loading schedules, ship's parameters like strength and stability, and factors like longitudinal moment, trim, and metacentric height are taken into consideration. Solution procedures and computational test examples are proposed.

Dubrovsky et al. [37] use a GA for solving the stowage planning problem of minimizing the number of container movements. Search space is significantly reduced by a compact and efficient encoding scheme. Ship stability criteria are reflected by appropriate constraints. Simulation runs demonstrate the efficiency and flexibility of the GA-based approach.

Simulation and online optimization in stowage planning is considered in [205, 204,182]. Especially in online settings as they are encountered in practice, waiting times of the cranes as well as congestions of transport means below the cranes have to be minimized to avoid productivity reduction. Winter [204] presents an integrated just-in-time scheduling model and algorithms for combined stowage and transport planning. In the first step a crane split is computed, based on the shipping company's stowage plan and a resulting loading sequence of bay positions and container types, respectively. The overall loading process is then optimized by flexibly assigning containers to straddle carriers fulfilling stowage criteria and minimizing late arrivals at the quay cranes. The assignment is based on container attributes instead of container numbers. Precedence constraints and transportation times depending on different travel distances (yard – quay) are considered, too. The model and the algorithms are tested with real-world data showing suitability for real-time planning with its special difficulties like delays of containers or incomplete information. A shorter version is given in [182].

Giemsch and Jellinghaus [57] present a MIP model for the stowage problem, based mainly on [9,8] and [202]. The basic model is extended with additional constraints and solution methods are modified. Results show improvements in comparison with [9].

For further references on stowage planning see, e.g., [86].

3.1.3 Crane split. The third step of ship planning is the allocation of quay cranes to ships and the ships' sections – the crane split (scheduling). Depending on the ship's size commonly three to five cranes operate at one oversea vessel. Feeder ships are operated with one to two cranes. In practice, crane to ship allocation has to reflect several constraints – especially technical data of cranes and ships and the accessibility of cranes at a berth. Because terminals are historically grown, in general different types of cranes exist at real terminals. The number of cranes operable at one berth in general is restricted because not every crane can be driven to every berth.

Crane split allocates a respective number of cranes to a ship and its sections (bays) on hold and deck and decides on which schedule the bays have to be operated. It not only reflects one ship but several ships – in neighboured berths and principally all ships moored at a terminal in a given period. There is no unique objective for optimization. Minimization of the sum of the delays of all ships can be an objective while maximization of one ship's performance or a well-balanced or economic utilization of the cranes can be others. In real terminal practice it depends on the actual terminal situation and the terminal's goal. In addition to the crane split, crane allocation decides on the mode how a ship and the ship's bays are loaded. A bay can be loaded either horizontally or vertically, starting at the quay or at the waterside, resulting in four different modes of loading. There are additional modes but these are the main ones. Stowage plan, crane split, and mode of loading together result in a working instruction which defines the loading sequence for every container of a bay. As mentioned earlier, the sequence for the landside transport has to match this loading sequence.

Literature review

Daganzo [32] shows a MIP for a static crane allocation problem with no additional ships arriving during the planning horizon. It is exactly solved for small problem instances (i.e. small number of ships), and a heuristic procedure for larger problems is proposed. In addition, the dynamic problem is considered. In both models the berth length is assumed to be unlimited.

Peterkofsky and Daganzo [155] study a branch and bound method for minimizing delay costs. Exact solutions for problems described in [32] are given in order to speed up the time-intensive and, therefore, cost-intensive (un)loading process.

Gambardella et al. [52] present a solution for the hierarchical problems of resource allocation – namely the allocation of quay cranes for (un)loading vessels and yard cranes for stack operations – and scheduling of equipment (i.e. (un)loading lists for each crane). Simulation results show reduction of equipment conflicts and of waiting times for truck queues. (See also related earlier papers of members of the same group of authors: [54, 129, 210, 166, 165, 53].)

The crane split as part of an integrated stowage and transport planning problem is discussed in [204,182] as mentioned in Section 3.1.2.

Bish [12] develops a heuristic method for minimizing the maximum turnaround time of a set of ships in the so called 'multiple-crane-constrained vehicle scheduling and location problem (MVSL)'. The problem is threefold: determination of a storage location in the yard for unloaded containers, dispatching vehicles to containers and scheduling of (un)loading operations to cranes.

Park and Kim [154] discuss an integer programming model for scheduling berth and quay cranes and propose a two-phase solution procedure. A first near-optimal solution for finding a berth place and time for each vessel and assigning the number of cranes is refined by a detailed schedule for each quay crane.

3.2 Storage and stacking logistics

Stacking logistics has become a field of increasing importance because more and more containers have to be stored in ports as container traffic grows continuously and space is becoming a scarce resource. Generally containers are stacked on the ground in several levels or tiers and the whole storage area is separated into blocks. A container's position in the storage area (or yard) is then addressed by the block, the bay, the row and the tier. The maximum number of tiers depends on the stacking equipment, either straddle carriers or gantry cranes. According to operational needs the storage area is commonly separated into different areas. There are different areas for import and export containers, special areas for reefer, dangerous goods or damaged containers. The average daily yard utilization of large container terminals in Europe is about 15.000–20.000 containers resulting in about 15.000 movements per day. The dwell time of containers in the yard is in the range of 3–5 days at an average.

A storage planning or stacking decision system has to decide which block and slot has to be selected for a container to be stored. Because containers are piled up, not every one is in direct access to the stacking equipment. Containers that are placed on top of the required one have to be removed first. Reshuffles (or rehandles) may occur due to several reasons; the most important ones result if data of containers to be stacked are wrong or incomplete. At European terminals 30–40 % of the export containers arrive at the terminal lacking accurate data for the respective vessel, the discharge port, or container weight – data which are necessary to make a good storage decision. Even after arrival, vessel and discharge port can be changed by the shipping line. For import containers unloaded from ships the situation is even worse: the landside transport mode is known in at most 10–15 % of all cases at the time of unloading a ship, e.g., when a location has to be selected in the yard.

To ease the situation and to ensure a high performance of ship, train and truck operation, containers sometimes are pre-stowed near to the loading place and in such an order that it fits the loading sequence. This is done after the stowage plan is finished and before ship loading starts. Because pre-stowage needs extra transportation, it is cost extensive and terminals normally try to avoid it by optimizing the yard stacking, but it is executed when ship loading has to be as fast as possible. Storage and stacking logistics are becoming more complex and sophisticated; they play an important role for the terminals' overall performance.

Two classes of storage logistics can be distinguished. In storage or yard planning systems, stack areas and storage capacities are allocated to a ship's arrival in advance according to the number of import and export containers expected. An appropriate number of slots in blocks and rows are reserved for a special ship. Depending on the planning strategy, the reservation for export containers can be split for discharge port, container type/length, and container weight. A common strategy for export planning is to reserve slots within a row for containers of the same type and discharge port while heavier containers are stacked on lighter ones assuming that they are loaded first because of the ship stability. For import containers only a reservation of yard capacity of respective size is done without further differentiation. This is because data and transport means of delivery generally are unknown at the time of discharge. If the transport mode is known, import areas can be subdivided according to them. Common strategies for import containers are either selecting any location in the import area or piling containers of the same storage date.

Yard or storage planning seldom matches the real delivery because container delivery is a stochastic process not exactly to be foreseen. The quality of this yard concept mainly depends on the strategy how to determine a good stack configuration and a good forecast of the container delivery distribution. Both factors are hard to solve, the result is a comparatively high amount of yard reshuffles. In addition, the reservation of yard locations occupies stack capacity.

Because of these disadvantages some terminals installed an alternative stacking concept, called scattered stacking. In scattered stacking, yard areas are no longer assigned to a specific ship's arrival but only once to a berthing place. On arrival of a container the computer system selects the berthing place of the ship from the ships schedule and automatically searches for a good stack location within the area assigned to the berth. A stack position is selected in real-time and containers with the same categories – ship, type/length, discharge port, and weight – are piled up one on top of the other. Containers for one ship are stochastically scattered over the respective stack area; reservation of yard slots is no longer necessary. This concept results in a higher yard utilization – because no slots are reserved – and a remarkable lower amount of reshuffles – because the stacking criteria merge the ship's stowage criteria.

Although the container attributes play a major role in yard stacking concepts, additional parameters have to be taken into account for improving logistic processes. Evidently, containers have to be stacked near to the future loading place, e.g., the transport distance should be minimal to ensure a high performance of the future operation. The performance of quay cranes is a multitude higher than the performance of stacking and transport equipment. Therefore, containers with the same categories have to be distributed over several blocks and rows to avoid congestions and unnecessary waiting times of vehicles. The actual workload of a gantry crane or other stacking equipment also has to be considered because allocating additional jobs to highly utilized equipment provokes waiting times. All these factors can be integrated into an algorithm while the weight of each factor is measured by parameters. The objective of yard optimization is to minimize the number of reshuffles and to maximize the storage utilization.

Literature review

Cao and Uebe [22] propose a tabu search based algorithm for solving the transportation problem with nonlinear side constraints – a general form of the problem of assignment of storage positions for containers with minimized searching and/or loading costs and satisfaction of limited space and other constraints.

Kim [89] investigates various stack configurations and their influence on expected number of rehandles in a scenario of loading import containers onto outside trucks with a single transfer crane. For easy estimation regression equations are proposed.

Kim and Bae [90] propose a methodology to convert a current order of export containers in the yard into a bay layout which is best from the point of view of operations for loading a vessel. The goal is to find the fewest possible number of containers and/or shortest possible travel distance in order to minimize the total turn-around time of a vessel in a port. The problem is decomposed, mathematical models (dynamic programming, transportation problem) for the three subproblems are suggested, and a numerical example is given. The authors demand heuristic algorithms due to time consuming computations.

Kim and Kim [92–94] discuss the determination of optimal amount of storage space and optimal number of transfer cranes for import containers. The decision is based on a cost model including fixed investment costs and variable operation costs. A simple solution procedure and sensitivity analysis is illustrated with a numerical example. Two different objectives are considered: minimization of the costs of only the terminal operator and minimization of these costs combined with the costs of the customers. Deterministic and stochastic models and simple solution methods are provided and illustrated using numerical examples. In [93] the authors focus on strategies for storage space allocation. Cases with constant, cyclic and dynamic arrival rates of import containers are analyzed. The objective is minimization of the expected total number of rehandles. Mathematical models and solution procedures are shown and illustrated by numerical examples.

Kim et al. [100] formulate a dynamic programming model for determination of the storage location of export containers in order to minimize the number of reshuffles expected for loading movements. The configuration of the container stack, the weight distribution of containers in the yard, and the weight of an arriving container are considered. For real-time decisions a fast decision tree is derived from the set of optimal solutions provided by dynamic programming.

A GA-based approach for minimizing the turnaround time of container vessels is described by Preston and Kozan [157]. The problem is formulated as an NP-hard MIP-model for determining the optimal storage strategy for various schedules of container handling (random, first-come-first-served, last-come-first-served). Computational experiments show that the type of schedule has no effect on transfer time if a good storage layout is used. Changes of storage area utilization in the range of 10–50 % result in linear changes of transfer time.

Kim and Park [99] focus on export containers and show a dynamic space allocation method in order to utilize storage space efficiently and to increase efficiency of loading operations. A basic MIP-model is formulated. Two heuristic algorithms – a myopic (least-duration-of-stay) rule and a sub-gradient optimization technique – are compared in computational experiments. Results are in 'almost the same level of objective values', but the decision rule is much faster. Effects of changing values of several model parameters are also analyzed.

Zhang et al. [211] study the storage space allocation problem in a complex terminal yard (with inbound, outbound and transit containers mixed). In each planning period of a rolling-horizon approach the problem is decomposed into two levels and mathematical models. The workload among blocks is balanced at the first level. The total number of containers associated with each vessel and allocated to each block is a result of the second step which minimizes the total distance to transport containers between blocks and vessels. Numerical experiments show significant reduction of workload imbalances and, therefore, possible bottlenecks.

As mentioned in Section 2, empty containers are often stored separately from loaded containers due to the possibility of using different equipment to store them higher than loaded containers. While methods for storage and stacking of empty containers do not differ from the above described approaches, the distribution of empty containers to ports has been considered as a separate problem deserving specialized approaches (see, e.g., [31,175,29]).

Additional references for storage and stacking logistics are, e.g., [183,21,24, 27,76,95,113].

3.3 Transport optimization

Two types of transport at a container terminal can be distinguished: the horizontal transport and the stacking transport carried out by gantry cranes. The horizontal transport itself subdivides into the quayside and the landside transport serving ships or trucks and trains, respectively. Trucks, multi-trailers, AGVs, manned or automatic straddle carriers can be used for the transport.

3.3.1 The quayside transport. For ship loading and unloading containers have to be transported from stack to ship and vice versa. Transport optimization at the quayside not only means to reduce transport times but also to synchronize the transports with the loading and unloading activity of the quay cranes. A general aim is to enhance crane productivity. Crane productivity does not only depend on the technical data of the cranes (50–60 boxes/h). The real performance at operation is much lower (in the range of 22–30 boxes/h). The reduction is caused by unproductive times like pauses and breaks during shifts, moves of hatch covers and lashing equipment, technical or operational disturbances and congestions occurring for the horizontal transport. Additionally, more transport vehicles provoke further costs and ship operation then is less economic.

Concerning logistics, a gain in ship productivity cannot be necessarily achieved by enhancing the number or the speed of transport vehicles operating at the quayside. This is because the possibility of congestions at the cranes and in the yard increases more than proportionally with the number of vehicles or their speed. Therefore, developing an optimization system also has to cope with the minimization of congestions.

Different modes of transport and strategies to allocate vehicles to cranes occur at the quayside. In single-cycle mode the vehicles serve only one crane. According to the crane's cycle they either transport discharged containers from the quay to the yard or export containers from the yard to the crane. In dual-cycle mode the transport vehicles serve several cranes which are in the loading or unloading cycle, respectively, thus combining the transports of export and import containers. Transport vehicles can either be allocated exclusively to one crane (gang structure) or to several cranes and ships (pooling).

In single-cycle mode no potential for the optimization of the import cycle exists. Optimization for discharged containers is restricted to the selection of optimal yard positions which is a task of the yard planning module (see above). As import containers have to be transported to the pre-selected stack locations, empty travels cannot be reduced. Travel distances can only be reduced if locations near to the quay are selected.

For export loading, however, there is a potential for optimization. In general the transport sequence is not identical to the loading sequence of the ship. The loading sequence is determined by the stowage plan, the crane split and the crane's loading strategy. The transport sequence, however, has to reflect different distances, yard reshuffles and special containers. The latter ones sometimes need special equipment which has to be provided before they can be transported. All effects result in additional transportation times. Therefore, the transport sequence has to be altered to ensure the right order of the loading sequence. Idle times of the cranes and vehicle congestions at cranes and stacks have to be avoided because both reduce productivity.

The dual-cycle is more complex. The dual-cycle mode combines the transports of export and import containers to/from cranes operating at the same ship or at neighbouring ships. The fixed allocation of transport vehicles to cranes is given up, vehicles operate in a pool serving several cranes in alternative modes (loading or discharging). Empty distances and transportation times are reduced in dual-cycle mode. This mode is more efficient but harder to organize because of the higher complexity. The possibility of crane waiting times can be reduced if containers can be buffered under the crane's portal.

In terminal practice, automatic transport vehicles like AGVs are always pooled while manned equipment like straddle carriers or trucks commonly operate at one crane (fixed allocation). If the loading capacity exceeds one container a multiple load mode is possible. Multiple load for AGVs contains potential for optimization, but it rarely occurs in practice because it is hard to organize. If unmanned equipment like AGVs or ALVs for transportation and automated gantry cranes for stacking are used, a main task of the control system is to synchronize the equipment in a way that the containers arrive 'in-time' at the interfaces (of the equipment such as, e.g., cranes and AGVs) and the idle times (of the cranes) are minimized.

Ship operation in practice is dynamic and, therefore, demands online optimization. For import containers, e.g., the precise yard location cannot be selected before the container is unloaded and its data and condition is physically checked. Disturbances occurring during ship operation often force to alter the loading or unloading sequence immediately. Such disturbances are: interruption of crane operation because of operational or technical problems, change of (un)loading sequences decided by the crane driver because of ship stability reasons or problems occurring during the horizontal transport. Such reasons force (re)calculating sequences only for few containers. The objective of optimization in any case is to minimize the lateness of container deliveries for the cranes and the travel times of the transport vehicles.

Literature review

A literature review regarding quayside transport is almost a dime a dozen and may be distinguished mainly based on the means of transport, i.e., AGVs, straddle carriers, etc. Even within the first category (AGVs) the number of references is enormous as AGVs are commonly used in warehouse operations and flexible manufacturing systems (see, e.g., [162] for a survey). In the sequel we first provide a wealth of references regarding AGVs before we are considering other means of transport.

Evers and Koppers [48] focus on movements of AGVs over the physical infrastructure with their model of an AGV traffic control system and the so-called semaphore technique.

Bruno et al. [18] focus on the control problem of dynamic determination of waiting positions for idle AGVs in order to reach good overall performance of the system (the paper deals with general material handling systems). Two fast effective heuristic algorithms are discussed and tested in real-world scenarios. The shown approach (without taking into account any information about future events) has better results than the traditional point-of-release-positioning rule.

Gademann and van de Velde [51] determine the waiting locations for idle AGVs in a loop layout with uni- or bidirectional flow system. The problem is restricted to a static setting, in which all AGVs are assumed to be idle at the same time. Objective functions are functions of travel times from the nearest waiting location of an AGV to a pickup point.

Wallace [200] presents an agent based AGV controller in order to provide effective flow even in complex designs. Agents allow AGVs to allocate only small possible segments or points on their paths. The agent approaches are tested in computational experiments with two layouts and are compared with an 'AutoMod' simulation. Results show higher efficiency without any deadlock situation.

Van der Heijden et al. [72] develop rules for management of empty AGVs in (general) automated transportation systems. Their performance (in terms of service levels, AGV requirements and empty travel distances) is evaluated by simulation. Look-ahead rules outperform the simple first-come-first-served rule.

Leong [119] develops an efficient dynamic deployment algorithm scheme for AGVs, that dispatches AGVs to containers in order to minimize the (un)loading time for a vessel. A deadlock prediction and avoidance algorithm – developed in [137] and also discussed later in [138] – is integrated. The new scheme is compared with the current scheme (used at a terminal in Singapore) in a simulation experiment. Analysis of results shows improvements, since the throughput is increased by the new scheme.

In a similar paper concerning the same project as in [119], Chan [25] models a network flow in order to develop an efficient dispatching strategy for AGVs. Constraints describe disparate instances of AGVs carrying one container or two containers. The performance of the proposed heuristic algorithms is tested and – in case of single load – compared with the current deployment strategy, that is outperformed by the new one.

Reveliotis [164] proposes a robust conflict resolution strategy for flexible operations on arbitrarily structured path networks. A dynamic closed-loop control scheme is developed, which organizes dispatching and routing of AGVs on basis of real-time feedback on the system traffic. Although the paper does not focus on automated container terminals, results may be transferred to this field.

Qiu and Hsu [158–161] address scheduling and routing problems for AGVs. They develop conflict-free routing algorithms for two different path topologies and two scheduling strategies. The methods are applied together in a case study.

Qiu et al. [162] provide a survey of scheduling and routing algorithms for AGVs. They show similarities and differences between scheduling and routing of AGVs and related problems like the vehicle routing problem, the shortest path problem, scheduling problems or others. They classify algorithms in groups for general path topology (static/time-window based/dynamic methods), for path optimization (0-1-integer-programming model, intersection graph method, integer LP model), for specific path topologies (linear/loop/mesh topology) and dedicated scheduling algorithms.

Grunow et al. [58, 59] focus on dispatching multi-load AGVs. A flexible priority rule based approach is proposed and compared to an alternative MIP formulation in different scenarios. Reduction of AGVs' lateness in case of multi-load mode is shown and an improvement of the terminal's overall performance is expected. In addition, a MIP is developed that allows determining optimal solutions for small problem instances. For real applications a hybrid approach using the MIP combined with fast heuristics on some special dispatching requests is suggested. A different MIP formulation can be found in [172].

Hanafi et al. [67] extend the simple multi-load case to the following problem related to container terminal logistics. Given a pool of containers, the container assignment problem consists of determining on which barges containers have to be loaded to minimize the total number of barges used while satisfying a number of side constraints. Different models and methods are compared on data provided by the Port of Lille.

Hartmann [68] develops a general scheduling model consisting of assignment of jobs to resources and (temporal) arrangement of the jobs with consideration of constraints. This model can be applied for scheduling of AGVs, straddle carriers, gantry cranes and even workers. A heuristic method based on priority rules and a GA for solving the problem are discussed and compared in a computational experiment, that shows promising results for the GA.

Yang et al. [207] analyze an increase of terminal productivity due to using ALVs rather than AGVs – based on the observation of unproductive and costly waiting of AGVs under quay cranes and in the blocks compared to that of ALVs. By means of a simulation model it is demonstrated, 'that the ALV is superior to the AGV in both productivity and economical efficiency principally because the ALV eliminates the waiting time in the buffer zone'. Similar findings are reported by [195].

Lim et al. [123] do not especially focus on container terminals, but suggest an auction algorithm as dispatching method for AGVs in a general context. The method implements a distributed decision process with communication among related vehicles and machines for matching multiple tasks with multiple vehicles. Future events are taken into account as well. Outperformance is shown by a simulation study.

Ulusoy et al. [188] address the problem of simultaneous scheduling of machines and a number of identical AGVs in a flexible manufacturing system in order to minimize the makespan. The discussed ideas and the GA may be transferred to problems arising at container terminals, especially the simultaneous scheduling of RMGs and AGVs.

Routing of straddle carriers for loading export containers is discussed by Kim and Kim [96]. The objective is the minimization of total travel distance of straddle carriers in the yard. The routing problem is composed of the container allocation problem – formulated as a transportation problem – and a carrier routing problem with given sequence of yard-bays to be visited by a carrier. The routing problem is solved by a beam search algorithm, that is evaluated in numerical tests. In [103] the number of containers picked up by a straddle carrier at each bay and the sequence of bay visits are determined in order to minimize total travel distance/time of the carrier. The proposed integer programming model is solved by a two-phase procedure. Sequencing of individual containers is not studied.

Böse et al. [15] investigate different dispatching strategies for straddle carriers to gantry cranes in order to reduce vessel's turnaround time at port by maximizing productivity of gantry cranes achieved by an efficient schedule of given straddle carriers. The potential of evolutionary algorithms for solving the considered allocation problem is shown in computational experiments based on real data (without taking stochastic influence into account). Different vehicle assignment strategies are suggested. The first approach suspends the static binding of carriers to gantry cranes using a dynamic strategy where a predetermined number of carriers perform container transports for several gantry cranes (straddle carrier pooling). Depending on the number of loading and discharging processes (structure of the waterside transshipment process), the carriers can be used in a double-cycle mode such that empty runnings are replaced by jobs for other gantry cranes. Two different cases of straddle carrier pooling are considered: semi-dynamic assignment (a fixed number of straddle carriers is assigned to the gantry cranes of one vessel) and dynamic assignment (a fixed number of straddle carriers can perform transports for all gantry cranes). Considering an online optimization setting, numerical results for real data may show that the influence of the number of sequenced containers need not have a large influence when the carriers operate in double cycle mode [128].

Li and Vairaktarakis [121] address the problem of minimizing the (un)loading time for a vessel at a container terminal with fixed number of internal trucks (not shared among different vessels). An optimal algorithm and some heuristic algorithms are developed for the case of a single quay crane. Effectiveness of the heuristics is shown by analysis and computational experiments. The case with multiple identical quay cranes is not solved, but the complexity is analyzed.

Bish et al. [13] focus on the NP-hard vehicle-scheduling-location problem of assigning a yard location to each import container and dispatching vehicles to the containers in order to minimize the total time for unloading a vessel. A heuristic algorithm based on an assignment problem formulation is presented. The algorithm's performance is tested in computational experiments.

Meersmans and Wagelmans [134] consider the problem of integrated scheduling of AGVs, quay cranes and RMGs at automated terminals. They present a branch and bound algorithm and a heuristic beam search algorithm in order to minimize the makespan of the schedule. Near optimal solutions are obtained in a reasonable time. In [133] a beam search algorithm and several dispatching rules are compared in a computational study under different scenarios with similar results. The study also indicates 'that it is more important to base a planning on a long horizon with inaccurate data, than to update the planning often in order to take newly available information into account'. These results are also included in the PhD-thesis of Meersmans [131].

Carrascosa et al. [23] present a multi-agent system architecture to solve the automatic allocation problem in container terminals in order to minimize the ships' docking time. The paper focuses on the management of gantry cranes by a 'transtainer agent'. This work is framed into a project to the integral management of the containers terminal of an actual port. The independence of subsystems obtained from a multi-agent approach is emphasized. (The approach is also described by the same group of authors in [163].)

Kim et al. [91] discuss the load sequencing problem for export containers in terminals with transfer cranes and yard trucks. They introduce various objectives and constraints. A flexible beam search algorithm for minimizing total handling time of cranes and trucks is suggested. Comparison of performance with other approaches shows high quality of the proposed algorithm.

3.3.2 The landside transport. The landside transport is split into the rail operation, the truck operation and the internal transports. A common means of operation is to allocate a given number of vehicles to each sphere of operation appropriate to the workload expected. A more advanced strategy is to pool the vehicles for all these working areas.

Trains are commonly loaded and unloaded by gantry cranes while the transports between the stack and the railhead are performed by straddle carriers, trucks and trailers or similar equipment. Containers are then buffered alongside the railhead or directly on trailers. Sometimes pure straddle carrier operation exists where straddle carriers drive over the wagons to pick up and drop containers.

Operation at the railhead is analogous to the quayside operation. A loading plan describes on which wagon a container has to be placed. The wagon position of a container depends on its destination, type and weight, the maximum load of the wagon and the wagon's position in the train sequence. A loading plan is either produced by the railway company and sent by EDI to the terminal operator or by the terminal operator himself. The aim of the rail operator is to minimize shunting activities during train transport while the aim of the terminal operator is to minimize the number of yard reshuffles, to minimize the crane waiting times and the empty transport distances of cranes and transport vehicles. Optimization at the railhead is facilitated if only a stowage instruction is sent to the terminal operator which indicates the wagon position for container attributes instead of specific positions for each container. The yard situation then can be reflected. Transport and crane activities have to be synchronized to avoid unnecessary crane waiting times or movements. Single- and dual-cycle mode exist depending on whether one or several trains are loaded and unloaded in parallel.

Trucks arrive at the terminal's in-gate where the data of the containers have to be checked and filed into the computer system or actualized in case of pre-advice. Trucks then drive to transition points where the containers are loaded or unloaded by internal equipment. Large container terminals serve some thousand trucks a day. Transition points are located either at the stack crane or inside the yard in case of straddle carrier operation. A truck driving schedule prescribes which points have to be accessed in which sequence. The arrival time of the trucks at the transition points cannot be precisely foreseen, i.e., transport jobs for the internal equipment cannot be released until the truck arrives at the transition point. Because of the permanently changing traffic volume, optimization has to be very flexible and fast. Online optimization is demanded for. Minimizing empty distances and travel times are the objectives of optimization at the truck operation area. Empty distances can be minimized if transports of export containers from the transition point to the yard are combined with transports of import containers from the yard to the interchange point.

Internal movements occur because of different reasons. If sheds or depots for empty containers exist at a terminal additional transports have to be performed: Import containers to be stripped have to be driven to the respective shed while packed containers have to be driven to the export stock. Empty containers are needed at the sheds for stuffing purposes while unpacked containers have to be stored in the empty depot or in the yard. Because of imbalances, empty containers are needed for ship, train and truck loading and have to be transported to the respective yard or transition area. Additional transports occur when containers assigned for a ship's departure are left back because of ship's overbooking. A reorganization of the yard then has to be performed. Characteristic for these types of transports is that sequences of jobs have to be performed. Sometimes time-windows have to be kept.

In general these kinds of transports are not as time critical as those for the ship or truck operation. Therefore, terminals try to execute them at times of less workload. The objective is to minimize empty and loaded travel times.

Literature review

Powell and Carvalho [156] propose a dynamic model for real-time optimization of the flow of flatcars considering constraints for assignment of trailers and containers to a flatcar. A smaller flatcar fleet is possible due to useful information for decision makers provided by the developed global logistics queueing network model.

Steenken [178] investigates methods to optimize the straddle carrier operation at the truck working area. The problem of assigning jobs to straddle carriers is solved with linear assignment procedures combining movements for export und import containers. Steenken et al. [181] deal with the optimization for the rail operation and internal moves. Different algorithmic approaches are used to solve the routing problems, as they can be found in machine scheduling, for solving the travelling salesman problem, the rural postman problem, etc. Both solutions were implemented in a real time environment and resulted in considerable gains of productivity. Results and architecture of implementation are presented in [180].

Kim et al. [97] discuss approaches and decision rules for sequencing pickup and delivery operations for yard cranes and outside trucks, respectively. Their goal is to maximize the service level of trucks by minimizing the turnaround time of them, both for automated and conventional terminals. A dynamic programming model for a static case (all arrivals of trucks are known in advance) is suggested. For a dynamic case (new trucks arrive continuously) a learning-based method for deriving decision rules is proposed besides several heuristic rules. The performances of the methods are compared in a simulation study. The rule of serving the truck with the shortest transfer time (sum of travel time and time for transferring the corresponding container to and from the truck, including occurring rehandling time) shows good, robust performance in various situations, whereas the learned rules outperformed other methods in case of non-uniform distribution of containers' arrival locations. The authors conclude that their single crane based approaches can be extended to the multiple crane case.

Koo et al. [109] present a two-phase fleet sizing and vehicle routing procedure for container ports with several yards. The goal is to find the smallest required fleet size and a route for each vehicle to fulfill all transportation requirements within a static planning horizon. A computational study shows solutions of good quality in comparison with two other existing methods.

3.3.3 Crane transport optimization. Another field of application of optimization methods are the transports of gantry cranes operating in stacks. The transport requirements do not differ from those of the horizontal transport described above. Sequences of jobs have to be calculated and jobs have to be assigned to the respective crane. Commonly the location of a container to be positioned in the stack is calculated by the yard module. This is also true for the containers which have to be reshuffled. Therefore, transport optimization for stack cranes reduces to the same requirements as for the horizontal transport and comparative algorithms can be applied. Priority of jobs have to be taken into account – as is the case for the horizontal transport. The objective of optimization is to minimize the waiting times of the transport vehicles at the stack interfaces and the travel times of the stacking cranes. Because the traffic at the interfaces changes rapidly online optimization is demanded for and job sequences have to be recalculated whenever a new job arises.

Literature review

Due to interdependencies of crane operations and quayside, landside and stack operations, references regarding crane transport optimization may be reviewed in either section as we have done above; see, e.g., [155, 92–94, 12, 97].

Kim and Kim [102] present a routing algorithm for a single gantry crane loading export containers out of the stack onto waiting vehicles. The objective is to minimize the crane's total transfer time including set-up and travel times. The model's solution determines the sequence of bay visits for pick-up operations and the number of containers to be picked up at each bay simultaneously. The developed algorithm is named 'efficient' and shows solutions to problems of practical size 'within seconds'. In a more detailed paper [95] the same algorithm is used for solving the MIP of a 'practical problem of a moderate size'. The load sequence of individual containers within a specific bay remains undetermined.

Kim and Kim [104] extend their problem shown in [102] and [95] to general yard-side equipment, such as gantry cranes or straddle carriers. Experiments show that the proposed beam search algorithm outperforms a GA. The pick-up sequence for individual containers in a bay remains undetermined as in [95].

Lin [124] deals with the problem of scheduling movements of RTGs among different storage blocks in order to balance the workload and minimize the total unfinished workload at the end of each time period. The complexity of the MIP is analyzed. Besides the Lagrangian decomposition solution procedure, a new approach ('successive piecewise-linear approximation') is discussed. This solution method can be applied to large size problem instances since computational experiments show efficiency and effectiveness. The same results are published later by Cheung et al. [30].

Narasimhan and Palekar [144] consider the minimization of a yard gantry crane's handling time for executing a given load plan with a given bay plan for export containers. An exact branch-and-bound based algorithm and a heuristic method are developed and tested by computational experiments on randomly generated problem instances. Besides the algorithmic approaches the authors provide a mathematical programming formulation and also consider some complexity issues.

Zhang et al. [212] describe the dynamic RTG deployment problem with forecasted workload per block per planning period (4 hours). The objective is to find times and routes of RTG movements among blocks with minimization of total delayed workload in the yard. For safety reasons a maximum of two RTGs per block is allowed. Only one transfer of a RTG in and out of a block can occur. The problem is formulated as a MIP model and is solved by a modified Lagrangian relaxation with excellent results.

A similar group of authors [125] solve this RTG deployment problem in a different way. The size of the problem is reduced by sorting blocks into categories like 'sink block' (needs and can take additional help), 'source block' (can spend capacity of RTGs) and 'neither block' (needs help but cannot take help, because two RTGs currently work in the block, or it does not need help). Neither blocks are excluded in the model. A pre-sort step identifies eligible RTGs and sink blocks, a following deployment step (formulated as MIP model) results in the optimal RTGs' deployment plan for source and sink blocks. The approach is tested with a set of real operation data (Hong Kong). Results demonstrate 'an excellent capability and potential of the model in minimizing the crane workload overflow'.

Routing and/or scheduling algorithms for multiple cranes are hardly addressed in literature. A simulation study on operational rules for Double-RMGs is shortly discussed by Kim et al. [101]. Crane dispatching rules with and without different roles for the different cranes and sequencing roles are tested. A second simulation study focuses on determining the storage location of arriving containers.

In [46] we consider the case of Double-RMGs and develop possible solution approaches for specific sequencing and scheduling problems in order to take advantage of using two cranes – which can overtake each other – instead of one crane and increase the terminal's throughput.

3.4 Simulation systems

In recent years, simulation has become an important tool to improve terminal operation and performance. Three types of simulation can be distinguished: strategical, operational and tactical simulation.

Strategical simulation is applied to study and compare different types of terminal layout and handling equipment in respect to efficiency and costs expected. It is mainly used if new terminals are planned or the layout or the equipment of existing terminals has to be altered. Strategical simulation systems allow for easy design of different terminal layouts and employment of different types of handling equipment. The chief goal of strategical simulation is to decide on terminal layout and handling equipment which promises high performance and low costs. To match reality, simulation systems allow to design realistic scenarios or to import data of existing terminals.

Operational simulation is applied to test different kinds of terminal logistics and optimization methods. It has achieved growing acceptance at least at large terminals. Terminal operation and logistics at large terminals are already very complex and the effect of alternative logistics or optimization methods has to be tested with objective methods. Therefore, optimization methods are tested in a simulation environment before they are implemented in real terminal control and steering systems.

Tactical simulation means integration of simulation systems into the terminal's operation system. Variants of operation shall be simulated parallel to the operation and advices for handling alternatives shall be given especially if disturbances occur in real operation. Real data of operation then have to be imported and analyzed synchronously to the operation. Because of this ambitious requirement, tactical simulation is seldom or only partially installed at container terminals.

Literature review

Simulation results provide valuable decision support information for terminal planners, operators, and managers (see, e.g., [56,70,107,139,77,186,135,145,185, 169,170]). Therefore, various groups have worked in simulation systems for container terminals; see, e.g., [54,129,166] or work based on [15].

A group of authors [42,114,150,149,38,151,193,41,39,44,40,43,152,194] demonstrate the usage of simulation for development of an automated container terminal by example of Rotterdam. The performance of (sub)systems with AGVs, ALVs, multi-trailers/manned trucks is tested. Valuable insights can be obtained about optimal stack height, optimal number of AGVs and other variables.

Bruzzone et al. [20] demonstrate effectiveness of simulation for supporting complex container port management. Presented application examples and experimental results show benefits in reusability, flexibility, modelling time, and performance estimation of a simulation approach.

Gehlsen and Page [55] present a framework (written in Java) for simulation projects including heuristic optimization procedures (GA) in a parallel distributed environment.

Liu et al. [126] use future demand scenarios to design the characteristics of different terminals in terms of configuration, equipment and operations. A microscopic simulation model is developed and used to investigate each terminal system for the same operational scenario and evaluate its performance. Moreover, a cost model is developed evaluating the cost associated with each terminal concept. Results indicate that automation could improve the performance of conventional terminals substantially at a considerably lower cost.

Nam et al. [143] examine optimal number of berths and quay cranes for a terminal in Busan (Korea). Different operational patterns are represented in four scenarios for performance evaluation by simulation experiments. Results reveal that 'sharing quay cranes with adjacent berths can increase productivity, and that the more berths per operator, the higher the productivity achieved'. Terminal development and operation policy implications are considered. Topics for further studies are given.

Shabayek and Yeung [174] describe a simulation model to simulate the Kwai Chung container terminals in Hong Kong. They investigate accuracy of prediction of actual terminal operations and conclude with good results.

Kia et al. [88] describe the role of simulation for evaluating the performance of a terminal's equipment and capacity. Performance criteria and interesting model parameters are discussed.

Hartmann [69] develops an approach for generating realistic scenario data of port container terminals as input for simulation models and for test of optimization algorithms. A scenario consists of data concerning arrivals of ships, trains and trucks within a time horizon and information about containers being delivered or picked up. Users can control various typical parameters.

Yun and Choi [208, 209] propose an object-oriented simulation model for analysis of container terminals consisting of gate, container yard, berth and equipment like transfer cranes, gantry cranes, trailers, and yard tractors. Output of resource statistics can be used for analysis of capacity and operational efficiency of an existing container terminal.

Saanen et al. [169, 168] use simulation models to account for cost values of different types of equipment to be installed at a terminal. Examples are based on the layout of terminals in Hamburg and Rotterdam where using straddle carriers versus AGVs or ALVs is compared with respect to productivity values. One of the major results is that at a certain point adding further equipment can no longer increase productivity (or even lead to decreasing productivity, e.g., if too many vehicles are blocking each other). Similar results are presented by Steenken [180].

Vis and Harika [195] study the performance of AGVs and ALVs. A simulation experiment shows effects on unloading times of a vessel using different equipment. A sensitivity analysis is performed. Results show that the optimal type of equipment and fleet size depend on the terminal's design and technical aspects of quay cranes. Investigations regarding the number of AGVs can also be found in [197].

Analytical approaches that use modern queuing techniques instead of discrete-event simulation in order to evaluate terminal allocation and layout planning problems can be found in, e.g., [110, 71].

4 Conclusions and outlook

The increasing number of publications in the last decade indicates the importance of operations research methods in the field of optimizing logistic operations at a container terminal. Until now the focus is not on optimizing the transport chain as a whole but on optimizing several separate parts of the chain. A tendency from relatively theoretical publications to more practical ones can be seen. Furthermore, operations research methods are applied more and more in real terminals. One of the drivers in this respect is an increased availability of modern information and communication technology that only allows the application of these methods.

High operating costs for ships and container terminals and also high capitalization of ships, containers and port equipment demand a reduction of unproductive times at port. Therefore, the potential for cost savings is high. A key to efficiency is the automation of in-yard transportation, storing and stacking to increase the terminal throughput and decrease ship turnaround time at the terminal. Due to severe competition the increasing pressure on container terminals to cut costs of operation and to increase productivity enforces the usage of optimization methods.

At terminals which already apply operations research methods to optimize transport and stacking processes, the need for 'integrated' optimization is becoming more and more relevant. The transport process between quay and yard or between hinterland and yard is broken into separate phases because different types of equipment are engaged for the whole transport chain. Additionally, containers have to be buffered in respective handshaking areas. In practice, optimization commonly is restricted to the partial phases of the whole transport or to rules (heuristics) for the handshaking. Thus not all sources of optimization are exploited, but high performing operations ask for it. An example shall be given which explains the problem: A solution for the crane split can allow that two (or more) cranes operate very close together at a ship. This can be optimal for the crane operation, but it will not be for the horizontal transport because then the cranes are not easily accessible by the vehicles and congestions are provoked. An integrated optimization of both the crane split and the horizontal transport is demanded for. Similar problems can be found for every transport or stowage process at container terminals.

Up to now there are only a few studies on such 'integrated problems' – e.g., in [134] or in [74,73], presenting a multi-agent system approach with several agents (agents for ship, berth, yard, and gate and utility agents for quay crane, gantry crane and transport) – although they are important for enhanced terminal performance. Therefore, 'integrated optimization' should be a field of increased investigation.

Besides the major research needs regarding the topics online optimization as well as integration, additional topics may become important. Operations research approaches for container terminals usually apply simulation when it comes to consideration of stochasticity. However, the area of stochastic optimization and scenario based planning may be applied, too. For instance, vehicle routing problems with time windows and stochastic travel times or with stochastic customers (see, e.g., [206,11]) may be important areas worth considering for container terminal operations.

Finally, a new challenge is given by advanced security issues. They will imply more versatile planning tools for optimization. Usage of techniques like, e.g., transponders and certain security procedures and their impact on the logistic chain have to be taken into account.

References

1. Alicke K (1999) Modellierung und Optimierung von mehrstufigen Umschlagsystemen. PhD thesis, Universität Karlsruhe
2. Alicke K (2002) Modeling and optimization of the intermodal terminal Mega Hub. OR Spectrum 24: 1–17
3. Anonymous. Container Traffic of the top 100 Ports, 1970–2000. http://www.ci-online.co.uk – last check of address: Oct 13, 2003
4. Anonymous (2003). The Journal of Commerce, Jul 14–20
5. Asef-Vaziri A, Cadavid M, Dougherty E (2003) A combined container handling system in maritime terminals. Paper presented at Annual Meeting INFORMS Atlanta 2003, Oct 19–22, California State University, Northridge
6. Asef-Vaziri A, Khoshnevis B, Parsaei H (2003) Potentials for ASRS and AGVS in Maritime Container Terminals. Working paper, University of Houston
7. Avriel M, Penn M (1993) Exact and approximate solutions of the container ship stowage problem. Computers and Industrial Engineering 25(1–4): 271–274
8. Avriel M, Penn M, Shpirer N (2000) Container ship stowage problem: complexity and connection to the coloring of circle graphs. Discrete Applied Mathematics 103(1–3): 271–279
9. Avriel M, Penn M, Shpirer N, Witteboon S (1998) Stowage planning for container ships to reduce the number of shifts. Annals of Operations Research 76: 55–71

10. Ball M, Magazine M (1988) Sequencing of insertions in printed circuit board assembly. Operations Research 36: 192–201
11. Bent R W, van Hentenryck P (2002) Scenario-based planning for partially dynamic vehicle routing with stochastic customers. Technical Report, Department of Computer Science, Brown University, Providence, RI
12. Bish E K (2003) A multiple-crane-constrained scheduling problem in a container terminal. European Journal of Operational Research 144: 83–107
13. Bish E K, Leong T-Y, Li C-L, Ng J W C, Simchi-Levi D (2001) Analysis of a new vehicle scheduling and location problem. Naval Research Logistics 48: 363–385
14. Bodin L, Golden B, Assad A, Ball M (1983) Routing and scheduling of vehicles and crews. Computers & Operations Research 10: 63–211
15. Böse J, Reiners T, Steenken D, Voß S (2000) Vehicle dispatching at seaport container terminals using evolutionary algorithms. In: Sprague R H (ed) Proceedings of the 33rd Annual Hawaii International Conference on System Sciences, DTM-IT, pp 1–10. IEEE, Piscataway
16. Boyes J R C (ed) (1994) Containerisation international yearbook, ranking of containerports of the world. National Magazin, London
17. Boyes J R C (ed) (1997) Containerisation international yearbook, ranking of containerports of the world. National Magazin, London
18. Bruno G, Ghiani G, Improta G (2000) Dynamic positioning of idle automated guided vehicles. Journal of Intelligent Manufacturing 11: 209–215
19. Bruzzone A, Signorile R (1998) Simulation and genetic algorithms for ship planning and shipyard layout. Simulation 71: 74–83
20. Bruzzone A G, Giribone P, Revetria R (1999) Operative requirements and advances for the new generation simulators in multimodal container terminals. In: Farrington P A, Nembhard H B, Sturrock D T, Evans G W (eds) Proceedings of the 31st Conference on Winter Simulation, vol 2, pp 1243–1252. ACM Press, New York
21. Cao B, Uebe G (1993) An algorithm for solving capacitated multicommodity p-median transportation problems. Journal of the Operational Research Society 44: 259–269
22. Cao B, Uebe G (1995) Solving transportation problems with nonlinear side constraints with tabu search. Computers & Operations Research 22: 593–603
23. Carrascosa C, Rebollo M, Julian V, Botti V (2001) A MAS approach for port container terminal management: the transtainer agent. In: Actas de SCI'01, pp 1–5. International Institute of Informatics and Systemics, Orlando, FL
24. de Castilho B, Daganzo C F (1993) Handling strategies for import containers at marine terminals. Transportation Research-B 27: 151–166
25. Chan S H (2001) Dynamic AGV-container job deployment strategy. Master of science, National University of Singapore
26. Chen C S, Lee S M, Shen Q S (1995) An analytical model for the container loading problem. European Journal of Operational Research 80: 68–76
27. Chen T (1999) Yard operations in the container terminal – a study in the 'unproductive moves'. Maritime Policy and Management 26: 27–38
28. Cheu R L, Chew E P, Wee C L (2003) Estimating total distance for hauling import and export containers. Journal of Transport Engineering 129: 292–299
29. Cheung R K, Chen C Y (1998) A two-stage stochastic network model and solution methods for the dynamic empty container allocation problem. Transportation Science 32(2): 142–162
30. Cheung R K, Li C-L, Lin W (2002) Interblock crane deployment in container terminals. Transportation Science 36(1): 79–93
31. Crainic T G, Gendreau M, Dejax P (1993) Dynamic and stochastic models for the allocation of empty containers. Operations Research 41: 102–126

32. Daganzo C F (1989) The crane scheduling problem. Transportation Research-B 23B(3): 159–175
33. Davies A P, Bischoff E E (1999) Weight distribution considerations in container loading. European Journal of Operational Research 114: 509–527
34. Dell'Olmo P, Lulli G (2003) Planning activities in a network of logistic platforms with shared resources. Technical Report, Department of Statistics, Probability and Applied Statistics, University of Rome
35. Desrochers M, Lenstra J K, Savelsbergh M W P (1990) A classification scheme for vehicle routing and scheduling problems. European Journal of Operational Research 46: 322–332
36. Dror M (ed) (2000) Arc routing. Kluwer, Boston
37. Dubrovsky O, Levitin G, Penn M (2002) A genetic algorithm with a compact solution encoding for the container ship stowage problem. Journal of Heuristics 8: 585–599
38. Duinkerken M B, Evers J J M, Ottjes J A (1999) TRACES: Traffic control engineering system – a case-study on container terminal automation. In: Proceedings of the 1999 Summer Computer Simulation Conference (SCSC 1999), July 1999, Chicago [SCS], http://www.ocp.tudelft.nl/tt/users/duinkerk/papers/chi9907b.pdf – last check of address: July 30, 2003
39. Duinkerken M B, Evers J J M, Ottjes J A (2001) A simulation model for integrating quay transport and stacking policies on automated container terminals. In: Proceedings of the 15th European Simulation Multiconference (ESM2001), June 2001, Prague [SCS], http://www.ocp.tudelft.nl/tt/users/duinkerk/papers/pra0106a.pdf – last check of address: July 30, 2003
40. Duinkerken M B, Evers J J M, Ottjes J A (2002) Improving quay transport on automated container terminals. In: Proceedings of the IASTED International Conference Applied Simulation and Modelling (ASM 2002), Crete, June 2002, http://www.ocp.tudelft.nl/tt/users/duinkerk/papers/cre0206.pdf – last check of address: July 30, 2003
41. Duinkerken M B, Ottjes J A (2000) A simulation model for automated container terminals. In: Proceedings of the Business and Industry Simulation Symposium (ASTC 2000), April 2000, Washington, DC [SCS], pp 134–139, http://www.ocp.tudelft.nl/tt/users/duinkerk/papers/was0004a.pdf – last check of address: July 30, 2003
42. Duinkerken M B, Ottjes J A, Evers J J M, Kurstjens S, Dekker R, Dellaert N (1996) Simulation studies on inter terminal transport at the Maasvlakte. In: Proceedings 2nd TRAIL PhD Congress 1996 'Defence or attack', May 1996, Rotterdam [TRAIL], http://www.ocp.tudelft.nl/tt/users/duinkerk/papers/rot9605a.pdf – last check of address: July 30, 2003
43. Duinkerken M B, Ottjes J A, Lodewijks G (2002) The application of distributed simulation in TOMAS: redesigning a complex transportation model. In: Yücesan E, Chen C-H, Snowdon J L, Charnes J M (eds) Proceedings of the 2002 Winter Simulation Conference (WSC 2002), December 2002, San Diego, http://www.ocp.tudelft.nl/tt/users/duinkerk/papers/san0212.pdf – last check of address: July 30, 2003
44. Duinkerken M B, Terstegge M J (2001) TRAVIS: an engineering tool to animate and validate AGV-systems. In: Proceedings of the International Conference on Simulation and Multimedia in Engineering Education (WMC 2001), January 2001, Phoenix [SCS], http://www.ocp.tudelft.nl/tt/users/duinkerk/papers/pho0101.pdf – last check of address: July 30, 2003

45. Edmond E D, Maggs R P (1978) How useful are queue models in port investment decisions for container berths? Journal of the Operational Research Society 29: 741–750
46. Eisenberg R, Stahlbock R, Voß S, Steenken D (2003) Sequencing and scheduling of movements in an automated container yard using double rail-mounted gantry cranes. Working paper, University of Hamburg
47. Eley M (2003) A bottleneck assignment approach to the multiple container loading problem. OR Spectrum 25: 54–60
48. Evers J J M, Koppers S A J (1996) Automated guided vehicle traffic control at a container terminal. Transportation Research-A 30: 21–34
49. Fagerholt K (2000) Evaluating the trade-off between the level of costumer service and transportation costs in a ship scheduling problem. Maritime Policy and Management 27(2): 145–153
50. Fung M K (2002) Forecasting Hong Kong's container throughput: an error-correction model. Journal of Forecasting 21: 69–80
51. Gademann A J R M, van de Velde S L (2000) Positioning automated guided vehicles in a loop layout. European Journal of Operational Research 127(3): 565–573
52. Gambardella L M, Mastrolilli M, Rizzoli A E, Zaffalon M (2001) An optimization methodology for intermodal terminal management. Journal of Intelligent Manufacturing 12: 521–534
53. Gambardella L M, Rizzoli A E (2000) The role of simulation and optimisation in intermodal container terminals. Working paper, Istituto Dalle Molle di Studi sull'Intelligenza Artificiale, Manno-Lugano, Switzerland, http://www.idsia.ch/luca/abstracts/papers/ess2000.pdf – last check of address: May 15, 2003
54. Gambardella L M, Rizzoli A E, Zaffalon M (1998) Simulation and planning of an intermodal container terminal. Simulation 71(2): 107–116
55. Gehlsen B, Page B (2001) A framework for distributed simulation optimization. In: Proceedings of the 33rd Conference on Winter Simulation, pp 508–514. IEEE Computer Society, Washington, DC
56. Gibson R, Carpenter B, Seeburger S (1992) A flexible port traffic planning model. In: Proceedings of the 1992 Winter Simulation Conference, pp 1296–1306
57. Giemsch P, Jellinghaus A (2003). Organization models for the containership stowage problem. Paper presented at the annual international conference of the German Operations Research Society (OR 2003), Sep 3–5, Heidelberg, http://www.andor.uni-karlsruhe.de/fak/inst/andor/a10_www/Preprint/PaperGOR2003.pdf – last check of address: Oct 15, 2003
58. Grunow M, Günther H-O, Lehmann M (2004) Dispatching multi-load AGVs in highly automated seaport container terminals. OR Spectrum 26(2): to appear
59. Grunow M, Günther H-O, Lehmann M (2004) Online- versus Offline-Einsatzplanung von fahrerlosen Transportsystemen in Containerhäfen. In: Spengler T, Voß S, Kopfer H (eds) Logistik Management, pp 399–410. Springer, Berlin Heidelberg New York
60. Guan Y, Cheung R K (2004) The berth allocation problem: models and solution methods. OR Spectrum 26: 75–92
61. Guan Y, Xiao W-Q, Cheung R K, Li C-L (2002) A multiprocessor task scheduling model for berth allocation: heuristic and worst case analysis. Operations Research Letters 30: 343–350
62. Gupta Y P, Somers T M (1992) The measurement of manufacturing flexibility. European Journal of Operational Research 60: 166–182
63. Gutenschwager K, Böse J, Voß S (2003) Effiziente Prozesse im kombinierten Verkehr – Ein neuer Lösungsansatz zur Disposition von Portalkränen. Logistik Management 5(1): 62–73

64. Gutenschwager K, Niklaus C, Voß S (2004) Dispatching of an electric monorail system: applying meta-heuristics to an online pickup and delivery problem. Transportation Science (to appear)
65. Gutin G, Punnen A P (eds) (2002) The traveling salesman problem and its variations. Kluwer, Boston
66. Haghani A, Kaisar E I (2001) A model for designing container loading plans for containerships. Working paper, University of Maryland, presented at Transportation Research Board 2001 Annual Meeting
67. Hanafi S, Jesus J, Semet F (2003) The container assignment problem: a case study on the port of Lille. Working paper, University of Valenciennes, presented at ODYSSEUS 2003
68. Hartmann S (2004) A general framework for scheduling equipment and manpower on container terminals. OR Spectrum 26: 51–74
69. Hartmann S (2004) Generating scenarios for simulation and optimization of container terminal logistics. OR Spectrum 26(2): to appear
70. Hayuth Y, Pollatschek M A, Roll Y (1994) Building a port simulator. Simulation 63: 179–189
71. van Hee K M, Wijbrands R J (1988) Decision support system for container terminal planning. European Journal of Operational Research 34: 262–272
72. van der Heijden M, Ebben M, Gademann N, van Harten A (2002) Scheduling vehicles in automated transportation systems: algorithms and case study. OR Spectrum 24: 31–58
73. Henesey L, Törnquist J (2002) Enemy at the gates: introduction of multi-agents in a terminal information community. In: Proceedings of Ports and Marinas, Rhodos, Greece, http://www.assert.bth.se/publications/HeneseyTornquist.pdf – last check of address: July 30, 2003
74. Henesey L, Wernstedt F, Davidsson P (2002) A market based approach to container port terminal management. In: Proceedings of the 15th European Conference on Artificial Intelligence, Workshop – Agent Technologies in Logistics, Lyon, France, http://www.assert.bth.se/publications/Marketv1.pdf – last check of address: July 30, 2003
75. Hoffarth L, Voß S (1994) Liegeplatzdisposition auf einem Containerterminal – Ansätze zur Entwicklung eines entscheidungsunterstützenden Systems. In: Dyckhoff H, Derigs U, Salomon M, Tijms H (eds) Operations Research Proceedings 1993, pp 89–95. Springer, Berlin Heidelberg New York
76. Holguín-Veras J, Jara-Díaz S (1999) Optimal pricing for priority service and space allocation in container ports. Transportation Research-B 33: 81–106
77. Holguín-Veras J, Walton C (1996) On the development of a computer system to simulate port operations considering priorities. In: Proceedings of the 28th Conference on Winter Simulation, pp 1471–1478. ACM Press, New York
78. Hulten L A R (1997) Container logistics and its management. PhD thesis, Chalmers University of Technology: Department of Transportation and Logistics
79. ICC (International Chamber of Commerce) (2000) Incoterms 2000: ICC official rules for the interpretation of trade terms. Paris
80. Imai A, Nagaiwa K, Tat C W (1997) Efficient planning of berth allocation for container terminals in Asia. Journal of Advanced Transportation 31: 75–94
81. Imai A, Nishimura E, Papadimitriou S (2001) The dynamic berth allocation problem for a container port. Transportation Research-B 35(4): 401–417
82. Imai A, Nishimura E, Papadimitriou S (2003) Berth allocation with service priority. Transportation Research-B 37(5): 437–457

83. Ioannou P, Chassiakos A, Zhang J, Kanaris A, Unglaub R (2002) Automated container transport system between inland port and terminals. Project Report, University of Southern California, http://www.usc.edu/dept/ee/catt/2003/jianlong/02%20METRANS%20Final%20Report.pdf – last check of address: Oct 30, 2003
84. Ioannou P A, Jula H, Liu C-I, Vukadinovic K, Pourmohammadi H, Dougherty Jr E (2001) Advanced material handling: automated guided vehicles in agile ports. Final Report, University of Southern California, http://www.usc.edu/dept/ee/catt/2002/jula/Marine/FinalReport_CCDoTT_981.pdf – last check of address: Oct 30, 2003
85. Ioannou P A, Kosmatopoulos E B, Jula H, Collinge A, Liu C-I, Asef-Vaziri A, Dougherty Jr E (2000) Cargo handling technologies. Final Report, University of Southern California, http://www.usc.edu/dept/ee/catt/2002/jula/Marine/FinalReport_CCDoTT_97.pdf – last check of address: Oct 30, 2003
86. Kang J-G, Kim Y-D (2002) Stowage planning in maritime container transportation. Journal of the Operational Research Society 53: 415–426
87. Khoshnevis B, Asef-Vaziri A (2000) 3D Virtual and physical simulation of automated container terminal and analysis of impact on in land transportation. Research Report, University of Southern California, http://www.metrans.org/Research/Final_Report/99-14_Final.pdf – last check of address: Oct 30, 2003
88. Kia M, Shayan E, Ghotb F (2002) Investigation of port capacity under a new approach by computer simulation. Computers & Industrial Engineering 42: 533–540
89. Kim K H (1997) Evaluation of the number of rehandles in container yards. Computers & Industrial Engineering 32: 701–711
90. Kim K H, Bae J W (1998) Re-marshaling export containers in port container terminals. Computers & Industrial Engineering 35: 655–658
91. Kim K H, Kang J S, Ryu K R (2004) A beam search algorithm for the load sequencing of outbound containers in port container terminals. OR Spectrum 26: 93–116
92. Kim K H, Kim H B (1998) The optimal determination of the space requirement and the number of transfer cranes for import containers. Computers & Industrial Engineering 35: 427–430
93. Kim K H, Kim H B (1999) Segregating space allocation models for container inventories in port container terminals. International Journal of Production Economics 59: 415–423
94. Kim K H, Kim H B (2002) The optimal sizing of the storage space and handling facilities for import containers. Transportation Research-B 36: 821–835
95. Kim K H, Kim K Y (1999) An optimal routing algorithm for a transfer crane in port container terminals. Transportation Science 33(1): 17–33
96. Kim K H, Kim K Y (1999) Routing straddle carriers for the loading operation of containers using a beam search algorithm. Computers & Industrial Engineering 36: 106–136
97. Kim K H, Lee K M, Hwang H (2003) Sequencing delivery and receiving operations for yard cranes in port container terminals. International Journal of Production Economics 84(3): 283–292
98. Kim K H, Moon K C (2003) Berth scheduling by simulated annealing. Transportation Research-B 37(6): 541–560
99. Kim K H, Park K T (2003) A note on a dynamic space-allocation method for outbound containers. European Journal of Operational Research 148(1): 92–101
100. Kim K H, Park Y M, Ryu K-R (2000) Deriving decision rules to locate export containers in container yards. European Journal of Operational Research 124: 89–101

101. Kim K H, Wang S J, Park Y-M, Yang C-H, Bae J W (2002). A simulation study on operation rules for automated container yards. Presentation/Proc. of the 7th Annual International Conference on Industrial Engineering
102. Kim K Y, Kim K H (1997) A routing algorithm for a single transfer crane to load export containers onto a containership. Computers & Industrial Engineering 33: 673–676
103. Kim K Y, Kim K H (1999) A routing algorithm for a single straddle carrier to load export containers onto a containership. International Journal of Production Economics 59: 425–433
104. Kim K Y, Kim K H (2003) Heuristic algorithms for routing yard-side equipment for minimizing loading times in container terminals. Naval Research Logistics 50: 498–514
105. Ko K-C, Egbelu P J (2003) Unidirectional AGV guidepath network design: a heuristic algorithm. International Journal of Production Research 41: 2325–2343
106. Koch J (1997) Die Entwicklung des Kombinierten Verkehrs. Gabler, Deutscher Universitäts-Verlag, Wiesbaden
107. Koh P-H, Goh J L K, Ng H-S, Ng H-C (1994) Using simulation to preview plans of a container port operations. In: Proceedings of the 26th Conference on Winter Simulation, pp 1109–1115. Society for Computer Simulation International, San Diego, CA
108. Konings J W (1996) Integrated centres for the transshipment, storage, collection and distribution of goods : a survey of the possibilities for a high-quality intermodal transport concept. Transport Policy 3(1–2): 3–11
109. Koo P-H, Lee W S, Jang D W (2004) Fleet sizing and vehicle routing for container transportation in a static environment. OR Spectrum 26(2): to appear
110. Kozan E (1997) Comparison of analytical and simulation planning models of seaport container terminals. Transportation Planning and Technology 20: 235–248
111. Kozan E (1997) Increasing the operational efficiency of container terminals in Australia. Journal of the Operational Research Society 48: 151–161
112. Kozan E (2000) Optimising container transfers at multimodal terminals. Mathematical and Computer Modelling 31: 235–243
113. Kozan E, Preston P (1999) Genetic algorithms to schedule container transfers at multimodal terminals. International Transactions of Operational Research 6: 311–329
114. Kurstjens S T G L, Dekker R, Dellaert N P, Duinkerken M B, Ottjes J A, Evers J J M (1996) Planning of inter terminal transport at the Maasvlakte. In: Proceedings 2nd TRAIL PhD Congress 1996 'Defence or attack', May 1996, Rotterdam [TRAIL], http://www.ocp.tudelft.nl/tt/users/duinkerk/papers/rot9605b.pdf – last check of address: July 30, 2003
115. Lai K K, Shih K (1992) A study of container berth allocation. Journal of Advanced Transportation 26: 45–60
116. Lawler E L, Lenstra J K, Rinnoy Kan A H G, Shmoys D B (eds) (1985) The traveling salesman problem – a guided tour of combinatorial optimization. Wiley, Chichester
117. Legato P, Mazza R M (2001) Berth planning and resources optimisation at a container terminal via discrete event simulation. European Journal of Operational Research 133: 537–547
118. Lemper B (2003). Containerschifffahrt und Welthandel – eine 'Symbiose'. http://www.gdv.de/download/VortragLemper,pdf – last check of address: July 15, 2003
119. Leong C Y (2001) Simulation study of dynamic AGV-container job deployment scheme. Master of science, National University of Singapore
120. Li C-L, Cai X, Lee C-Y (1998) Scheduling with multiple-job-on-one-processor pattern. IIE Transactions 30: 433–445

121. Li C-L, Vairaktarakis G L (2001) Loading and unloading operations in container terminals. Technical Memorandum 745, Case Western Reserve University, Weatherhead School of Management, Department of Operations, http://www.weatherhead.cwru.edu/orom/research/technicalReports/Technical%20Memorandum%20Number%20745.pdf – last check of address: July 30, 2003
122. Lim A (1998) The berth planning problem. Operations Research Letters 22: 105–110
123. Lim J K, Kim K H, Yoshimoto K, Lee J H (2003) A dispatching method for automated guided vehicles by using a bidding concept. OR Spectrum 25: 25–44
124. Lin W (2000) On dynamic crane deployment in container terminals. Master of philosophy in industrial engineering and engineering management, University of Science and Technology, http://www.isye.gatech.edu/ linwq/mthesis.pdf – last check of address: July 30, 2003
125. Linn R J, Liu J-y, Wan Y-w, Zhang C, Murty K G (2003) Rubber tired gantry crane deployment for container yard operation. Computers & Industrial Engineering 45: 429–442
126. Liu C-I, Jula H, Ioannou P A (2002) Design, simulation, and evaluation of automated container terminals. IEEE Transactions on Intelligent Transportation Systems 3(1): 12–26
127. Mahoney J H (1985) Intermodal freight transportation. Eno Foundation for Transportation, Westport, CN
128. Martinssen D, Steenken D, Wölfer F, Reiners T, Voß S (2001) Einsatz bioanaloger Verfahren bei der Optimierung des wasserseitigen Containerumschlags. In: Sebastian H, Grünert T (eds) Logistik management – supply chain management und e-business, pp 377–388. Teubner, Stuttgart
129. Mastrolilli M, Fornara N, Gambardella L M, Rizzoli A E, Zaffalon M (1998) Simulation for policy evaluation, planning and decision support in an intermodal container terminal. In: Merkuryev Y, Bruzzone A, Novitsky L (eds) Proceedings of the International Workshop 'Modeling and Simulation within a Maritime Environment', Sep 6–8, pp 33–38. Society for Computer Simulation International, Riga, Latvia
130. Mattfeld D (2003). The management of transshipment terminals. Habilitation thesis, Universität Bremen
131. Meersmans P J M (2002) Optimization of container handling systems. Thela Thesis, Amsterdam
132. Meersmans P J M, Dekker R (2001) Operations research supports container handling. Technical Report EI 2001-22, Erasmus University Rotterdam, Econometric Institute, http://www.eur.nl/WebDOC/doc/econometrie/feweco20011102151222.pdf – last check of address: Oct 30, 2003
133. Meersmans P J M, Wagelmans A P M (2001) Dynamic scheduling of handling equipment at automated container terminals. Technical Report EI 2001-33, Erasmus University Rotterdam, Econometric Institute, http://www.eur.nl/WebDOC/doc/econometrie/feweco20011128140514.pdf – last check of address: Oct 30, 2003
134. Meersmans P J M, Wagelmans A P M (2001) Effective algorithms for integrated scheduling of handling equipment at automated container terminals. Technical Report EI 2001-19, Erasmus University Rotterdam, Econometric Institute, http://www.eur.nl/WebDOC/doc/econometrie/feweco20010621101333.pdf – last check of address: Oct 30, 2003
135. Merkuryev Y, Tolujew J, Blümel E, Novitsky L, Ginters E, Viktorova E, Merkuryeva G, Pronins J (1998) A modelling and simulation methodology for managing the Riga harbour container terminal. Simulation 71: 84–95

136. Moon K C (2000) A mathematical model and a heuristic algorithm for berth planning. PhD thesis, Pusan National University, http://logistics.ie.pusan.ac.kr/bk21/pdf/kcMoon.pdf – last check of address: July 30, 2003
137. Moorthy R L, Hock-Guan W (2000) Deadlock prediction and avoidance in an AGV system. Master of science, Sri Ramakrishna Engineering College, National University of Singapore
138. Moorthy R L, Hock-Guan W, Wing-Cheong N, Chung-Piaw T (2003) Cyclic deadlock prediction and avoidance for zone-controlled AGV system. International Journal of Production Economics 83(3): 309–324
139. Mosca R, Giribone P, Bruzzone A (1994) Simulation and automatic parking in a training system for container terminal yard management. In: Proceedings of ITEC, pp 65–72
140. Muller G (1995) Intermodal freight transportation, 3rd edn. Eno Foundation for Transportation, Westport, CN
141. Murty K G, Liu J, Wan Y-w, Linn R J (2003) A DSS (Decision Support System) for operations in a container terminal. Working paper, University of Michigan, USA, http://www-personal.engin.umich.edu/~murty/terminal-10.pdf – last check of address: Oct 13, 2003
142. Nam K-C, Ha W-I (2001) Evaluation of handling systems for container terminals. Journal of Waterway, Port, Coastal and Ocean Engineering 127(3): 171–175
143. Nam K-C, Kwak K-S, Yu M-S (2002) Simulation study of container terminal performance. Journal of Waterway, Port, Coastal and Ocean Engineering 128(3): 126–132
144. Narasimhan A, Palekar U S (2002) Analysis and algorithms for the transtainer routing problem in container port operations. Transportation Science 36(1): 63–78
145. Nevins M R, Macal C M, Love R J, Bragen M J (1998) Simulation, animation and visualization of seaport operations. Simulation 71: 96–106
146. Nishimura E, Imai A, Papadimitriou S (2001) Berth allocation planning in the public berth system by genetic algorithms. European Journal of Operational Research 131: 282–292
147. NKM Noell Special Cranes (2003) Crane construction. http://www.nkmnoell.com/ – last check of address: July 30, 2003
148. Orient Overseas Container Line – World's Top 50 Container Ports 2002 vs 2001. http://www.oocl.com/trade_news/20030715.htm – last check of address: Oct 13, 2003
149. Ottjes J A, Duinkerken M B, Evers J J M, Dekker R (1996) Robotised inter terminal transport of containers: a simulation study at the Rotterdam port area. In: Proceedings of the 8th European Simulation Symposium (ESS 1996), October 1996, Genua [SCS], http://www.ocp.tudelft.nl/tt/users/duinkerk/papers/gen9610b.pdf – last check of address: July 30, 2003
150. Ottjes J A, Hogedoorn F P A (1996) Design and control of multi-AGV systems : reuse of simulation software. In: Proceedings of the 8th European Simulation Symposium (ESS 1996), October 1996, Genua [SCS], http://www.ocp.tudelft.nl/tt/users/duinkerk/papers/gen9610a.pdf – last check of address: July 30, 2003
151. Ottjes J A, Veeke H P M (1999) Simulation of a new port ship interface concept for inter modal transport. In: Proceedings of the 11th European Simulation Symposium (ESS 1999), October 1999, Erlangen [SCS], http://www.ocp.tudelft.nl/tt/users/duinkerk/papers/erl9910b.pdf – last check of address: July 30, 2003
152. Ottjes J A, Veeke H P M, Duinkerken M B (2002) Simulation studies of robotized multi terminal systems. In: Proceedings International Congress on Freight

Transport Automation and Multimodality (FTAM), May 2002, Delft [TRAIL], http://www.ocp.tudelft.nl/tt/users/duinkerk/papers/del0205.pdf – last check of address: July 30, 2003

153. Park K T, Kim K H (2002) Berth scheduling for container terminals by using a subgradient optimization technique. Journal of the Operational Research Society 53: 1054–1062
154. Park Y-M, Kim K H (2003) A scheduling method for berth and quay cranes. OR Spectrum 25: 1–23
155. Peterkofsky R I, Daganzo C F (1990) A branch and bound solution method for the crane scheduling problem. Transportation Research-B 24B: 159–172
156. Powell W B, Carvalho T A (1998) Real-time optimization of containers and flatcars for intermodal operations. Transportation Science 32(2): 110–126
157. Preston P, Kozan E (2001) An approach to determine storage locations of containers at seaport terminals. Computers & Operations Research 28: 983–995
158. Qiu L, Hsu W-J (2000) Adapting sorting algorithms for routing AGVs on a mesh-like path topology. Technical Report CAIS-TR-00-28, Nanyang Technological University, School of Applied Science, Centre for Advanced Information Systems
159. Qiu L, Hsu W-J (2000) Conflict-free AGV routing in a bi-directional path layout. In: Proceedings of the 5th International Conference on Computer Integrated Manufacturing (ICCIM 2000), March 28–30, vol 1, pp 392–403. Gintic Institute of Manufacturing Technology, Singapore
160. Qiu L, Hsu W-J (2001) Scheduling of AGVs in a mesh-like path topology. Technical Report CAIS-TR-01-34, Nanyang Technological University, School of Computer Engineering, Centre for Advanced Information Systems
161. Qiu L, Hsu W-J (2001) Scheduling of AGVs in a mesh-like path topology (II): a case study in a container terminal. Technical Report CAIS-TR-01-35, Nanyang Technological University, School of Computer Engineering, Centre for Advanced Information Systems
162. Qiu L, Hsu W-J, Huang S Y, Wang H (2002) Scheduling and routing algorithms for AGVs: a survey. International Journal of Production Research 40: 745–760
163. Rebollo M, Julian V, Carrascosa C, Botti V (2000) A MAS approach for port container terminal management. In: Proceedings of the 3rd Iberoamerican workshop on DAI and MAS, 2000, pp 83–94. Atibaia, Sao Paulo, Brasil
164. Reveliotis S A (2000) Conflict resolution in AGV systems. IIE Transactions 32: 647–659
165. Rizzoli A E, Fornara N, Gambardella L M (1999) A simulation tool for combined rail-road transport in intermodal terminals. In: Proceedings of the conference MODSIM 1999, Modelling and Simulation Society of Australia and New Zealand, ftp://ftp.idsia.ch/pub/luca/papers/modsim99.ps.gz – last check of address: July 30, 2003
166. Rizzoli A E, Gambardella L M, Zaffalon M, Mastrolilli M (1999) Simulation for the evaluation of optimised operations policies in a container terminal. In: HMS99, Maritime & Industrial Logistics Modelling and Simulation, Sep 16–18, Genoa, Italy, ftp://ftp.idsia.ch/pub/luca/papers/hms99.ps.gz – last check of address: July 30, 2003
167. Roach P A, Wilson I D (2002) A genetic algorithm approach to strategic stowage planning for container-ships. Working paper, University of Glamorgan, UK
168. Saanen Y, van Meel J, Verbraeck A (2003) The next generation automated container terminals. Technical Report, TBA Nederland/Delft University of Technology
169. Saanen Y A (2000) Examining the potential for adapting simulation software to enable short-term tactical decision making for operational optimisation. Technical Report, TBA Nederland/Delft University of Technology

170. Saanen Y A, Verbraeck A, Rijsenbrij J C (2000) The application of advanced simulations for the engineering of logistic control systems. In: Mertins K, Rabe M (eds) The new simulation in production and logistics – prospects, views and attitudes. Proc. 9. ASIM Fachtagung 'Simulation in Produktion und Logistik', pp 217–231. IPK, Berlin
171. Scheithauer G (1999) LP-based bounds for the container and multi-container loading problem. International Transactions in Operational Research 6: 199–213
172. Schneidereit G, Voß S, Parra A, Steenken D (2003) A general pickup and delivery problem for automated guided vehicles with multiloads: a case study. Working paper, University of Hamburg
173. Sculli D, Hui C F (1988) Three dimensional stacking of containers. Omega 16(6): 585–594
174. Shabayek A A, Yeung W W (2002) A simulation for the Kwai Chung container terminal in Hong Kong. European Journal of Operational Research 40: 1–11
175. Shen W S, Khoong C M (1995) A DSS for empty container distribution planning. Decision Support Systems 15: 75–82
176. Shields J J (1984) Container stowage: a computer aided preplanning system. Marine Technology 21: 370–383
177. Sideris A, Boilé M P, Spasovic L N (2002) Using on-line information to estimate container movements for day-to-day marine terminal operation. Working Paper (submitted for consideration for publication in the Journal of Advanced Transportation), New Jersey Institute of Technology, http://www.transportation.njit.edu/nctip/publications/forecastTRB.pdf – last check of address: Oct 30, 2003
178. Steenken D (1992) Fahrwegoptimierung am Containerterminal unter Echtzeitbedingungen. OR Spektrum 14: 161–168
179. Steenken D (1999) Latest operational experience with satellite based container positioning systems. In: Proceedings IIR-Conference on 'Container Handling, Automation and Technology', IIR, London, via http://www.hhla.de/de/Presse/index.jsp – last check of address: Oct 30, 2003
180. Steenken D (2003) Optimised vehicle routing at a seaport container terminal. ORbit 4: 8–14
181. Steenken D, Henning A, Freigang S, Voß S (1993) Routing of straddle carriers at a container terminal with the special aspect of internal moves. OR Spektrum 15: 167–172
182. Steenken D, Winter T, Zimmermann U T (2001) Stowage and transport optimization in ship planning. In: Grötschel M, Krumke S O, Rambau J (eds) Online optimization of large scale systems, pp 731–745. Springer, Berlin
183. Taleb-Ibrahimi M, de Castilho B, Daganzo C F (1993) Storage space vs handling work in container terminals. Transportation Research-B 27B: 13–32
184. The Rotterdam Municipal Port Management – The annual report 2001. http://www.portmanagement.com/Images/17_45237.pdf – last check of address: Oct 13, 2003
185. Thiers G F, Janssens G K (1998) A port simulation model as a permanent decision instrument. Simulation 71: 117–125
186. Tolujev J, Merkuryev Y, Blümel E, Nikitins M (1996) Port terminal simulation: state of the art – a survey. Technical Report, Riga: AMCAIRiga Technical University
187. Toth P, Vigo D (eds) (2002) The vehicle routing problem. Society for Industrial & Applied Mathematics, SIAM, Philadelphia

188. Ulusoy G U, Sivrikaya-Şerifoğlu F, Bilge Ü (1997) A genetic algorithm approach to the simultaneous scheduling of machines and automated guided vehicles. Computers & Operations Research 24: 335–351
189. United Nations Conference on Trade and Development – secretariat (1999) Review of maritime transport. UNCTAD/RMT(99)/1, United Nations Publication, http://www.unctad.org/en/docs/rmt1999_en.pdf – last check of address: Jul 8, 2003
190. United Nations Conference on Trade and Development – secretariat (2000) Review of maritime transport. UNCTAD/RMT(2000)/1, United Nations Publication, http://www.unctad.org/en/docs/rmt2000_en.pdf – last check of address: Jul 8, 2003
191. United Nations Conference on Trade and Development – secretariat (2001) Review of maritime transport. UNCTAD/RMT/2001, United Nations Publication, http://www.unctad.org/en/docs/rmt2001_en.pdf – last check of address: Jul 8, 2003
192. United Nations Conference on Trade and Development – secretariat (2002) Review of maritime transport. UNCTAD/RMT/2002, United Nations Publication, http://www.unctad.org/en/docs/rmt2002_en.pdf – last check of address: Jul 8, 2003
193. Veeke H P M, Ottjes J A (1999) Detailed simulation of the container flows for the IPSI concept. In: Proceedings of the 11th European Simulation Symposium (ESS 1999), October 1999, Erlangen [SCS], http://www.ocp.tudelft.nl/tt/users/duinkerk/papers/erl9910a.pdf – last check of address: July 30, 2003
194. Veeke H P M, Ottjes J A (2002) A generic simulation model for systems of container terminals. In: Proceedings of the 16th European Simulation Multiconference (ESM 2002), June 2002, Darmstadt [SCS], http://www.ocp.tudelft.nl/tt/users/duinkerk/papers/dar0206b.pdf – last check of address: July 30, 2003
195. Vis I F A, Harika I (2004) Comparison of vehicle types at an automated container terminal. OR Spectrum 26: 117–143
196. Vis I F A, de Koster R (2003) Transshipment of containers at a container terminal: an overview. European Journal of Operational Research 147: 1–16
197. Vis I F A, de Koster R, Roodbergen K J, Peeters L W P (2001) Determination of the number of automated guided vehicles required at a semi-automated container terminal. Journal of the Operational Research Society 52: 409–417
198. Volk B (2002). Growth factors in container shipping. http://maritimebusiness.amc.edu.au/papers/AMC3_GRO.pdf – last check of address: July 15, 2003
199. Voß S, Böse J W (2000) Innovationsentscheidungen bei logistischen Dienstleistern – Praktische Erfahrungen in der Seeverkehrswirtschaft. In: Dangelmaier W, Felser W (eds) Das reagible Unternehmen, pp 253–282. HNI, Paderborn
200. Wallace A (2001) Application of AI to AGV control – agent control of AGVs. International Journal of Production Research 39(4): 709–726
201. Wilson I D, Roach P A (1999) Principles of combinatorial optimization applied to container-ship stowage planning. Journal of Heuristics 5: 403–418
202. Wilson I D, Roach P A (2000) Container stowage planning: a methodology for generating computerised solutions. Journal of the Operational Research Society 51: 1248–1255
203. Wilson I D, Roach P A, Ware J A (2001) Container stowage pre-planning: using search to generate solutions, a case study. Knowledge-Based Systems 14(3–4): 137–145
204. Winter T (2000) Online and real-time dispatching problems. GCA-Verlag, Herdecke
205. Winter T, Zimmermann U (1999) Combinatorial online and real-time optimization. In: Fundamentals – foundations of computer science. Proceedings of the 15th IFIP World Computer Congress, pp 31–48, Österreichische Computer Gesellschaft, Vienna

206. Wong J C F, Leung J M Y (2002) On a vehicle routing problem with time windows and stochastic travel times. Technical Report, Department of Systems Engineering and Engineering Management, Chinese University of Hong Kong, Hong Kong, China
207. Yang C H, Choi Y S, Ha T Y (2004) Performance evaluation of transport vehicle at automated container terminal using simulation. OR Spectrum 26(2): to appear
208. Yun W Y, Choi Y S (1999) A simulation model for container-terminal operation analysis using an object-oriented approach. International Journal of Production Economics 59: 221–230
209. Yun W Y, Choi Y S (2003) Simulator for port container terminal using an object oriented approach. Working paper, Pusan National University
210. Zaffalon M, Rizzoli A E, Gambardella L M, Mastrolilli M (1998) Resource allocation and scheduling of operations in an intermodal terminal. In: ESS98, 10th European Simulation Symposium and Exhibition, Simulation in Industry, October 26–28th, pp 520–528, Nottingham, UK, ftp://ftp.idsia.ch/pub/luca/papers/ess98.ps.gz – last check of address: July 30, 2003
211. Zhang C, Liu J, Wan Y-w, Murty K G, Linn R J (2003) Storage space allocation in container terminals. Transportation Research-B 37: 883–903
212. Zhang C, Wan Y, Liu J, Linn R J (2002) Dynamic crane deployment in container storage yards. Transportation Research-B 36: 537–555

Comparison of vehicle types at an automated container terminal

Iris F. A. Vis[1] **and Ismael Harika**[2]

[1] Free University Amsterdam, School of Economics and Business Administration, De Boelelaan 1105, Room 3A-31, 1081 HV Amsterdam, The Netherlands (e-mail: ivis@feweb.vu.nl)

[2] Raab Karcher de Waardt Bouwstoffen, Staringlaan 8, P.O. Box 14, 2740 AA Waddinxveen, The Netherlands

Abstract. At automated container terminals, containers are transshipped from one mode of transportation to another. Automated vehicles transport containers from the stack to the ship and vice versa. Two different types of automated vehicles are studied in this paper, namely automated lifting vehicles and automated guided vehicles. An automated lifting vehicle is capable of lifting a container from the ground by itself. An automated guided vehicles needs a crane to receive and deliver a container.

In designing automated container terminals one have to consider the choice for a certain type of equipment. The choice for a certain type of equipment should be made by performing a feasibility and economic analysis on various types of equipment. In this paper, we examine effects of using automated guided vehicles and automated lifting vehicles on unloading times of a ship, with simulation studies. In choosing a certain type of equipment we have considered criteria such as unloading times of a ship, occupancy degrees and the number of vehicles required. 38% more AGVs need to be used than ALVs. From this specific study, we conclude that, by observing only purchasing costs of equipment, ALVs are a cheaper option than AGVs.

To obtain an accurate analysis we have performed a sensitivity analysis. It can be concluded that the design of the terminal and technical aspects of quay cranes impact the number of vehicles required and as a result the choice for a certain type of equipment.

Keywords: Container logistics – Simulation – AGVs and ALVs

Correspondence to: I. F. A. Vis

1 Introduction

Containers are large boxes used to transport goods from one destination to another. With containers a bulk unit can be created out of the individual pieces of freight. As a result, *containerisation* can be defined, according to *The Containerization Institute*, as the utilisation, grouping or consolidating of multiple units into a larger container for more efficient movement. Compared to conventional bulk, the use of containers has several advantages, namely less product packaging, less damaging and higher productivity (Agerschou et al., 1983). The dimensions of containers have been standarised. The term *TEU* (twenty-feet-equivalent-unit) is used to refer to one container with a length of twenty feet. A container of 40 feet is expressed by 2 TEU. Several transportation systems can be used to transport containers from one destination to another. Transport over sea is carried out by ships. On the other hand, trucks or trains can be used to transport containers over land. To transship containers from one mode of transportation to another, ports and terminals can be used. For example, at a container terminal, a container can be taken off a train and placed on a ship.

Containers were used for the first time in the mid-fifties. Through the years, the proportion of cargo handled with containers has steadily increased. As a result of the enormous growth, the capacity of ships has been extended from 400 TEU to 4000 TEU and more. An extensive overview of the history of containers is given in Rath (1973). Furthermore, the importance of ports and terminals has grown. With the introduction of larger ships, small terminals have changed into large terminals. To ensure a fast transshipment process at large terminals information technology and automated control technology can be used. A detailed description of the use of these technologies in container terminals can be found in Johansen (1999). To use these kinds of technologies large investments have to be made and ongoing database management is required. Wan et al. (1992) show that the application of information technology in the port of Singapore results in more efficiency and a higher performance. In Leeper (1988) it is concluded that, in order to achieve an improvement of productivity and reduction in investment costs, an advanced automated control technology is a necessary condition.

The process of unloading and loading a ship at an automated container terminal is illustrated in Figure 1 and may be described as follows: when a ship arrives at the port, the containers have to be taken off the ship. This is done by manned Quay Cranes (QCs), which take the containers from the ship's hold and the deck. Next, the QCs put the containers on vehicles, like automated guided vehicles (AGVs). After receiving a container, the AGV moves to the stack. This stack consists of a number of lanes where containers can be stored for a certain period. These lanes are served by, for example, automatically controlled Automated Stacking Cranes (ASCs). When an AGV arrives at a lane, the ASC takes the container off the AGV and stores it in the stack. After a certain period the containers are retrieved from the stack by the ASCs and transported by the AGVs to transportation modes such as barges, deep-sea ships, trucks or trains. This process is also be executed in reverse order, to load containers on a ship.

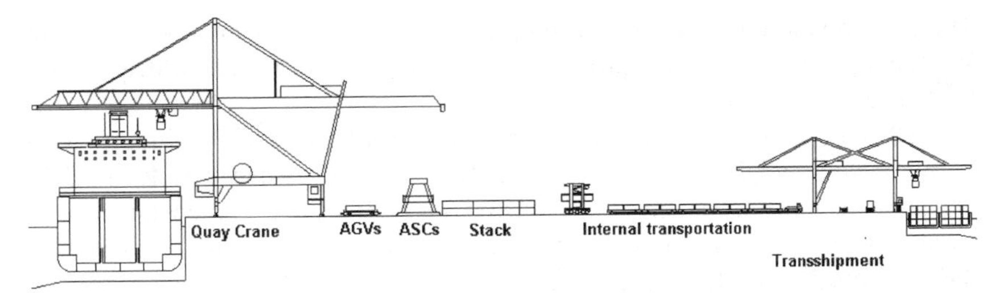

Fig. 1. An example of processes at an automated container terminal. Containers can also be transshipped to other modes of transportation, such as trucks or trains

We can distinguish between four planning and control levels in making decisions to obtain efficient processes and operations in material handling centres, namely the strategic level, the tactical level, the operational level and the real time level. At the strategic level it is, for example, decided at which location a container terminal is built. The time-horizon of decisions at this level covers one to several years. These decisions lead to the definition of a set of constraints under which the decisions at the tactical and operational level have to be made. At the tactical level, it is, for example, decided which types of equipments are used and which layout is chosen. The time-horizon of these decisions covers months to years. At the operational level decisions are made of which the time-horizon covers a day to months. Decisions at this level are, for example, the determination of the number of vehicles required to ensure an efficient transportation process and the determination of the number of QCs working on a certain ship. Finally, at the real-time level, real-time decisions are made each minute on, for example, the operation of the vehicle or crane.

Logistic activities that are potential candidates for improvement in container terminals are among other things: receiving and shipment of containers, internal transport of containers, storage and retrieval of containers. To perform these activities material handling equipments, like internal transport equipments and storage and retrieval systems can be used. Due to technological innovations, new types of equipments have been developed during the past decades, namely automated material handling equipment. Without human intervention, automated cranes and vehicles perform their tasks. Automated control technology is used for controlling automated material handling equipment. Automated material handling equipment demands high investments but little labour costs. For manned equipment, labour costs form a high percentage of the total costs. The difficulty of operational control problems can increase if the level of automation increases. Besides automated control technologies, information technology can be used in material handling centres, for example, for recognition of pallets, products and containers.

The choice for a certain type of equipment should be made by performing a feasibility and economic analysis on various types of equipments. Fisher et al. (1988) discuss heuristic rules for selecting equipment types which are both technically and economically appropriate for the movement of specified loads over specified types of paths. Furthermore, the choice of equipment depends on the design of the container terminal. For example, wide vehicles cannot travel through narrow lanes.

There is a relation between the terminal's layout and its equipment. For example, Welgama and Gibson (1996) propose an integrated methodology for determining both aspects simultaneously.

In this paper, we examine an automated container terminal. Two different types of automated vehicles can be used, namely automated guided vehicles (AGVs) and automated lifting vehicles (ALVs). AGVs and ALVs transport containers over fixed paths. An AGV receives a container from a crane and a crane is required to take the container off the vehicle. ALVs are capable of lifting a container from the ground by itself. To uncouple the unloading process and the transportation process, buffer areas are created where cranes position containers. After a while an ALV retrieves a container from a buffer area and transports it to its destination. At Patrick Terminals in Australia a proven prototype of a driverless straddle carrier has been developed. At one of their terminals a small number of these automated straddles is used during the unloading and loading process. At the port of Rotterdam in the Netherlands research is done on the way ALVs should be developed and to the usefulness of these types of equipment. The specifications of ALVs, which we use in this paper, are obatined from these studies (see Celen et al., 1997).

In this paper, we study the effect of using ALVs and AGVs on the unloading times of a ship by simulation studies. Section 2 provides more specific characteristics of both types of equipment. Furthermore, Section 2 presents the specifications of the simulation model for AGVs and the simulation model for ALVs. Section 3 describes how the models were implemented in the simulation software.

Several criteria can be used in choosing a certain type of equipment. In this paper, we concentrate on the several aspects, namely waiting times of cranes, occupancy degrees of vehicles, unloading times of a ship and number of vehicles required. Results of the various experiments are given in Section 4. To study the effect of a technical aspect of types of equipment (i.e. speed and capacity), human influences (i.e. capabilities of crane drivers) and layout aspects (i.e. buffer size) on the number of vehicles required we have performed a sensitivity analysis in Section 5. By examining the results of the simulation experiments in Section 4 and by examining the results of the sensitivity analysis in Section 5, we provide an advice on the choice of automated equipment in an automated container terminal in Section 6.

2 Specifications of the models

We observe two types of automated vehicles, namely automated guided vehicles (AGVs) and automated lifting vehicles (ALVs). These vehicles transport containers from the ship to the stack during the unloading process and from the stack to the ship during the loading process. The main difference between these two types of vehicles is the fact that automated lifting vehicles are capable of lifting containers from the ground by themselves. By using lifting vehicles the unloading and loading process at the cranes and the transportation process can be decoupled. A crane places a container on the ground and does not have to wait for a vehicle to place the container on. This last action is required in case non lifting vehicles are used. If no vehicle is present the crane has to wait with the container until a vehicle arrives.

Due to the decoupling of processes it might be expected that less ALVs than AGVs are required to minimise unloading times of the ship.

In this paper, we only consider the unloading process. In practice, Quay Cranes first unload a ship and thereafter these Quay Cranes load containers on the ship. Unloading and loading operations are only mixed during a very short period of time. The way of modelling unloading operations and loading operations are quite similar. In the AGV model containers are transported from the ship to the stack and vice versa in a fixed sequence. The container that is unloaded from the ship is placed directly on an AGV and transported to the stack. The loading containers have a fixed position in the ship and therefore the sequence in which these containers are transported from the stack to the ship is also fixed. In the ALV model, unloading containers are placed in a buffer area and can be transported in a random order. Also in the ALV model containers are loaded on the ship according to a loading plan. By using a buffer area, transportation and loading sequences might be different. However, to ensure an continously operating crane, the right container should be availabe, the moment the crane needs it. Therefore, it might be expected that the advantages of buffer areas are smaller during the loading operation than during the unloading operation. The number of AGVs and ALVs would probably be closer together in the loading case than in the unloading case. Thus, for both types of vehicles we only model the unloading process. By considering a similar situation, we can compare the results of both models.

In the next section, we describe the layout of a terminal. We have used this layout in both models. Sections 2.2 and 2.3 present the specifications of respectively the model for AGVs and the model for ALVs.

2.1 Layout of the terminal

Automated vehicles can travel along fixed guidepaths. A flowpath layout at a container terminal connects quay cranes and automated stacking cranes. This layout is usually represented by a directed network in which we can consider pick up and delivery points at cranes and intersections as nodes. Directed arcs indicate the direction of travel of vehicles. In the literature, several approaches exist for the design of flowpath layout. Examples, are unidirectional, bidirectional, multiple lane and loop layouts. Sinreich (1995) presents a literature overview on approaches for the design of flow networks. In practice different layouts are used at semi-automated container terminals. At the port of Rotterdam vehicles drive in a loop with one lane from ship to stack and back. At some ports in Germany, for example, Hamburg, and Australia (Patrick Terminals) vehicles drive on multiple lane paths.

In this paper, we consider a combination of these two types of layouts, namely a multiple lane layout and a loop layout. Figure 2 presents the layout of the terminal. In a multiple lane guidepath various flowpaths exist between nodes in the network. In a loop layout vehicles travel in a loop and they are controlled by simple traffic control algorithms. The loop is a fixed sequence of pick up and delivery points at QCs and ASCs. Disadvantages of loop layouts are, for example (see Sinreich, 1995), the fact that once a station is passed an AGV has to travel the complete loop before it reaches the station again and the fact that a vehicle which needs to stop

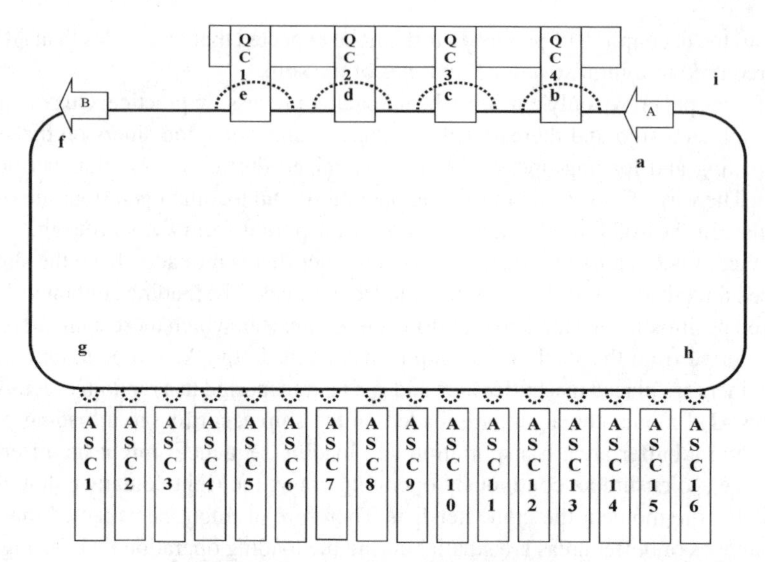

Fig. 2. Layout of a container terminal

to pick up or deliver a container, blocks the other vehicles. At a container terminal a vehicle starts during the unloading process at a quay crane and needs to travel full to the stack before it can travel back to the quay crane. Thus, in this case it is not a problem that a vehicle has to travel the complete loop before returing to the original station.

To solve the problem of vehicles blocking each other, we have used the principle of multiple lanes. At each crane a second lane exists at which vehicles can stop to pick up or deliver a container. In the ALV model, buffer areas are located on these extra lanes. The rest of the vehicles can continue their route on the main lane without losing time to wait for other vehicles. Of course some minor congestion might occur when a vehicle wants to reenter the main lane. By using right-of-way rules, collisions can be avoided. Examples of papers considering a terminal with a specific loop layout are Kim and Bae (1999), Van Hee and Wijbrands (1988), Meersmans (2002) and Vis (2002).

There is a relation between the terminal's layout and its equipment. The effects of layout aspects on the number of vehicles required could be studied. Therefore, we perform experiments in Sections 5.1 and 5.2 on the impact of the size of buffer areas on the number of vehicles required.

As indicated in Figure 2, vehicles drive counterclockwise in a loop from ship to stack to ship. Arrow A indicates the queue of vehicles waiting for transportation requests of quay cranes. Quay Cranes request vehicles waiting in this queue. Arrow B indicates the position where a vehicle is directed to the ASC which will store the container in the stack. The vehicle returns to arrow A after delivering the container to the stacking crane. Table 1 represents the various distances in the terminal, which we use in the simulation studies.

Table 1. Distances (in meters) in the terminal. An x indicates an infeasible path

	a	b	c	d	e	f	g	h	i
a	x	10	60	110	160	315	515	920	1120
b	x	x	50	100	150	305	505	910	1110
c	x	x	x	50	100	255	455	860	1060
d	x	x	x	x	50	205	405	810	1010
e	x	x	x	x	x	155	355	760	906
f	x	x	x	x	x	x	200	605	805
g	x	x	x	x	x	x	x	405	605
h	x	x	x	x	x	x	x	x	200
i	90	x	x	x	x	x	x	x	x

2.2 Automated guided vehicles

In this section, we discuss the unloading process for non lifting vehicles. We can divide the unloading process in three parts, namely

- processes at the QCs,
- transportation process,
- processes at the ASCs

We will discuss these three processes with their specific characteristics related to the use of non lifting vehicles. Furthermore, we present a schematic overview of the unloading process with non lifting vehicles.

Processes at the QC

A ship arrives at the terminal and berths at the quay. After executing some standard procedures, the Quay Crane starts to unload containers off the ship. The number of QCs working on a ship might differ per ship. Nowadays, four cranes are used for large ships. In the simulation study, we also used four QCs for the unloading process.

We assume that 2000 containers need to be unloaded off the ship. These containers are equally distributed over the ship. We do not distinguish between the various sizes of containers. Each of the four cranes needs to unload 500 containers.

Manned QCs unload containers from the deck and hold of the ship. This takes a certain cycle time. The cycle time of a container is the time required to unload a container of the ship and to position it on the vehicle. Cycle times depend, for example, on the capabilities of the crane driver, the type of ship, specifications of the crane and the position of the container in the ship. The time between two successive containers is smaller if the crane is capable of unloading very fast. In the simulation models, we use data which were used at a study at a large container terminal in the Netherlands (see Celen et al., 1997). The data used in this paper are, furthermore, obtained from interviews with logistics managers of Europe Combined Terminals (ECT) at the port of Rotterdam. ECT has a semi-automated container

Table 2. Empirical distribution of the cycle times of a QC

Fraction	Cycle time in seconds
0.05	30–40
0.15	40–50
0.25	50–60
0.20	60–70
0.17	70–80
0.11	80–90
0.04	90–120
0.02	120–150
0.01	150–180

terminal. Table 2 represents an empirical distribution of the cycle times. Possible large disturbances, such as breakdowns of equipment, are not taken into account.

With these data, the simulation program can generate cycle times for each container (see Fig. 3) as follows: a class of cycle times is randomly selected according to the empirical distribution. In this class each cycle time has an equal probability to be chosen. Within a class cycle times are assumed to follow a continuous uniform distribution.

The average cycle time of the distribution in Table 2 equals 65,9 seconds. On average, a crane can unload 55 containers per hour. Consequently, four cranes can unload on average 2000 containers in nine hours.

The moment a QC starts unloading a container, an AGV is requested from the queue (arrow A in Fig. 2). At the start of the simulation all vehicles are present in this queue. Vehicles are dispatched to cranes according to the principle of nearest vehicle first (see, for example, Egbelu and Tanchoco, 1984). The first vehicle in line is dispatched to the crane needing a vehicle. Other dispatching rules cannot be used in practice due to the fact that the vehicles travel in a loop without the possibility to pass each other. The AGV travels from this queue to the respective QC. At the QC the AGV waits for the container. The capacity under a crane equals one vehicle. The QC places the container on the AGV. It might also occur that the crane needs to wait for an AGV to arrive at the crane. After receiving the container the AGV starts transporting the container.

This part of the unloading process is given in a schematic way in Figure 3.

Transportation process

The AGV travels with a container to the stack. At arrow B (see Fig. 2) the container is dispatched to a stacking crane. We use the following rule: the container is dispatched to the crane at the farthest travel distance and with the least number of scheduled jobs. In this way congestion at the stack will be avoided. The speed of a full AGV equals 4 m/s.

Usually, 8 AGVs are used per Quay Crane. In the model, we use 32 AGVs. In Section 4, we will vary the number of AGVs to study the impact on the per-

Table 3. Characteristics of an AGV used in the model

	AGV
Speed of full vehicle	4 m/s
Speed of empty vehicle	5.5 m/s
Acceleration	0.5 m/s^2
Deceleration	0.5 m/s^2
Capacity vehicle	1 container
Number of vehicles	32
Maximum number of vehicles in queue	unlimited

Table 4. Empirical distribution of the cycle times of an ASC

Fraction	Cycle time in seconds
0.05	90–120
0.15	120–150
0.25	150–180
0.20	180–210
0.17	210–240
0.11	240–270
0.04	270–360
0.02	360–450
0.01	450–540

formance of the terminal during the unloading process. Table 3 represent various characteristics of an AGV.

The AGV travels to its stacking crane. If the ASC handles another vehicle or stores another container in the stack the AGV needs to wait. The ASC lifts the container off the AGV and drives into the stack to store the container. The empty AGV travels to the queue at the QCs. This part of the unloading process is illustrated in Figure 3.

Processes at the ASC

The stack consists of several blocks. A block consists of six rows with containers. These containers are positioned next to each other and on top of each other. An ASC serves one block. Such a block has a length of 300 meters and is 25 meters wide. Sixteen ASCs serve one ship. At the front of each block a pickup and delivery point is located where a container is transferred from an AGV to an ASC.

The cycle time of an ASC equals the time to lift a container from an AGV, store it in the stack and return to the pickup and delivery point. The value of the cycle time depends, for example, on the storage location of the container and specifications of the ASC. In practice, the cycle time of an ASC equals on average three times the cycle time of a QC. Table 4 represents the empirical distribution of the cycle times of an ASC. These data are obtained by multiplying the value of the intervals

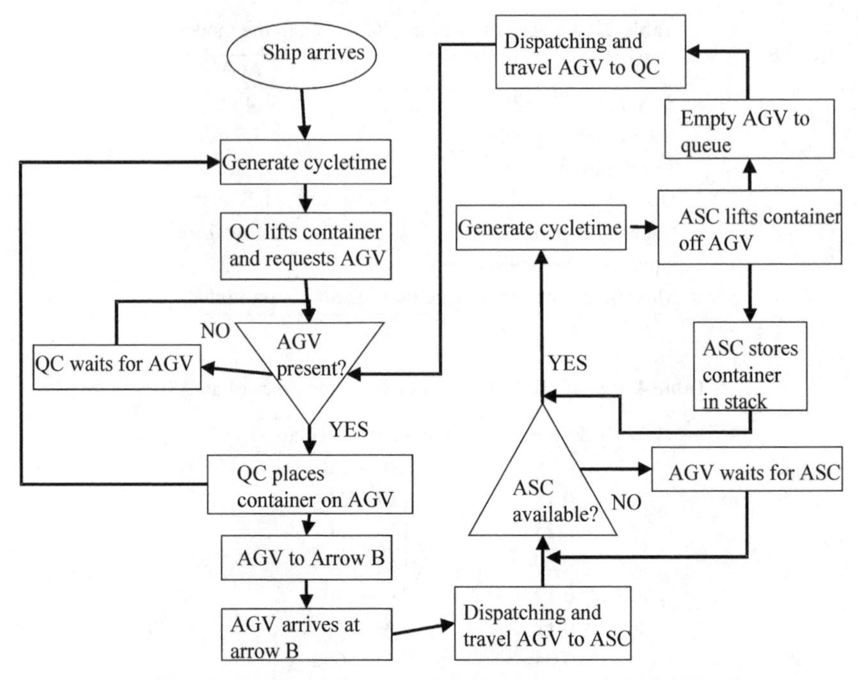

Fig. 3. Unloading process with an automated guided vehicle

of the classes of the distribution of the QC with a factor three. The cycle times are obtained as follows: randomly select a class of cycle times according to the empirical distribution. In this class each cycle time has an equal probability to be chosen. Within a class cycle times are assumed to follow a continuous uniform distribution. The average cycle time equals 197.7 seconds. On average, an ASC stores 18 containers per hour.

The ASC lifts a container from the AGV and stores it in the stack. After storing the container the ASC returns to the pickup and delivery point and is free to store the next container. Figure 3 represent also this part of the unloading process.

2.3 Automated lifting vehicles

In this section, we describe the unloading process for lifting vehicles. Also, in this case, we can divide the unloading process in three parts, namely processes at QCs, the transportation process and processes at the ASCs. We indicate the specific elements of these processes related to lifting vehicles. We also illustrate the processes in a schematic way in Figure 4.

Processes at the QC

A QC unloads a container from the ship and places it on the ground or on top of another container in a buffer area. A lifting vehicle is able to lift a container by

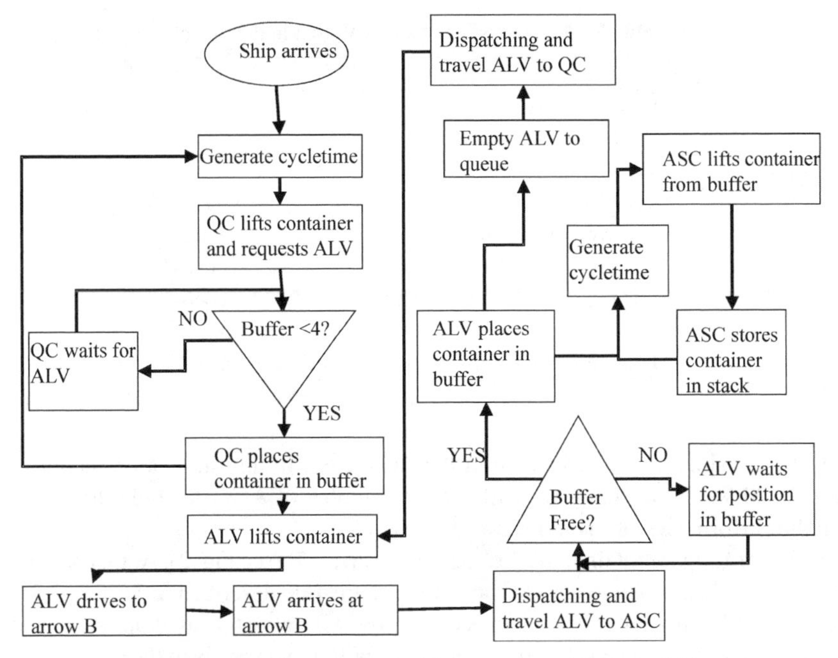

Fig. 4. Unloading processes with an automated lifting vehicle

itself. As a result, the QC does not needs to wait for an empty vehicle. After placing the container in a buffer area, the QC can unload another container. The buffer area is located under the crane. This area has a limited capacity due to restrictions in space. We assume that the maximum capacity equals 4 containers (2 in the length and 2 high). The cycle times follow the empirical distribution from Table 2 (see Celen et al., 1997). Celen et al. (1997) assume that the quay crane can easily deviate from its central axis to place containers into the buffer area without the need to be moved physically. As a result, it can be assumed that cycle times do not increase due to the fact that containers need to be positioned on top of, behind or next to another container.

An ALV is capable of travelling over 2 containers placed on the ground. Thus, no congestion will occur in the flow of vehicles with buffers areas of capacity 4 under the cranes. We will vary the size of the buffer in Section 5 to study the impact of the buffer size on the unloading times of a ship.

If the buffer area is full, the QC needs to wait for a vehicle, before it can place a new container in the buffer. Free ALVs are waiting in a waiting queue at arrow A (see Fig. 2) for a new transportation request. Dispatching of ALVs to containers is done according to the same dispatching rule, namely nearest vehicle first. This part of the unloading process is illustrated in Figure 4.

Transportation process

The processes at the QC and the transportation process are decoupled by using buffer areas. Van der Meer (2000) and Vis (2002) have shown that this decoupling

Table 5. Characteristics of an ALV used in the model

	ALV
Speed of full vehicle	6 m/s
Speed of empty vehicle	7 m/s
Acceleration	0.5 m/s^2
Deceleration	0.5 m/s^2
Capacity vehicle	1 container
Time required to lift a container	22 seconds
Time required to put a container down	22 seconds
Number of vehicles	20
Maximum number of vehicles in queue	unlimited

results in using less vehicles than in the AGV case. In this simulation model we use 20 ALVs. In Section 4 we will vary the number of ALVs to study the impact on the performance of the unloading process.

After lifting a container the ALV travels to arrow B (see Fig. 2). At this position the ALV is dispatched to an ASC according to the rule described in Section 2.2.

Each ASC has a buffer with capacity 1. We will relax this assumption in Section 5.2. The ALV places the container in this buffer and travels to the queue at the QCs. The ASC lifts the container and stores it in the stack. If the buffer is full, the vehicle needs to wait before it can deliver its container to the stack. Table 5 gives some specifications of an ALV. These specifications are obtained from Celen et al. (1997). To obtain these specifications Celen et al. (1997) have used a combination of the data of AGVs (automated vehicles) and straddle carriers (non automated lifting vehicles).

Figure 4 also illustrates this part of the unloading process.

Processes at the ASCs

The ASC lifts a container from the buffer and stores it in the stack. After storing the container, the ASC returns to the buffer to store the next container. Table 4 represents the empirical distribution of the related cycle time.

3 Implementation of the models

We have used Arena 3.5 as simulation software. In this section, we describe how we have implemented both models of Section 2. The schematic illustrations in Figures 3 and 4 are used in the modelling of respectively the AGV model and the ALV model. We will describe the following two parts of the model in more detail, namely:

- processes at QCs
- processes at ASCs

First, we describe these parts for the AGV model in Section 3.1. In Section 3.2 we describe the parts for the ALV model.

3.1 Automated guided vehicles

Processes at QCs

Figure 5 represents the implementation of the processes at a QC.

The block *create* generates containers. Each 30 seconds a container is generated. In this way, always a container is available to be unloaded. The generation process stops the moment 500 containers are generated. A QC is represented by the blocks *station*, *seize*, *delay* and *release*. *Station* represents the physical location of the QC. *Seize* activates the crane. The cycle time of a crane, drawn from the empirical distribution in Table 2 is incorporated in the block *delay*. An entity (i.e. container) is duplicated (*Duplicate*) to allow the QC to perform two tasks simultaneously. These tasks are the unloading of a container with a related cycle time and requesting an AGV from the queue (*request*).

A QC and AGV need to wait for each other if one of them is available earlier than the other. If an AGV arrives before the QC has finished removing the container from the ship, the AGV has to wait (*Wait*). This AGV receives a signal (*Signal*) if the crane is ready. With the block choose the crane examines the presence of an AGV. If no AGV is present, the crane needs to wait for the AGV (blocks *delay* and *count*). Each second the QC examines the presence of an AGV. With *count* we can measure the waiting times of the QC, namely the sum of all seconds.

After placing the container on the AGV, the crane is released (*Release*) and free to unload a new container. The AGV travels with the container in the direction of the stack (*Leave*). With *dispose* the duplicate is removed from the model.

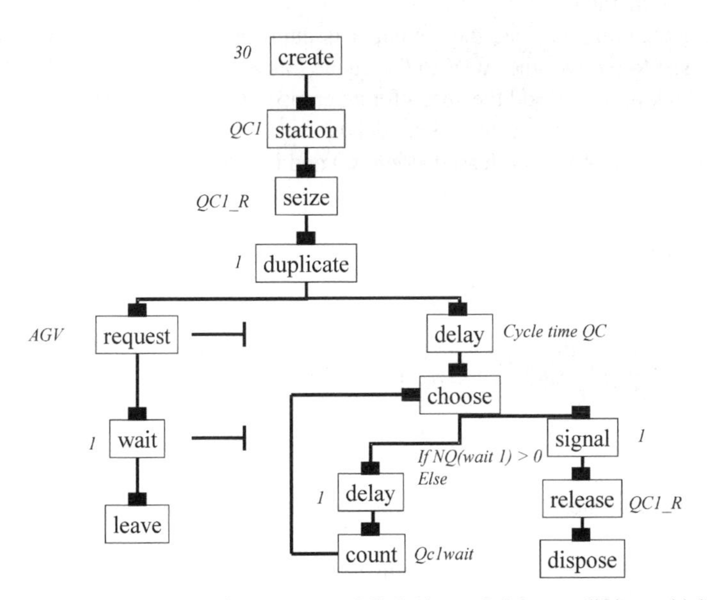

Fig. 5. Implementation of processes at QCs in Arena 3.5 for non lifting vehicles

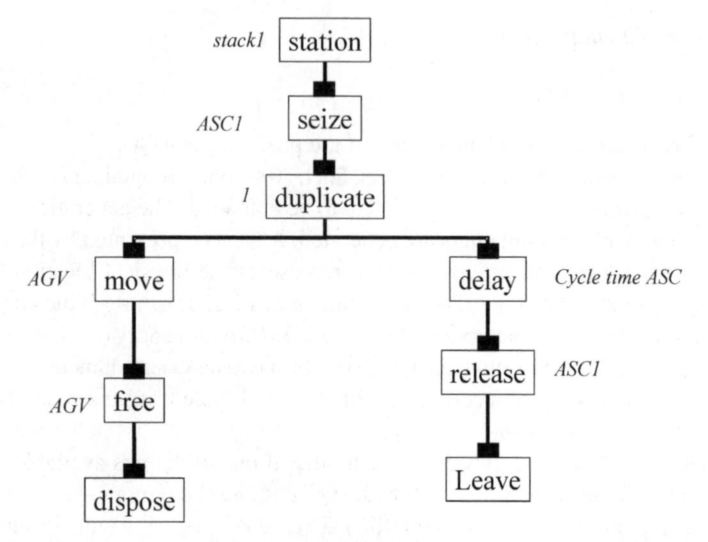

Fig. 6. Implementation of processes at ASCs in Arena 3.5 for non lifting vehicles

Processes at ASCs

The processes at the ASCs are given in Figure 6.

An ASC is represented by the blocks station, seize, delay and release. Station represents the physical location of the ASC. With seize an ASC is claimed to perform a storage request. In delay we incorporate the cycle times drawn from the distribution in Table 4.

After claiming an ASC the container is duplicated (*Duplicate*). The original one is used to redirect the AGV to the queue (*move*) and to free the AGV (*Free*). In the block move we add the speed for an empty AGV. The container is disposed (*Dispose*). The duplicate is used to model the storage of the container. After storing the container the ASC is released (*Release*) and is able to store a new container in the stack.

3.2 Automated lifting vehicles

In this section, we illustrate how to model the processes at the QCs and ASCs in the case that lifting vehicles are used.

Processes at QCs

In Figure 7 we indicate how we have modelled the processes at the QCs in the ALV model.

The processes at the QC are comparable to the processes at the QCs in the AGV model. In this model, it is a necessity to ensure that the QC places at most 4 containers in the buffer area. Therefore, we use the block *choose*. The QC needs to wait for a vehicle if the buffer area is full. By using the loop *delay* and *count*

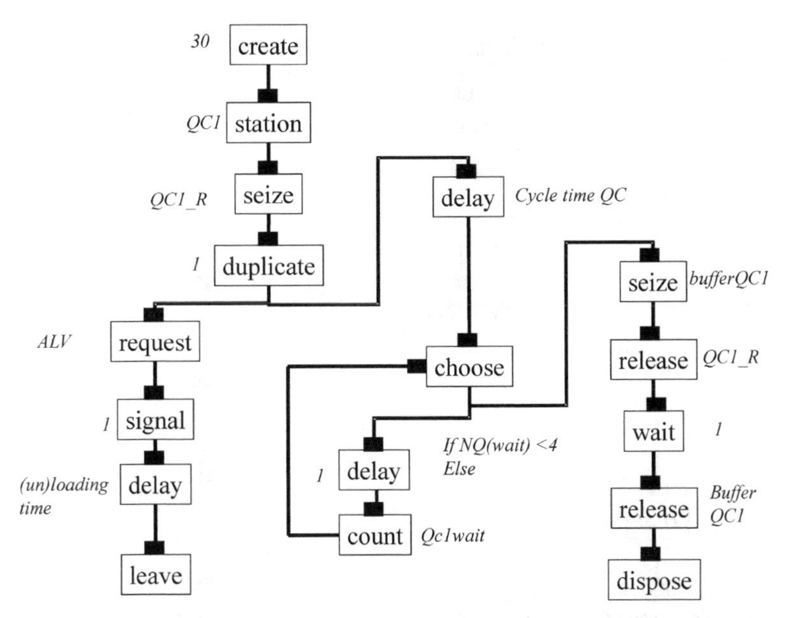

Fig. 7. Implementation of processes at QCs in Arena 3.5 for lifting vehicles

it is checked each second if there is a free place in the buffer. If there is a free location in the buffer, the crane claims this position (*seize bufferQC*). After placing the container in the buffer, the QC is released (*Release*) and is able to unload a new container. The container in the buffer needs to wait for an ALV (*Wait*). When the ALV arrives at the buffer, it gives a signal (*Signal*) and it lifts the container from the buffer. The block *delay* contains the related lifting time. After lifting the container, the buffer is released (*Release BufferQC*) and a location is available for a new container.

Processes at the ASCs

Figure 8 represents the modelling of the processes at the ASC in Arena.

When an ALV arrives at the ASC, it should first check if there is a position available in the buffer. An ALV has to wait if the buffer is full. Otherwise, the ALV seizes (*Seize*) the buffer and places the container in the buffer. The time required to place the container in the buffer is included in the block *delay*. The moment the ASC lifts a container from the buffer (*seize*, *delay*) the buffer is released and a new container can be placed in the buffer. The rest of the blocks is similar to the blocks in Figure 6.

4 Results

To obtain a high accuracy sufficient replications of each experiment should be simulated. By using the formula of Law and Kelton (1991), we have calculated

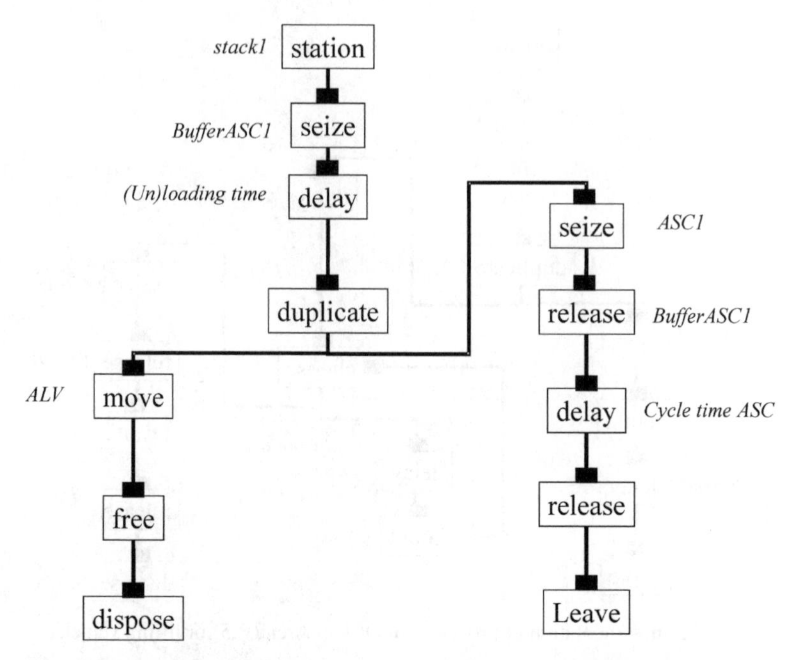

Fig. 8. Implementation of the processes at ASCs in Arena 3.5 for lifting vehicles

that a replication size of 100 is sufficient for all experiments in this paper. Thus, for every simulation experiment we generate 100 experiments and determine the average value among these 100 replications for, for example, unloading and waiting times. On average, the computation times for 100 replications on a pc with 350MHz are 30 minutes.

In this section, we compare the performance of both models. We simulate the models with their data as described in Sections 2 and 3. We examine several criteria, namely

- unloading times of a ship
- waiting times of quay cranes
- waiting times of automated guided vehicles
- occupancy degrees of vehicles and quay cranes

Thereafter, in Section 4.4 we determine the minimum number of vehicles required of both types of vehicles such that unloading times of a ship are minimised.

4.1 Unloading times of a ship

We define the unloading time of a ship as the time required to unload 2000 containers and store them into the stack. 32 AGVs and 20 ALVs are respectively used to unload the ship. The average unloading times are given in Table 6. Furthermore, we have incorporated the half width of a 95% confidence interval on the expected value of

Table 6. Average unloading times of a ship

Model	Unloading time (seconds)	95% half width
AGV	34800	81.8
ALV	34500	132.0

the unloading times. A 95% confidence interval can be calculated as follows:

$$\left[\bar{X}(n) - 1.96 * \sqrt{\frac{s^2(n)}{n}}, \bar{X}(n) + 1.96 * \sqrt{\frac{s^2(n)}{n}}\right],$$

where $\bar{X}(n)$ equals the average value, s the standard deviation and n the number of values. The value of the half width equals $1.96 * \sqrt{\frac{s^2(n)}{n}}$.

These average unloading times are comparable to the theoretical time, required to unload 2000 containers, of 9 hours (see Sect. 2).

4.2 Occupancy degrees of vehicles and quay cranes

Table 7 presents the various occupancy degrees for the AGV and ASC model.

The differences in occupancy degrees of ASCs and QCs can be explained as follows: theoretically an ASC can handle 18 containers per hour (see Sect. 2.2). Consequently, 16 ASCs can handle 288 containers per hour. A QC can unload 55 containers per hour (see Sect. 2.2). 4 QCs can handle 220 containers per hour. Therefore, we can conclude that in the unloading process the QCs are the bottleneck. Consequently, the occupancy degrees of the QCs are higher than the occupancy degrees of the various ASCs.

Due to higher waiting times of QC1, QC1 has the highest occupancy degree. From the occupancy degrees of the ASCs we can conclude firstly that it is possible to use less and longer stacking lanes. In this way, less ASCs can ensure an comparable efficient unloading process. Secondly, it becomes clear that the dispatching rule, which dispatches vehicles to the ASCs at the farthest travel distances (see Sect. 2.2), results in higher occupancy degrees for ASCs with higher numbers.

During the unloading process QCs need to wait with a container for AGVs or a free location in the buffer area. These waiting times are incorporated in the occupancy degrees of QCs. The waiting time of an empty ASC for a full AGV is not incorporated in the time an ASC is occupied with storing a container in the stack. Consequently, the periods of time that ASCs need to wait for AGVs equal the periods of time that ASCs are not occupied. Thus, contrary to QCs, these waiting times can be derived directly from the occupancy degrees of ASCs. Therefore, we only study the waiting times of QCs in the next section.

The occupancy degrees of the vehicles are relatively high. It might be expected that using some more vehicles might result in lower occupancy degrees and lower unloading times of a ship. In the Section 4.4, we will study the effect of the number of vehicles on unloading times of the ship. To ensure an efficient unloading process with small unloading times, AGVS and ALVs are inferior to the bottleneck in the

Table 7. Average occupancy degrees of QCs, ASCs and vehicles

	AGV model	ALV model
vehicle	0.970	0.966
QC1	0.988	0.983
QC2	0.961	0.962
QC3	0.948	0.957
QC4	0.948	0.958
ASC1	0.649	0.596
ASC2	0.643	0.621
ASC3	0.637	0.641
ASC4	0.627	0.659
ASC5	0.621	0.667
ASC6	0.614	0.678
ASC7	0.606	0.684
ASC8	0.604	0.690
ASC9	0.617	0.696
ASC10	0.655	0.701
ASC11	0.721	0.715
ASC12	0.792	0.734
ASC13	0.849	0.766
ASC14	0.889	0.813
ASC15	0.910	0.868
ASC16	0.922	0.931

Table 8. Average total waiting times of quay cranes in seconds

Model	QC1		QC2		QC3		QC4	
	time	half width	time	half width	time	half width	time	half width
AGV	1420	48	416	34	212	26	124	18
ALV	876	145	225	48	61	18	13	6

process the QCs. In other words, we prefer that AGVs and ALVs need to wait for a QC with a container instead of QCs that need to wait for AGVs or ALVs. Therefore, we study the waiting times of QCs for AGVs and ALVs in the next section.

4.3 Waiting times of quay cranes

Either, QCs need to wait for an empty AGV or for an empty position in the buffer area. Table 8 indicates the average waiting times of the four QCs.

It can be noticed that the waiting times for AGVs are much longer than for ALVs. By using buffers a QC only needs to wait for an ALV if the buffer is full. Thus, a QC only has to wait after positioning 4 container in the buffer without a vehicle arriving at the buffer. Furthermore, waiting times at QC1 are higher than for the other QCs. QC 1 is located farthest from the queue with vehicles. Vehicles have to travel for a longer period of time before they arrive at QC1. Furthermore, the chance of congestion is higher if a vehicle needs to drive a longer distance.

4.4 Minimum number of vehicles required

In this section, we determine the minimum number of vehicles required to minimise unloading times of the ship. Therefore, we have varied the number of vehicles used in both models. First, we study the impact of the number of automated guided vehicles on the unloading times of a ship. Figure 9 presents the relation between the number of AGVs and the unloading times.

From the results in Figure 9 we can conclude that the minimum number of vehicles required to minimise unloading times of a ship equals 37 AGVs. One can notice stabilisation in the number of vehicles required. Using more than 37 vehicles does not have any effect on unloading times of the ship. An explanation for this stabilisation is the fact that using more vehicles results in the occurrence of more congestion. This congestion might result in longer travel times from ship to stack and back to ship. These longer travel times might result in higher unloading times. This phenomenon can be compared with traffic jams on the highway during rush hours. We can conclude that we need 5 vehicles more than in the original model to obtain minimal unloading times of the ship.

Figure 10 presents the unloading times of a ship for a varying number of automated lifting vehicles.

From Figure 10 we can conclude that the unloading times of the ship stabilise by using 23 vehicles or more. Therefore, the minimum number of ALVs required to minimise unloading times equals 23.

We can conclude that there exist a difference in the number of vehicles of both types of 38%. By using 37 AGVs the unloading times of ship equal 34471 seconds. The unloading times of the ship are 33912 seconds by using 23 ALVs.

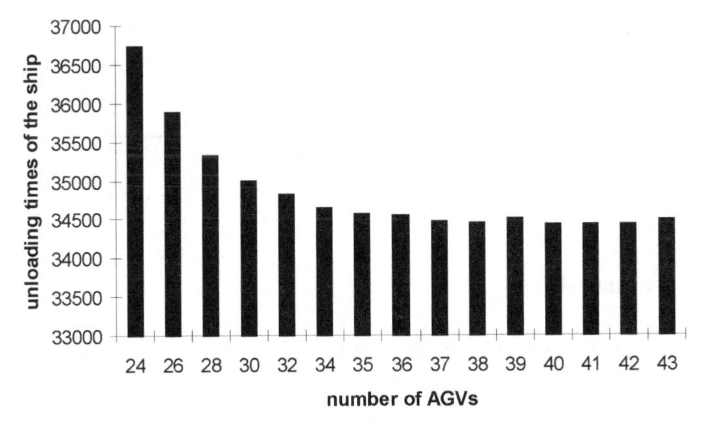

Fig. 9. Unloading times (in seconds) of a ship for a varying number of AGVs

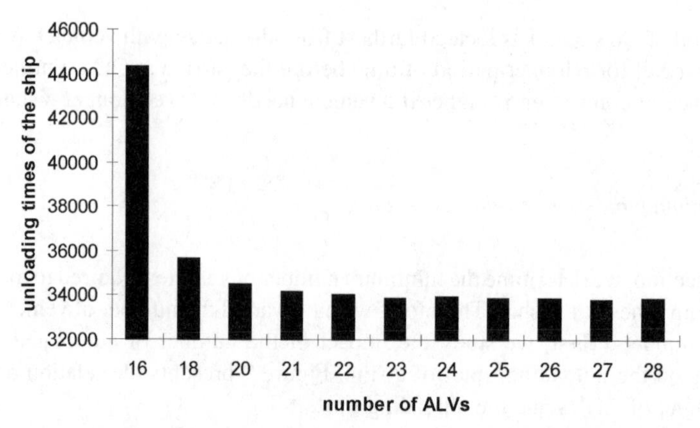

Fig. 10. Unloading times (in seconds) of a ship for a varying number of ALVs

4.5 Comparison of both models

In this section, we compare the results of both models. From the results, we can conclude that using ALVs or AGVs results in similar unloading times. However, the number of vehicles required to obtain these unloading times differ. In the case that we use AGVs we need to use 38% more vehicles than if we use ALVs.

The waiting times of QCs are higher in the case that AGVs are used. By using ALVs, a crane can place containers in a buffer. As a result, the unloading process is decoupled of the transportation process and the crane only needs to wait for a vehicle in the case the buffer area is full. The occupancy degrees of both types of vehicles and at the cranes are comparable.

Except for comparing these models on the above mentioned criteria, we can compare the costs of both models. The costs at the end of 2000 to buy an AGV were 318,000 euro. An ALV costs (at the end of 2000) 455,000 euro. These data are obtained from interviews with logistics managers of Europe Combined Terminals (ECT) in Rotterdam by the end of the year 2000. The total costs for 37 AGVs are 11,766,000 euro. The total costs for 23 ALVs are 10,465,000 euro. If we use the purchasing costs of equipment as the only criteria to decide on, we should choose for using ALVs to transport containers. However, in practice there are also costs for the layout related to the type of equipment and the costs for the related planning and control system. Furthermore, to compare both models accurately we have to study the effect of several factors on the number of vehicles. Therefore, we have performed a sensitivity analysis. We present the results of this analysis in Section 5.

5 Sensitivity analysis

In this section, we present the results of a sensitivity analysis. We have varied one of the variables in the model while the other variables remain the same. We have studied the effect on the number of vehicles required of both types of vehicles by varying a variable. The following variables have been studied:

- size of the various buffer areas in the ALV model
- speed of empty vehicles
- twin load capacity of AGVs
- cycle times of the quay cranes

These variables are of different categories influencing the decision process (see also Sect. 1). The size of the buffer area is related to the layout of the terminal. In terminals with severe restrictions on space, the buffer area might be smaller than in terminals with enough space. The speed of empty vehicles and the capacity of AGVs are technical aspects of types of equipment. The cycle times of quay cranes depend on the specifications of the crane and capabilities of the crane driver.

The results of the various studies are presented in the next sections.

5.1 Size of the buffer area at the QC

We vary the size of the buffer area from 1 to 4 containers. We study the impact of the number of lifting vehicles, in a terminal with these buffer sizes, on the unloading times of the ships. Similar to the patterns in Figures 9 and 10 we can notice a stabilisation in the number of ALVs. Table 9 represents the minimum number of vehicles required to minimise unloading times by a varying size of the buffer area. Furthermore, the related unloading times of the ship are given. We also consider the case of a buffer area with capacity 0, which is the AGV model.

From the data in Table 9 we can conclude that the size of the buffer influences the number of vehicles required to minimise unloading times of the ship. The larger the buffer area the smaller the number of vehicles required. A larger buffer can easier compensate for vehicles returning later to the buffer.

The layout of the terminal, therefore, influences the choice of equipment. If the buffer size is small due to severe restrictions in space one should keep in mind that the choice for lifting vehicles results in the use of more vehicles than in a terminal with more space for buffer areas. In the case that the buffer area has capacity 1 or 2, 26 vehicles are required. The related purchasing costs are 11,830,000 euro. These costs are higher than purchasing 37 AGVs.

5.2 Size of the buffer area at the ASC

We vary the size of the buffer area at the ASC from 1 (original case) to 4 containers. We study the impact of the buffer size at the ASCs on the number of ALVs required

Table 9. Number of vehicles (#) required by varying buffer size at the QCs and related average unloading times (in seconds)

Model	buffer=0		buffer=1		buffer=2		buffer=3		buffer=4	
	time	#	time	#	time	#	time	#	time	#
AGV	34471	37								
ALV			34121	26	33951	26	33962	25	33912	23

Table 10. Number of vehicles (#) required by varying buffer size at the ASCs and related average unloading times (in seconds)

Model	buffer=0		buffer=1		buffer=2		buffer=3		buffer=4	
	time	#	time	#	time	#	time	#	time	#
AGV	34471	37								
ALV			33912	23	33876	23	33862	23	33836	23

to minimise unloading times. Again we can notice a stabilisation in the number of ALVs. Table 10 represents the results, unloading times and number of ALVs, from this experiment. We have also added the results for the case in which the buffer area has a capacity of zero (AGV model).

From the data in Table 10 we can conclude that the size of the buffer area does not influence the number of vehicles required. This can be explained as follows: from the occupancy degrees in Table 7, one can conclude that the occupancy degrees of ASCs are low compared to the occupancy degrees of the QCs. In Section 4.2 we explained that the QCs are the bottleneck in the unloading process. QCs are almost continously occupied with unloading containers. To ensure that ALVs do not have to wait for QCs buffer areas can be used. Clearly, in that case the larger the buffer areas, the less vehicles are required (see Sect. 5.1). From the occupancy degrees of the ASCs it can be concluded that, already with a buffer area of capacity 1, the ASCs have much free time and are not continously in operation. Therefore, the effect of increasing the buffer area does not result in a more efficient transportation process with less vehicles. From the unloading times in Table 10, we can conclude that a very small decrease in unloading times (76 seconds) can be obtained by increasing the capacity of the buffer area from 1 to 4. Probably, a small number of vehicles need to wait a little shorter at the ASCs to deliver the container. This small decrease in unloading times is outweighed by the costs of extra capacity that is required at the ASCs to create larger buffer areas.

5.3 Speed of empty vehicles

In this section, we study the impact of the speed of empty vehicles on the unloading times of a ship and the number of vehicles required. In the models we vary the empty speed. The remaining variables remain the same. Thus, the capacity of the buffer area at the QCs equals again 4 and at the ASCs 1.

One might expect that we need more vehicles if empty speeds decrease. Vehicles arrive later at the ship. Consequently, more vehicles need to be used to acquire similar minimal unloading times of a ship. The other way round, we expect that increasing speeds of vehicles results in using less vehicles.

We have varied the empty speed of AGVs from 4.5 m/s, 5.5 m/s (original case), 6.5 m/s to 7.5 m/s. Also, in this situation the number of AGVs required stabilise. In Table 11, we give the number of vehicles required by varying empty speeds and related unloading times in seconds.

Table 11. Number of AGVs (#) required by varying empty travel speeds and related average unloading times (in seconds)

Model	4.5 m/s		5.5 m/s		6.5 m/s		7.5 m/s	
	time	#	time	#	time	#	time	#
AGV	34474	38	34471	37	34529	37	34462	37

Table 12. Number of ALVs (#) required by varying empty travel speeds and related average unloading times (in seconds)

Model	6 m/s		7 m/s		8 m/s		9 m/s	
	time	#	time	#	time	#	time	#
ALV	33877	23	33912	23	33902	23	33861	23

From the results in Table 11, we can conclude that decreasing the speed with 1 m/s results in an increase in the number of vehicles required with 1. Increasing the speed does not result in a decrease of the number of vehicles required. Only a small decrease in unloading times can be noticed, if we have an empty travel speed of 7.5 m/s.

We have varied the empty speed of ALVs from 6 m/s, 7 m/s (original case), 8 m/s to 9 m/s. Again, we can notice a stabilisation in the number of vehicles required to minimise unloading times. Table 12 presents the number of vehicles required by varying empty travel speeds of ALVs and related unloading times.

The minimum number of ALVs required to minimise unloading is not influenced by the empty travel speed of the vehicle.

It can be concluded that this specific technical aspect of the equipment does not influence the number of vehicles. In other words, this aspect does not influence the choice for a certain type of equipment in the case that the purchasing costs are used as decision criteria.

5.4 Twinload capacity of AGVs

From a technical perspective it is also interesting to study the impact of transporting two containers simultaneously by one AGV. Currently, no information is available concerning the technical feasibility of twin load capacity at ALVs. Therefore, we did not perform simulation studies with ALVs with capacity two. However, it might be expected that the consequences for the number of AGVs are similar to the consequences for the number of ALVs, due to the fact that all other variables remain the same.

After receiving the first container the AGV needs to wait until the moment the QC has placed the second container on the AGV. In the period of time the QC is unloading the first container, the AGV travels a little further to allow the QC to place the container on the back position. We study the effect of increasing the capacity of AGVs from one to two on the number of AGVs required such that unloading times are minimised. Similar to the patterns in Figure 9 we can notice a stabilisation in the number of AGVs. The minimum number of AGVs with capacity two required to minimise the unloading time (34161 seconds) equals 35. Consequently, by introducing AGVs with capacity two the number of vehicles required decreases with two.

This small decrease in the number of AGVs can be explained as follows: on average each 65.9 seconds a QC unloads a container and places it on an AGV. On average 131.8 seconds are required to position two containers on an AGV with capacity two. By using the travel distance and the maximum empty and full speed of a vehicle, we can derive the theoretical minimum driving time of an AGV. This driving time equals 283 seconds. At the stack the ASC lifts the first container from the AGV and stores it in the stack. This takes on average 197.7 seconds. The AGV needs to wait at the ASC until the ASC has returned from storing the container. After storing the first container, the ASC is able to lift the second one. The AGV needs to wait on average 197.7 seconds to leave the stack and to travel back to the QCs. The total time between leaving and arriving again at the QCs equals on average 480.7 seconds. Using these data the theoretical average number of vehicles required equals 4 AGVs per quay crane. By performing a similar calculation for AGVs with capacity one, the theoretical average number of vehicles required equals 5 AGVs per quay crane. Consequently, theoretically a difference of four vehicles can be obtained by using AGVs with capacity two.

The total costs for 37 AGVs are 11,766,000 euro (see Sect. 4.5). The total costs for 35 AGVs with capacity one equal 11,130,000. Consequently, the extra costs for producing AGVs with twin load capacity should be smaller than 18171 euro per AGV to obtain advantage from the reduction in the number of vehicles required.

5.5 Cycle times of the quay cranes

The cycle times depend on the specifications of the cranes and the capabilities of the crane drivers. In this section, we examine the effects of decreasing cycle times. To increase the performance of a terminal, the management wants to improve all various processes in a terminal. Improvement in the processes at a quay crane can be realised by decreasing cycle times and not by increasing cycle times. Therefore, we only examine decreasing cycle times.

We decrease the intervals of the cycle times (see Table 2) with 5, 10 and 15 seconds. We study the effect of decreasing cycle times on the number of vehicles required to minimise unloading times. Table 13 presents the results of this experiment.

We can notice several facts from the results in Table 13. First, the decrease in unloading times as a result of decreasing cycle times from −10 to −15 seconds is smaller than decreasing cycle times from −5 to −10 seconds. This can be explained as follows: by unloading faster, containers are available for transport earlier. As a result, automated stacking cranes need to store more containers in the same period of time. Cycle times of ASCs remain the same and as a result, vehicles have to wait longer at the stack before they can return to the ship. Consequently, QCs need to

Table 13. Number of AGVs and ALVs (#) required by varying cycle times and related average unloading times (in seconds)

Model	0 seconds		−5 seconds		−10 seconds		−15 seconds	
	time	#	time	#	time	#	time	#
AGV	34471	37	32928	39	31794	42	31421	44
ALV	33912	23	32132	26	30213	32	29671	34

wait relatively longer for vehicles. Therefore, the decrease in unloading times is relatively smaller when cycle times decrease with 15 seconds.

Secondly, we can notice that the number of AGVs required increases with 7 vehicles in the case that cycle times decrease with 15 seconds. The minimum number of ALVs required increases with 10 vehicles in the case that cycle times decrease with 15 seconds. The impact of decreasing cycle times on the number of ALVs required is higher than on the number of AGVs required. By unloading faster, the pressure on the buffer increases. More ALVs are required to transport containers from the buffer such that waiting times of the cranes are restricted. The increase in waiting times of QCs is, therefore, larger in the case that we use lifting vehicles. Quay cranes already needed to wait for non lifting vehicles and as a result the number of AGVs required extra is smaller to minimise unloading times.

By unloading on average 15 seconds faster, 23% more AGVs are required than ALVs. The purchasing costs of 44 AGVs are 13,992,000 euro. The purchasing costs for 34 ALVs are 15,470,000 euro. The purchasing costs for ALVs are higher than for AGVs. We can conclude that the decrease of cycle times influences the decision concerning the type of equipment required, if we use the purchasing costs as decision criterion.

6 Conclusions

At automated container terminals containers are transshipped from one mode of transportation to another. The internal transport of containers from the ship to the stack (and vice versa) can be done by automated vehicles. We consider two different types of vehicles, namely automated guided vehicles (AGVs) and automated lifting vehicles (ALVs). AGVs and ALVs transport containers over fixed paths. An ALV is able to lift a container from the ground by itself. On the contrary, an AGV needs a crane to receive a container and a crane is required to take a container off the vehicle. We have studied the effect of using AGVs and ALVs on the unloading times of a ship by using simulation models. First, we have described characteristics of both types of equipment and the way of modelling in Section 2. As simulation software we have used Arena 3.5. Section 3 describes how we have implemented both models in Arena.

In choosing a certain type of equipment at a container terminal, several criteria can be used. We have focussed on waiting times of cranes, occupancy degrees of QCs, ASCs and vehicles, unloading times of the ship and the number of vehicles required at the terminal. In Section 4 we describe the results of the various experiments. It can be concluded that using either AGVs or ALVs does not impact the unloading times of a ship. However, we can notice a difference in the number of vehicles required. 38% more AGVs need to be used than ALVs. If we use purchasing costs of vehicles as a criteria in choosing a certain type of equipment, ALVs should be chosen to transport containers. However, for an accurate comparison we have also performed a sensitivity analysis in Section 5.

Successively, we have varied the size of the various buffer areas, the speed of empty vehicles, the capacity of AGVs and the cycle times of quay cranes. First, we conclude that the size of the buffer area at the QCs influences the number of vehicles required to minimise unloading times of the ship. A larger buffer area can compensate more easily for vehicles returning later. As a result, less vehicles are required when a larger buffer is used. Due to the low occupancy degrees of the ASCs in the original case (buffer area with capacity 1), it can be seen that enlargement of buffer areas at the ASCs does not impact the number of ALVs required to minimise unloading times.

Varying the empty speed of a vehicle does not impact the number of vehicles required to minimise unloading times. Using AGVs with capacity two instead of capacity one results in a small decrease in the number of AGVs required. Finally, we have varied the cycle times of quay cranes by decreasing them with 5, 10 and 15 seconds. It can be seen that a decrease in the time between two successive containers results in the use of more vehicles. Consequently, the layout of the terminal (buffer size at the quay crane) and technical aspects of equipment (quay crane and vehicle) influence the choice of equipment, when we consider purchasing costs as a decision criteria.

References

Agerschou H, Lundgren H, Sørensen T, Ernst T, Korsgaard J, Schmidt LR, Chi WK (1983) Planning and design of ports and marine terminals. Wiley, Chichester

Celen HP, Slegtenhorst RJW, van der Ham RT, Nagel A, de Vos Burchart, R, Berg J, van den Evers JJM, Lindeijer DG, Dekker R, Meersmans PJM, de Koster MBM, van der Meer R, Carlebur AFC, Nooijen FJAM (1997) FAMAS NewCon. Concept TRAIL report. Technical University of Delft and Erasmus University Rotterdam

Egbelu PJ, Tanchoco, JMA (1984) Characterization of automatic guided vehicle dispatching rules. International Journal of Production Research 22(3): 359–374

Fisher EL, Farber JB, Kay MG (1988) MATHES: an expert system for material handling equipment selection. Engineering Costs and Production Economics 14: 297–310

Johansen RS (1999) Gate solutions. Paper presented at the Containerport and Terminal Performance Conference, Amsterdam

Kim KH, Bae JW (1999) A dispatching method for automated guided vehicles to minimize delays of containership operations. International Journal of Management Science 5(1): 1–25

Law AM, Kelton WD (1991) Simulation modeling and analysis, 2. edn. McGraw-Hill, New York

Leeper JH (1988) Integrated automated terminal operations. Transportation Research Circular 33: 23–28

Meersmans PJM (2002) Optimization of container handling systems. Ph.D. Thesis, Tinbergen Institute 271, Erasmus University Rotterdam

Rath E (1973) Container systems. Wiley, New York

Sinriech D (1995) Network design models for discrete material flow systems: a literature review. International Journal of Advanced Manufacturing Technology 10: 277–291

Van der Meer JR (2000) Operational control of internal transport. ERIM Ph.D. Series Research in Management 1. Erasmus University Rotterdam

Van Hee KM, Wijbrands RJ (1988) Decision support systems for container terminal planning. European Journal of Operational Research 34(3): 262–272

Vis IFA (2002) Planning and control concepts for material handling systems. ERIM Ph.D. Series Research in Management 14. Erasmus University Rotterdam

Wan TB, Wah ELC, Meng LC (1992) The use of information technology by the port of Singapore authority. World Development 20 (12): 1785–1795

Welgama PS, Gibson PR (1996) An integrated methodology for automating the determination of layout and materials handling system. International Journal of Production Research 34 (8): 2247–2264

Simulation-based performance evaluation of transport vehicles at automated container terminals

Chang Ho Yang, Yong Seok Choi, and Tae Young Ha

Shipping, Logistics and Port Research Center, Korea Maritime Institute,
NFFC B/D, 11-6, Shinchun-Dong, Songpa-Ku, Seoul, 138-730, Korea
(e-mail: {yang;drasto;haty}@kmi.re.kr)

Abstract. Significant unproductive and costly waiting occurs during AGV (Automated Guided Vehicle) use, both under the CC (Container Crane) and in the blocks compared to that of a manual yard tractor. A possible solution to this problem is that, in the design of ACT (Automated Container Terminals), ALV (Automated Lifting Vehicles), which can load and unload their own containers, be considered as an alternative. In this paper, the objective is to analyze how increases in the use of ALVs rather than AGVs affects the productivity of ACTs. We derived four inferences regarding the cycle time of vehicles and verified their validity in a simulation. A simulation model of an ACT with perpendicular layout was developed and is described in this paper. From the results of the simulation analysis, we determined the savings effect by cycle time and the required number of vehicles. We demonstrated that the ALV is superior to the AGV in both productivity and efficiency principally because the ALV eliminates the waiting time in the buffer zone.

Keywords: AGV (Automated Guided Vehicle) – ALV (Automated Lifting Vehicle) – ACT (Automated Container Terminal) – Simulation model – Productivity

1 Introduction

The age of automated container handling systems has commenced with them having been adopted in several ports and harbors in the world, such as the ECT (Europe Combined Terminal) in Rotterdam, Thamesport in London, the CTA (Container Terminal Altenwerder) in Hamburg, and the PPT (Pasir Panjang Terminal) in Singapore. Furthermore, management engaged in container operations all over the world has begun to take a keen interest in automated container handling systems.

Correspondence to: Y. S. Choi

To solve problems such as the increase in operation time due to larger and wider vessels, high personnel expenses, lack of qualified manpower, and for the higher efficiency of land utilization, modern port facilities generally and ACT (Automated Container Terminals) specifically have become the foci of interest around the world. In some advanced countries, ideas promoting efficiency in response to the above maladies, have been practically implemented. ECT, the most modern ACT in the world, honed the concept of ACT 10 years ago; in 1997, they started operating the second generation ACT while currently they are testing technologies for the third generation ACT.

Van der Meer has studied the performance of several well-known on-line dispatching rules and some case-specific dispatching rules for container transshipments in terms of pre-arrival information [9]. Bish theoretically analyzed operation problems in container terminals such as the vehicle dispatching problem, the vehicle scheduling location problem, and the vehicle routing problem [2].

Evers and Koppers proposed the distributed traffic control technique for AGVs (Automated Guided Vehicles), which led to the development of the conceptual model [4]. Vis et al. discussed a method for determining the minimum number of AGVs needed for completing a given set of delivery tasks without causing a delay in semi-automated container terminals. To determine the minimum number of AGVs, they suggested a dispatching method that utilized the maximum flow problem technique [10]. Kim and Bae discussed a dispatching method for AGV to minimize delays during containership operations [7]. Lim et al. proposed a dispatching method for AGVs based on a bidding concept and discussed the theoretical rationale behind the distributed dispatching method. And the performance of the method is compared with that of a popular dispatching rule using simulation [8]. Grunow et al. discussed a priority rule based algorithm for the dispatching of multi-load vehicles in automated container terminals [5].

Most research has focused on equipment allocation and dispatching problems and its results set limits on the particular operational situation based upon port properties, and it is not sufficient to analyze the vehicle operations of a specific ACT equipped with newly sophisticated container handling equipment. Regarding vehicle dispatching in container terminals, in studies to minimize the delays of ship operations, the works of Bish, Van der Meer, and Vis et al. failed to explore the characteristics of loading/unloading operations or the strategies of equipment allocation.

The objective of this study is to determine the extent to which an increased number of ALVs (Automated Lifting Vehicles), compared with AGVs, improves the productivity of ACTs. To determine the number of transport vehicles needed to transport containers between seaside (berth apron) and landside (yard) in an ACT, we compare the required number of ALVs and AGVs at a given service level and their impact on cycle time. This study investigates both the seaside and landside operations and tries to synchronize the goal productivity of CCs (Container Cranes) and the delivery tasks of these transport vehicles.

Because a simulation analysis reflects the characteristics of a system more precisely than a mathematical analysis does, we analyze the effect of vehicle operations using a simulation. Therefore, we develop the simulation model taking into con-

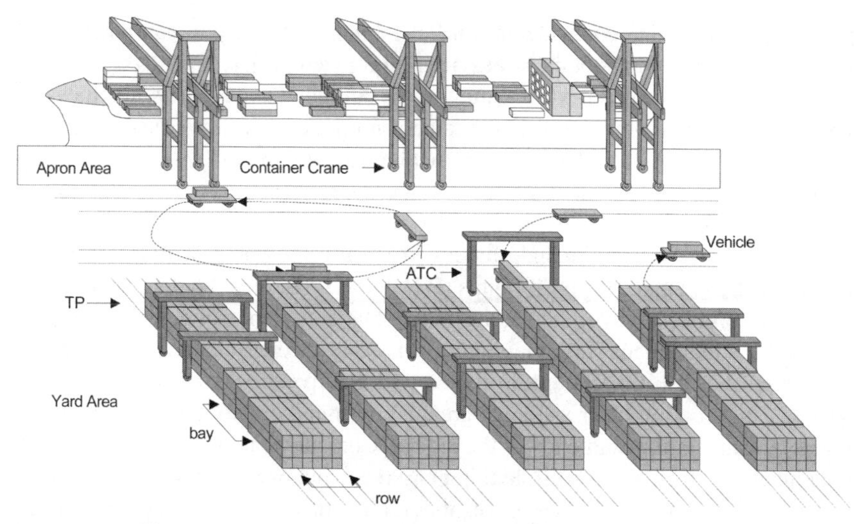

Fig. 1. Container operations in an automated container terminal

sideration the characteristics of an ACT. In this simulation model, we assume that the cycle time of transport vehicles consists of the moving time between apron and yard, the waiting time in buffer zones, both apron and yard, and the waiting time for loading and unloading by the CC and the ATC (Automated Transfer Crane). These time elements are used to define the state transition model and are utilized as performance measures.

2 Automated container terminal

The container handling operations in an ACT, generally, consist of seaside (or apron) and landside (or yard) operations, as shown in Figure 1.

Seaside operations are divided into three kinds of operations, that is to say, container handling with ships at the apron, container shifting between apron and yard and container handling in the yard. The first operation, container handling with ships at the apron, is performed by CCs, excepting roll on/roll off operations. AGVs are used for the shifting of containers between apron and yard. Container handling in the yard is performed by ATCs.

The work to shift containers between apron and yard has been automated by using AGVs. But, special devices, such as a chassis-loader system used at ECT, are required for the container handling duties of the AGV in cooperation with CCs at seaside.

The proposed operation system of the Kwangyang ACT in Korea is partially modified from the system adopted by ECT and CTA. In flow planning, the access of trucks to the yard has been minimized and the intersecting of trucks and AGVs has also been prohibited in the yard. AGVs are used for the transport of containers between apron and yard, and ATCs are used for any work done in the yard. By providing for rails that go inside the yard, the amount of transport equipment used for bringing in and out containers has been minimized [11].

A yard that is located next to a quay stores outbound containers while they are waiting for a ship or other inbound containers, until a vehicle arrives and transports them. Containers are stored in a yard formation, called a "block". A stack is a group of containers on top of each other. A bay is a group of container side by side. A tier is a layer of containers.

The layouts of ACTs include blocks in a yard, the paths of AGVs, and the locations of pickup and drop-off points (P/D points). Terminal layouts can be categorized according to perpendicular and parallel layouts, depending on the direction of an ATC's movement. Terminal layouts can vary in their characteristics and in their requirements of operation. In the case of a perpendicular layout, AGVs and outside trucks do not enter a storage area in a terminal due to their TPs (Transfer Points) located at the both ends of the blocks. Containers are placed end to end in long blocks. In the case of a discharging or a receiving operation, an ATC receives containers at the TPs, and then moves them to the designated storage position. In the case of a loading or a delivery operation, an ATC brings containers from storage, then transfers them to AGVs or outside trucks at the TPs. The flow of containers and the operation of equipment in a terminal are simple, but ATCs are expensive to purchase and maintain [1].

In this study, it is assumed that the container terminal is automated, in which the yard crane is an ATC such as an automated stacking crane, as exists at the ECT, and the prime mover is an AGV or an ALV. The terminal layout used in the simulation is a perpendicular layout such as the ECT and CTA.

3 Transport equipment model

3.1 Vehicle model

A vehicle as transport equipment is a component that can transport containers from the loading point (apron/yard) to their destination (yard/apron). Every vehicle possesses data relating to its speed and states such as the loaded state and the empty state, its pickup and delivery points of origin and destination, and its load. In some cases, vehicle speed may be considered a decision variable in the design of the transport equipment, and the vehicle speed is specified by the user of the simulation model. A vehicle has a process description in which the trip between two types of cranes and its activations are defined. In this study, we consider two types of vehicles: AGV and ALV. We assume that the AGV must wait for lifting to be performed by the crane, but the ALV can load and unload its own containers.

The ALV system combines the best of both worlds, and which aims to enable the full potential of CC productivity to be exploited. The ALV system is essentially a small transporter SC (Shuttle Carrier), an integral part of port operation, which has been in existence since the 1950's, but now updated with modern technology and performance to find its place in high throughput container terminals and to more efficiently bridge the gap between CCs and ATCs [3].

ALVs are used for transporting containers between apron and yard and for loading and unloading trucks. In this case, an ATC is used only for stacking. Therefore, the ALV improves the productivity of CCs and the utilization of the buffer zone under a CC. It also improves the productivity of the ATC, and TPs under the ATC are used as buffer zones both at the seaside and at the landside of the yard [6].

The ALV is a vehicle that can both load and unload containers and travels from the loading point to its destination under its own power. In this study, the control activates an idle ALV. The ALV loads a container at its origin, then it travels to the destination of the container, unloads the container and is idle again. In this case the idle ALV is sent to a destination unloaded.

The ALV has an independent work cycle from that of the CC and does not need to wait for a transport vehicle. This factor can reduce both the cycle time and the number of transport vehicles. The loading and unloading of a ship's containers operate under a CC portal within a buffer zone, but when using an AGV it operates under a backreach with no buffer. Therefore, it can reduce the work cycle of the CC. Through these improvements with the ALV, ship turnaround time can be reduced.

3.2 State transition model

An AGV is a vehicle that is loaded and unloaded by both a CC and an ATC, but a vehicle that can also travel from the loading point to its destination under its own power. In this study, the control assigns an idle AGV to respond to the needs of the loading/unloading equipment such as the CC or the ATC, which load the AGV. Then the crane activates the AGV, which moves to the destination with a user-determined speed. At the destination, the AGV activates the crane and waits until unloading is completed. Then the AGV is empty again and the empty AGV is sent to a destination unloaded.

The AGV model utilizes state transition, a system consisting of six states, which is shown in Figure 2; it also contains two conditions in order to check the availability of the buffer zone. At the end of the AGV's task, the state of AGV transitions from a moving state to an idle state. The time interval between the start time and the end time of a relevant event is defined as the transition time of a state.

The ALV model also utilizes state transition, as shown in Figure 3. It includes eight conditions: the upper four conditions represent situations of the yard and the lower four conditions represent situations of the apron. From those conditions, we know that the ALV model is different from the AGV model. In addition, the ALV model divides the waiting time of the operation into loading and unloading periods.

From Figure 2 and Figure 3, Table 1 summarizes the state-transition systems of the AGV and the ALV mentioned above, and shows the relationship between these and the elements of cycle time proposed in this study.

In the case of the AGV model, the basic element is moving time $(\overline{G_3}+\overline{G_6})$ without considering crashing, the fixed element is waiting time for loading/unloading $(\overline{G_2}+\overline{G_5})$, and the reducible element is waiting time in the buffer zone $(\overline{G_1}+\overline{G_4})$. In case of the ALV model, similarly, the basic element is moving time $(\overline{L_3}+\overline{L_6})$ without considering crashing, the fixed element is loading/unloading time $(\overline{L_2}+\overline{L_5})$,

Table 1. Organization of state transition for transport vehicles

AGV		ALV	
$\overline{G_1}$	Mean waiting time in buffer of apron	$\overline{L_1}$	Mean waiting time in buffer of apron
$\overline{G_2}$	Mean waiting time for loading/unloading by CC	$\overline{L_2}$	Mean loading/unloading time by CC or itself
$\overline{G_3}$	Mean moving time from apron to yard	$\overline{L_3}$	Mean moving time from apron to yard
$\overline{G_4}$	Mean waiting time in buffer of yard	$\overline{L_4}$	Mean waiting time in buffer of yard
$\overline{G_5}$	Mean waiting time for loading/unloading by ATC	$\overline{L_5}$	Mean loading/unloading time by ATC or itself
$\overline{G_6}$	Mean moving time from yard to apron	$\overline{L_6}$	Mean moving time from yard to apron
CT_G(Cycle Time) = $\sum_{i=1}^{6} \overline{G_i} / N_G$		CT_L(Cycle Time) = $\sum_{j=1}^{6} \overline{L_j} / N_L$	
N_G : Average number of transported containers per AGV		N_L : Average number of transported containers per ALV	
$\overline{G_1} + \overline{G_4} =$ Mean waiting time in buffer zone		$\overline{L_1} + \overline{L_4} =$ Mean waiting time in buffer zone	
$\overline{G_2} + \overline{G_5} =$ Mean waiting time for loading/unloading		$\overline{L_2} + \overline{L_5} =$ Mean loading/unloading time	
$\overline{G_3} + \overline{G_6} =$ Mean moving time between apron and yard		$\overline{L_3} + \overline{L_6} =$ Mean moving time between apron and yard	

Table 2. Patterns of traffic lane selection

Task \ Traffic area			Traveling lane from apron to yard			Block	Traveling lane from yard to apron		
No	Origin	Destination	Apron	Change	Yard	TP	Yard	Change	Apron
1	CC1	A block	A1	C1→C1	Y1	assigned TP	Y1	C1→C1	A1
2	CC2	C block	A2	C2→C3	Y3	assigned TP	Y1	C1→C2	A2
3	CC3	E block	A3	C3→C5	Y5	assigned TP	Y1	C1→C3	A3

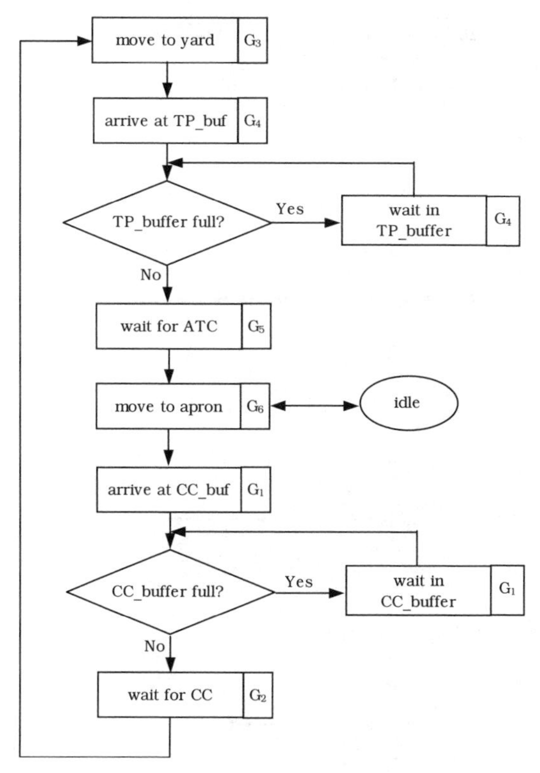

Fig. 2. AGV model by state transition

and the reducible element is waiting time in the buffer zone $(\overline{L_1} + \overline{L_4})$. The loading/unloading time of the ALV is selected according to a small value between the CC work time and the ALV work time. The sum of the cycle time of the AGV, $\sum_{i=1}^{6} \overline{G_i}$ means that the sum of the state transition times (i.e. $\overline{G_i}$) is equal to the completion time of all the tasks, while the sum of the cycle of the ALV, $\sum_{j=1}^{6} \overline{L_j}$ includes the six state- transition times.

We expect that the cycle time of the AGV is longer than that of the ALV and that the waiting time of the ALV is shorter than that of the AGV under the same conditions because the buffer zone of the ALV is more flexible. We assume that the speed of the transport vehicle is identical in both cases; however, it may be different due to the performance of the CC and the ATC.

Therefore, we derive the following four inferences.

1) $\overline{G_1} + \overline{G_4} \quad \leq \quad \overline{L_1} + \overline{L_4}$
   The ALV can lift a container by itself without the help of cranes, and the loading/unloading time is reduced and the length of waiting time in the buffer is also reduced. Therefore, the ALV arrives early at the next buffer where it may encounter more downtime, waiting in the buffer.

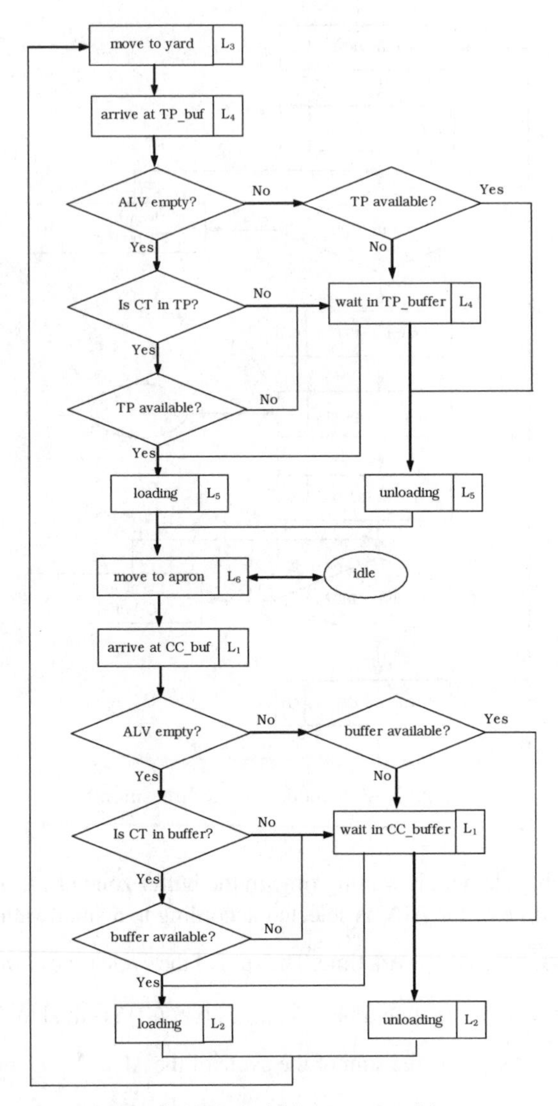

Fig. 3. ALV model by state transition

2) $\overline{G_2} + \overline{G_5} \quad \geq \quad \overline{L_2} + \overline{L_5}$
   In most cases, the loading and unloading of containers is performed by the CC and the ATC. However since the ALV performs the loading/unloading of containers by itself, this time is reduced.
3) $\overline{G_3} + \overline{G_6} = \overline{L_3} + \overline{L_6}$
   If vehicles are assigned the same task, the travel distance is equal, both for the AGV and the ALV. Therefore, an equality of moving time is realized.

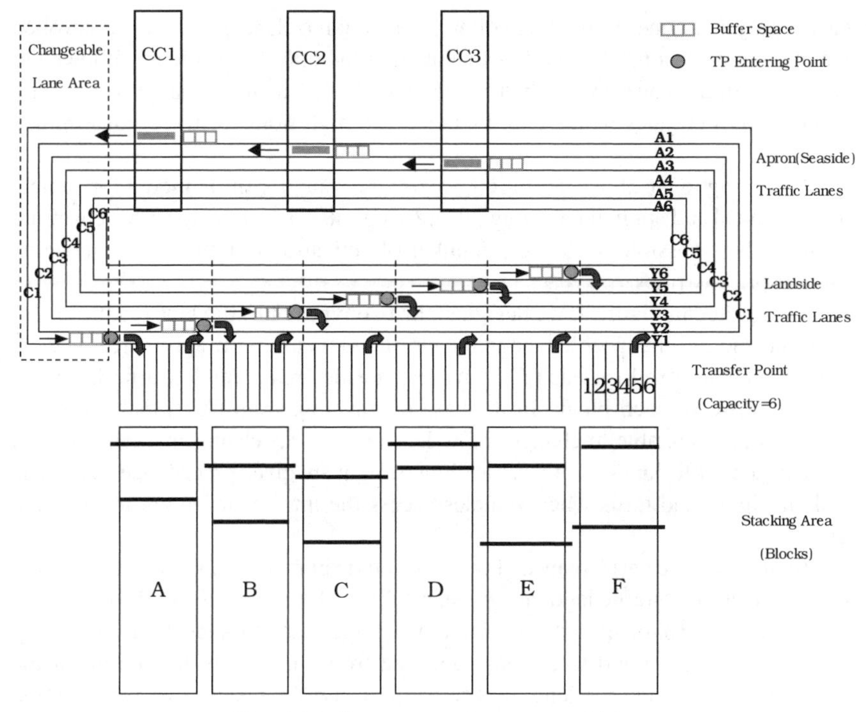

Fig. 4. Traffic patterns of transport vehicles

4) $\sum_{i=1}^{6} \overline{G_i} \geq \sum_{j=1}^{6} \overline{L_j}$

It is possible to reduce the waiting time and the loading/unloading time of the AGV. After all, the cycle time of the ALV is smaller than that of the AGV. If $\sum_{i=1}^{6} \overline{G_i}$ dose not equal $\sum_{j=1}^{6} \overline{L_j}$, then the difference, $\sum_{i=1}^{6} \overline{G_i} - \sum_{j=1}^{6} \overline{L_j}$, represents a savings effect that reduces the number of ALVs.

The above four inferences show the effectiveness of analysis through simulation results.

3.3 Vehicle traffic model

For the setup of traffic control systems, two concepts have to be taken into consideration: central traffic control and distributed traffic control. Although there are many forms and combinations of control in use today, the most popular form of zone control is that in which the zones are individually controlled for movements within the zones and centrally controlled to interface with other zones in the system. The general rule is that only one vehicle can occupy a zone [12]. In this paper, a setup is described in which several vehicles may access an intersection area at the same time so the priority in the area has to be that no vehicles collide within the area. A basic entity is a zone (in the form of an intersection of lanes), which is controlled via mutual exclusion. This implies that at any time only one vehicle is allowed to pass through the zone, even when the layout of a terminal may show zones with several adjacent lanes which could guarantee a high usage of the predetermined capacity.

According to the above definition, we use the state transition model for vehicle traffic control and apply the waiting and moving rules or criteria to activate waiting vehicles. It is possible to define a number of activation criteria, like sequencing rules or priority rules.

As depicted in Figure 4, we designed the two types of traffic lanes as the exclusive lane and the changeable lane in the terminal layout. In the apron area, vehicle traffic lanes are fixed for each CC. In the landside area, vehicle traffic lanes are fixed in terms of each block. In the case of traveling from CC (block) to block (CC), vehicles are able to change traffic lanes within the changeable lane area. In the landside traffic lanes, vehicles traveling straight are given precedence over those making right-hand turns when vehicles access the intersection point at the same time.

To illustrate the performance of the task of transporting a container, the patterns for the selection of traffic lanes are given in Table 2. A task consists of the origin and the destination. To travel to the destination, vehicles have to select both a traveling lane from the apron and to the yard and one from the yard to the apron for the given task. To change the lane, it is possible to use the changeable lanes. In the landside area, the waiting area consists of a buffer space from which to enter the transfer point, and the transfer point has a capacity of 6. In front of the transfer point, vehicles enter in an assigned order. Therefore, Task No. 3 of Table 3 performs the sequence of A3-C3-C5-Y5- assigned TP-Y1-C1-C3-A3 in one cycle.

4 Simulation and analysis

4.1 Simulation model

The simulation model was developed using Visual BASIC, a general-purpose language. Figure 5 shows the configuration of our simulation model as displayed by user input.

Figure 5 describes the behavior of the ACT in the simulation model. Before the simulation run, the user can input parameters and information through the user interface to construct an experimental simulation model. During the simulation, the model can interface with the state transition model and the vehicle traffic model, and displays a 2D animation.

We used a reduced-size model, which represents a part of the ACT, to show an example of model building by simulation, because the real container terminal requires a massive amount of data for terminal operation and planning. However, this model considers the values of the various parameters of facility operations and some measures are used to evaluate the effectiveness of the transport vehicle model.

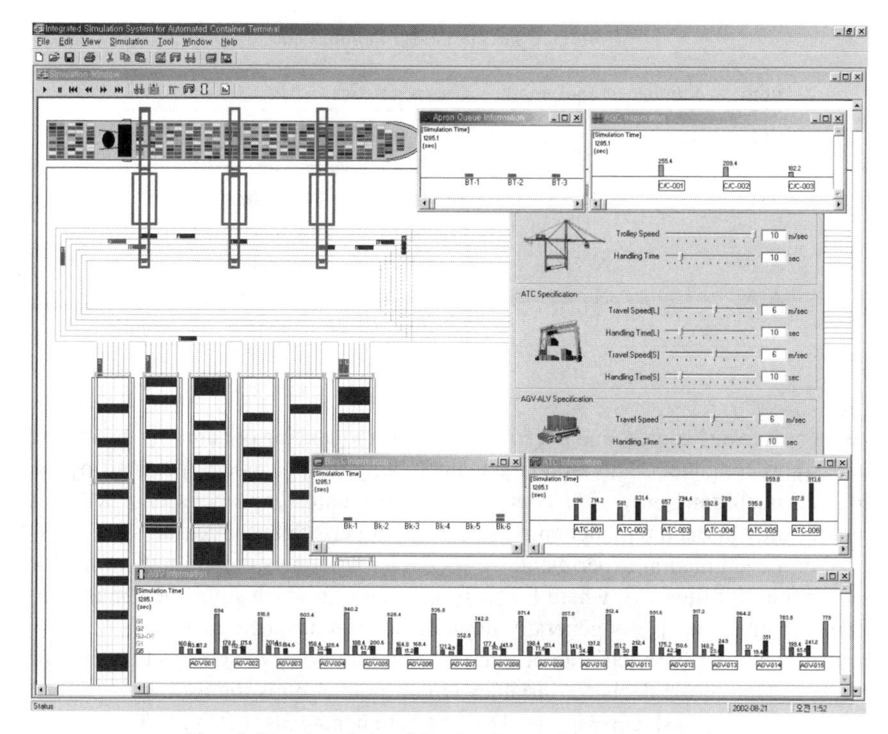

Fig. 5. User interface of the developed simulation model

4.2 Experiment design

The scope of our experiment's model is as follows. The container yard has 6 blocks and 2 ATCs per block. A block includes 40 bays, each bay consisting of 10 rows by 5 tiers. The berth has a quay and 3 CCs.

Table 3 shows the equipment characteristics. The operation of cranes such as the CC and the ATC involves trolley speed and loading/unloading time. Similarly, the ALVs have travel speed and loading/unloading time as characteristics, but the AGVs have only travel speed. The loading/unloading time of the ALVs is shorter than that of cranes. The number of cranes is fixed at 3 CCs and 12 ATCs. The number of vehicles is a decision variable determining the level of more efficient performance.

We start the simulation exercise by generating the ship arrivals at the berth. Upon berthing, equipment (CC, ATC, and AGV or ALV) is assigned to the ships according to the experiment conditions, and the discharging operations start at time 0. Loading operations follow the discharging operations. When all the container operations (discharging and loading) are finished, we terminate the simulation run. In our model, we complete all the discharging operations before starting the loading operations. In reality, however, the discharging and loading operations may be alternated to suit the stowage plan. It should be realized that the total time for container operations is mostly determined by the number of import and export

Table 3. Characteristics of equipment

Item	Number or equipment	Characteristics
CC	3	Trolley speed = 3m/second, Loading/unloading time = 30 second
ATC	12	Travel speed = 2m/second, Loading/unloading time = 30 second
AGV	N	Travel speed = 3m/second
ALV	N	Travel speed = 3m/second, Loading/unloading time = 20 second

containers to be handled, and not so much by the sequencing of handling the import and export containers. Therefore, the main assumption in a model (i.e., starting the loading operations only after completing all the unloading operations) will not significantly affect the model outputs in terms of performance measures.

For the simulation experiment, we used a terminating simulation that runs for the duration of time T_E, where E is a specified event which stops the simulation. The stopping time T_E is generally unpredictable in advance, and, in fact, T_E is probably the response variable of interest, as it represents the completion time of the task. One of the decision variables in our simulation is the estimated $E(T_E)$, the mean time to task completion. The task is to handle 300 lifts per CC, 900 lifts in total including the number of import and export containers.

The two alternative models have the same equipment and the same operation flows except for the transport vehicle. Therefore, the operation strategies used the same parameters as input.

4.3 Experiment results

Since one of the important factors that affect the turn-around time of ships is the productivity of the CCs at the apron, we use productivity of the CCs as an evaluation measure and determine the vehicle speed through a preliminary test.

Until the simulation terminates, we use a given travel speed for vehicles. The results produced by simulation, Table 4 and Table 5, are provided. The productivity limit of the CCs is 27.60 lifts/hr as shown in Table 4 and Table 5. We consider that 27.60 lifts/hr is the productivity limit of a CC.

In Table 4 and Table 5, the gray areas show that the feasible solutions are satisfied by the goal productivity of the CCs. In the case of 1m/second travel speed, a high productivity of a CC working with an AGV is infeasible. The savings effect of the vehicles does not effect the efficiency at the speed of over 4m/second. In most cases, when the travel speeds of the two vehicles are the same, fewer ALVs than AGVs are required.

With constraints on the productivity of the CCs, the speed of the vehicle is a necessary determinant in order to maximize the difference of the number of assigned vehicles and to minimize the number of allocated vehicles. Due to the fact that both

Table 4. CC productivity by AGV travel speed

Number of AGVs	AGV travel speed						
	1m/second	2m/second	3m/second	4m/second	5m/second	6m/second	7m/second
3	5.24	9.59	13.16	16.14	18.61	21.07	22.40
6	10.47	18.92	25.59	27.21	27.23	27.38	27.48
9	15.65	26.68	27.57	27.60[a]	27.60[a]	27.60[a]	27.60[a]
12	20.59	27.59	27.60[a]	27.60	27.60	27.60	27.60
15	25.18	27.60[a]	27.60	27.60	27.60	27.60	27.60
18	27.55	27.60	27.60	27.60	27.60	27.60	27.60
21	27.57	27.60	27.60	27.60	27.60	27.60	27.60

[a] The minimum number of vehicles among feasible solutions

Table 5. CC productivity by ALV travel speed

Number of ALVs	ALV travel speed						
	1m/second	2m/second	3m/second	4m/second	5m/second	6m/second	7m/second
3	5.39	10.12	14.26	17.87	21.03	24.20	26.52
6	10.77	20.20	27.60[a]	27.60[a]	27.60[a]	27.60[a]	27.60[a]
9	16.12	27.60[a]	27.60	27.60	27.60	27.60	27.60
12	21.46	27.60	27.60	27.60	27.60	27.60	27.60
15	26.77	27.60	27.60	27.60	27.60	27.60	27.60
18	27.60[a]	27.60	27.60	27.60	27.60	27.60	27.60
21	27.60	27.60	27.60	27.60	27.60	27.60	27.60

[a] The minimum number of vehicles among feasible solutions.

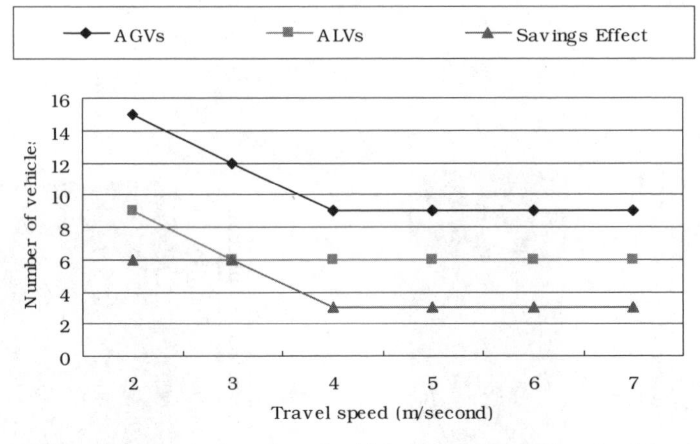

Fig. 6. The required number of vehicles for a given CC productivity

the 2m/second and 3m/second speeds are suitable for the purpose of determining the difference between the two types of transport vehicles, both speeds are possible, as shown in Figure 6. Here we select the 3m/second travel speed to show the savings effect of the vehicles due to the higher-percent decrease in the number of required vehicles (i.e., 2m/second: 15→9 (40%), 3m/second: 12→6 (50%)).

From the results, we determine the number of vehicles with the completion time constraint. We also consider the goal productivity of the CC for 27.60 lifts/hr at a 3m/second travel speed. Consequently, Table 6 shows that 6 ALVs (2 ALVs per CC) and 12 AGVs (4 AGVs per CC) satisfy constraints including the completion time, the goal productivity of the CCs, and the travel speed of vehicles, at the same time.

4.4 Results implementation

Cause analysis of cycle time The completion time, $\sum_{i=1}^{6} \overline{G_i}$ (or $\sum_{j=1}^{6} \overline{L_j}$), consists of 6 $\overline{G_i}$(or $\overline{L_i}$) and the cycle time of a vehicle, $\sum_{i=1}^{6} \overline{G_i}/N_G$ (or $\sum_{j=1}^{6} \overline{L_j}/N_L$), consists of each $\overline{G_i}/N_G$ (or $\overline{L_i}/N_L$). Using the cycle time, it is not easy to analyze the effect of the differences. So, we analyze it using $\sum_{i=1}^{6} \overline{G_i}$ and $\sum_{j=1}^{6} \overline{L_j}$(=T$_E$), but they include all the elements of the cycle time.

The simulation results of the AGV and the ALV cases are presented in Tables 7 and 8, respectively. In Tables 7 and 8, positive values of $\overline{G_1}$ and $\overline{L_1}$ are possible due to the buffer size of the apron being 1, but $\overline{G_4}$ and $\overline{L_4}$ are 0 because the buffer size in a block is sufficient at six. We expect a savings effect by the reduction of the waiting time. In fact, as must happen, $\overline{G_4}$ and $\overline{L_4}$ are 0 due to sufficient buffer size, and $\overline{G_3} + \overline{G_6}$ is equal to $\overline{L_3} + \overline{L_6}$.

Table 6. Comparison of CC productivity between AGVs and ALVs

Number of vehicles	AGV (3m/second)			ALV (3m/second)		
	Completion time (sec.)	CC Waiting time (sec.)	CC productivity (lifts/hr)	Completion time (sec.)	CC waiting time (sec.)	CC productivity (lifts/hr)
3	82,060.00	42,752.00	13.16	75,760.00	36,281.33	14.26
6	42,204.00	3,501.67	25.59	39,130.00	178.33	27.60[a]
9	39,168.00	135.33	27.57	39,130.00	0.00	27.60
12	39,130.00	9.00	27.60[a]	39,130.00	0.00	27.60
15	39,130.00	9.00	27.60	39,130.00	0.00	27.60
18	39,130.00	8.67	27.60	39,130.00	0.00	27.60
21	39,130.00	8.67	27.60	39,130.00	0.00	27.60

[a] The minimum number of vehicles among feasible solutions.

Table 7. Cycle time of AGVs

Number of vehicles	AGV (3m/second)					
	$\overline{G_1}$	$\overline{G_2}$	$\overline{G_3}+\overline{G_6}$	$\overline{G_4}$	$\overline{G_5}$	$\sum_{i=1}^{6}\overline{G_i}$ (=T_E)
3	0.00	8,986.00	63,200.33	0.00	9,873.67	82,060.00
6	40.50	5,338.00	31,149.83	0.00	5,675.67	42,204.00
9	1,684.33	12,308.22	21,012.33	0.00	4,163.11	39,167.99
12[a]	10,458.33	9,757.33	15,816.17	0.00	3,098.17	39,130.00
15	16,165.93	7,809.07	12,676.40	0.00	2,478.60	39,130.00
18	19,978.50	6,507.78	10,577.61	0.00	2,066.11	39,130.00
21	22,699.19	5,578.10	9,083.14	0.00	1,769.57	39,130.00

[a] The minimum number of vehicles among feasible solutions.

Table 8. Cycle time of ALVs

Number of vehicles	ALV (3m/second)					
	$\overline{L_1}$	$\overline{L_2}$	$\overline{L_3}+\overline{L_6}$	$\overline{L_4}$	$\overline{L_5}$	$\sum_{j=1}^{6}\overline{L_j}$ (=T_E)
3	0.00	6,218.00	63,348.67	0.00	6,193.33	75,760.00
6[a]	8.50	4,703.33	31,360.50	0.00	3,057.67	39,130.00
9	2,998.44	13,012.89	21,068.00	0.00	2,050.67	39,130.00
12	12,008.59	9,761.33	15,822.08	0.00	1,538.00	39,130.00
15	17,409.46	7,809.07	12,681.07	0.00	1,230.40	39,130.00
18	21,015.33	6,507.78	10,581.56	0.00	1,025.33	39,130.00
21	23,586.66	5,578.10	9,086.38	0.00	878.86	39,130.00

[a] The minimum number of vehicles among feasible solutions.

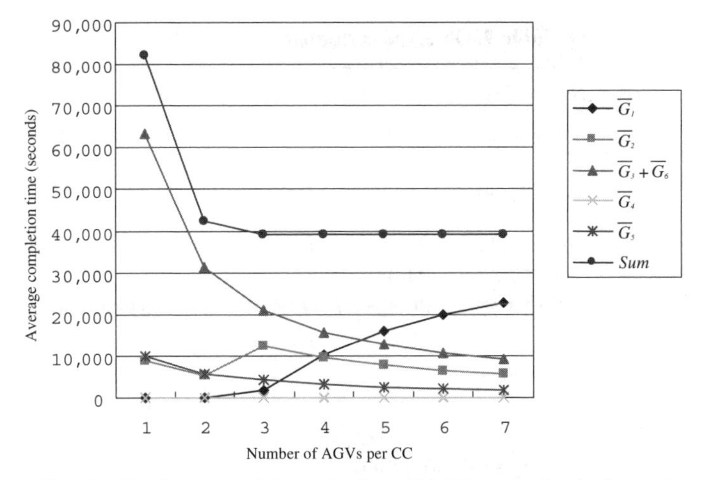

Fig. 7. The elements of the cycle time of AGVs at 3m/second speed

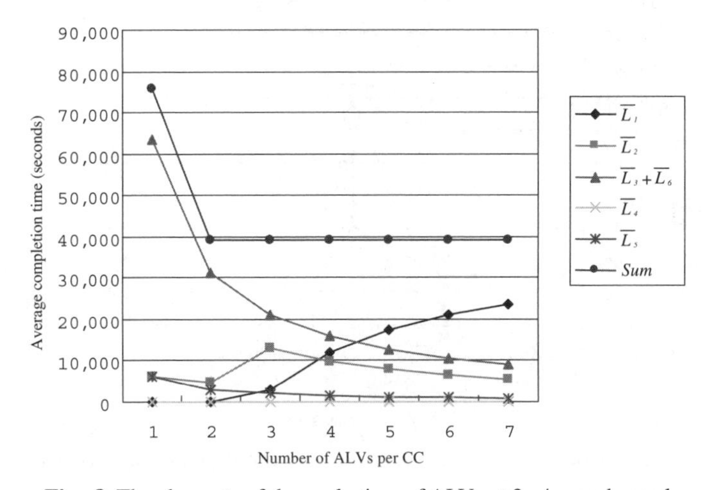

Fig. 8. The elements of the cycle time of ALVs at 3m/second speed

From Tables 7 and 8, we validate the above-mentioned four inferences within the feasible solutions.

Though these inferences are obvious, they each include all the variable effects. An illustration of the cycle time for AGVs in Table 7 is shown in Figure 7. And an illustration of Table 8 is shown in Figure 8. Figures 7 and 8 show how to compose time elements related to the cycle time for vehicles based on the assigned number of vehicles per CC.

The performance measures for vehicles after the termination of the simulation time are listed in Table 9. The $\overline{G_1}$ of the AGVs during the T_E is more than 26.73% compared with the ALVs because, occasionally, the AGVs have to wait for the CCs. The $\overline{G_2}$ of the AGVs based on the mean waiting time for loading and unloading by CC is on the increase. The difference between $\overline{G_3}+\overline{G_6}$ and $\overline{L_3}+\overline{L_6}$ is due to the assigned number of vehicles, and in the case of the ALVs, the total routing distance

Table 9. Observed performance of vehicles

AGV(No = 12)			ALV(No = 6)		
Measures	Time (unit: second)	% cycle	Measures	Time (unit: second)	% cycle
$\sum_{i=1}^{6} \overline{G_i}$	39,130.00	100.00	$\sum_{j=1}^{6} \overline{L_j}$	39,130.00	100.00
$\overline{G_1}$	10,458.33	26.73	$\overline{L_1}$	8.50	0.02
$\overline{G_2}$	9,757.33	24.94	$\overline{L_2}$	4,703.33	12.02
$\overline{G_3} + \overline{G_6}$	15,816.17	40.42	$\overline{L_3} + \overline{L_6}$	31,360.50	80.14
$\overline{G_4}$	0.00	0.00	$\overline{L_4}$	0.00	0.00
$\overline{G_5}$	3,098.17	7.92	$\overline{L_5}$	3,057.67	7.81

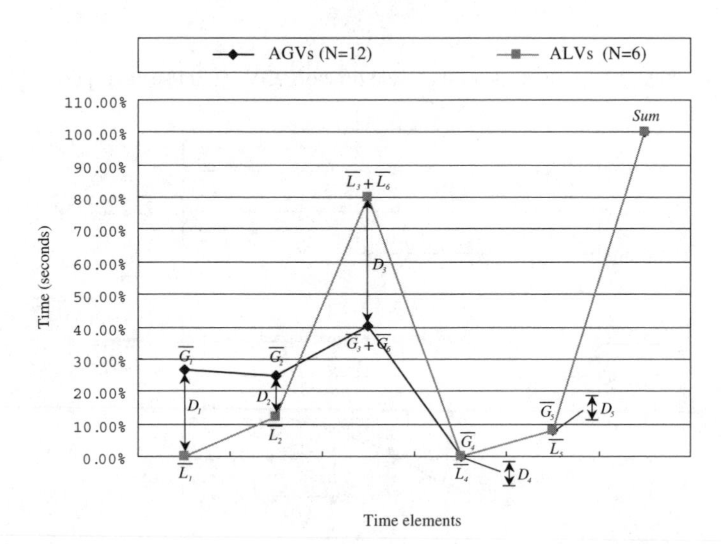

Fig. 9. Comparison of time elements between 12 AGVs and 6 ALVs

is increased by a factor of 2. The mean waiting time for loading and unloading by ATCs, $\overline{G_5}$ and $\overline{L_5}$, provides almost the same results. The zero value of $\overline{G_4}$ and $\overline{L_4}$ means that the waiting time in the buffer was not due to the sufficient buffer size of the block. In the case of the ALVs, 80% of the cycle time was spent on moving because the ALV has a shorter waiting time and a longer travel distance compared with the AGV.

In Figure 9, we know that the difference between $(\overline{L_3} + \overline{L_6})$ and $(\overline{G_3} + \overline{G_6})$ can cover the sum of $(\overline{G_1} - \overline{L_1})$ and $(\overline{G_2} - \overline{L_2})$. Therefore, it is possible to reduce the number of ALVs to 6.

5 Replacement range

We discover the measure to determine the replacement range by savings effect.

Table 10. Savings effect of D'

AGV \ ALV	6	9	12	15	18	21
12	81.0	2,095.0	3,211.0	3,736.9	4,147.9	4,440.5
15	−1,158.1	855.9	1,822.9	2,497.7	2,908.8	3,201.4
18	−1,983.1	30.9	1,057.9	1,672.8	2,083.8	2,376.4
21	−2,576.2	−562.2	464.8	1,079.7	1,490.7	1,783.3

Gray area : positive savings effect for waiting at seaside.

Using the above-mentioned four inferences, the notations are defined as follows:

$$D_1 = \overline{G_1} - \overline{L_1} \tag{1}$$
$$D_2 = \overline{G_2} - \overline{L_2} \tag{2}$$
$$D_3 = (\overline{L_3} + \overline{L_6}) - (\overline{G_3} + \overline{G_6}) \tag{3}$$
$$D_4 = \overline{G_4} - \overline{L_4} \tag{4}$$
$$D_5 = \overline{G_5} - \overline{L_5} \tag{5}$$
$$D' = D_3 - (D_1 + D_2) + (D_4 + D_5) \tag{6}$$
$$D'' = D_3 + (D_1 + D_2) - (D_4 + D_5) \tag{7}$$

In notations (1)–(5), D_i indicates the different effects of an AGV and an ALV on the set of time elements. In notations (6) and (7), we let D' denote the savings effect of waiting at seaside and D'' denote the savings effect of waiting at landside. Let $D' \times D''$ denote the savings effect of waiting in the buffers. By using $D' \times D''$, it is a rather simple calculation to derive an expression for the savings effect of assigning vehicles.

Comparing Table 10 with Table 11, a common savings effect, $D' \times D''$, can be determined, as shown in Table 12. Note that positive values are used to determine the savings effect in order to be able to assign between an AGV and an ALV.

In reference to the above notation, we see that if $D' \times D'' > 0$, then there is a positive savings effect and the number of vehicles is reducible until $D' \times D''=0$. Hence, $D' \times D''$represents a savings effect determining the required number of vehicles. But if $D' \times D'' < 0$, then there is a negative savings effect and the number of vehicles is irreducible.

Figure 10 shows that the contour of the savings effect was divided into 4 degrees. From the above-mentioned results, we know that the gray area in Figure 9 indicates a positive savings effect. Within the positive savings effect, it is possible to assign any combination of AGVs and ALVs while maintaining the balance of productivity. For example, as shown in Figures 10, 9 ALVs can replace 18 AGVs.

The results of the application of this simulation model have been encouraging. Our model has demonstrated how the use of a replaceable range can reduce the number of vehicles by assigning ALVs in the places of AGVs.

Table 11. Savings effect of D''

AGV \ ALV	6	9	12	15	18	21
12	31,007.7	8,408.7	−3110.2	−10,007.1	−14,617.1	−17,900.1
15	38,526.7	15,927.3	4,408.5	−2,488.4	−7,098.4	−10,381.4
18	43,548.9	20,949.9	9,431.1	2,534.2	−2,075.9	−5,358.9
21	47,130.9	24,531.9	13,013.1	6,116.2	1,506.1	−1,776.9

Gray area : positive savings effect for waiting at landside.

Table 12. Savings effect of $D' \times D''$

AGV \ ALV	6	9	12	15	18	21
12	2.51E+06	1.76E+07	−9.7E+06	−3.74E+07	−6.06E+07	−7.95E+07
15	−4.46E+07	1.36E+07	8.30E+06	−6.22E+06	−2.06E+07	−3.32E+07
18	−9.64E+07	6.47E+05	9.98E+06	4.24E+06	−4.33E+06	−1.27E+07
21	−1.21E+08	−1.38E+07	6.05E+06	6.60E+06	2.25E+06	−3.17E+06

Gray area : positive savings effect between AGV and ALV.

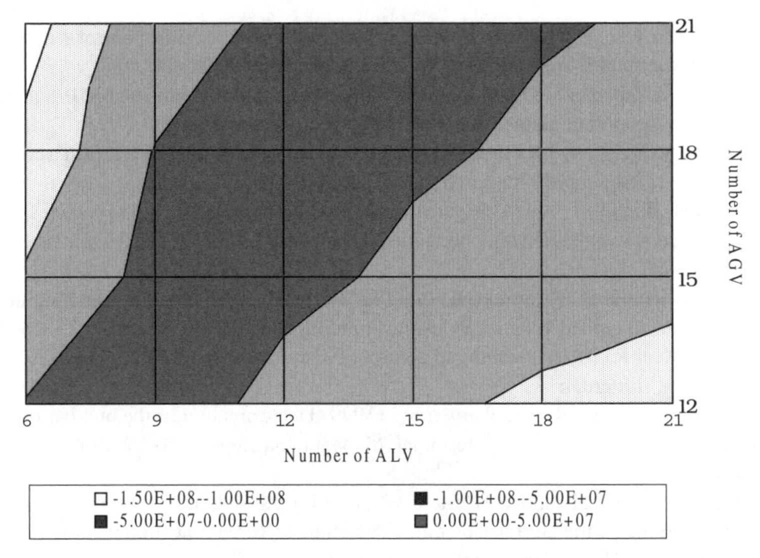

Fig. 10. Contour of savings effect comparing the AGV and the ALV

6 Conclusion

This paper presents a simulation model and a procedure governing transport vehicles of ACTs; the state transition model and the traffic model were proposed for the purpose of vehicle modeling. Using the simulation model developed for an ACT, we analyzed the travel speed of the vehicle with constraints on the productivity of the CC, and obtained the required number of vehicles and a savings effect by cycle time. Using a vehicle speed of 3m/second, we found that the number of ALVs is reducible while maintaining the same service level. This was due to the AGVs spending more time than the ALVs waiting in the ATC and CC buffer zones, respectively. This means that the ALV applied the wasteful waiting time of an AGV to its additional moving time.

As for the results, we demonstrated that the ALV is superior to the AGV in productivity because it reduces the waiting time in the buffer zones. In addition, we know that there are savings effects by assigning a vehicle mix between AGVs and ALVs.

References

1. Bae J W (2001) Operational methods for yard equipment in automated container terminals. Ph.D. Thesis, Pusan National University, Korea
2. Bish E K (1999) Theoretical analysis and practical algorithm for operational problems in container terminals. Ph.D. Thesis, Northwestern University
3. Duinkerken M B, Ottjes J A, Evers J J M, Kurstjens, S T G L, Dekker R N, Dellaert P (1996) Simulation studies on inter terminal transport at the Maasvlakte. TRAIL Research School

4. Evers J J M, Koppers S A J (1996) Automated guided vehicle traffic control at automated container terminal. Transportation Research Part A 30 (1): 21–34
5. Grunow M, Günther H O, Lehmann M (2004) Dispatching multi-load AGVs in highly automated seaport container terminals. OR Spectrum 26: 211–235
6. Joseph J M E, Stijn A J K (1996) Automated guided vehicle traffic control at a container terminal. Transportation Research Part A 30 (1): 21–34
7. Kim K H, Bae J W (1999) A dispatching method for automated guided vehicles to minimize delays of containership operations. International Journal of Management Science 15 (1): 1–26
8. Lim J K, Kim K H, Yoshimoto K, Lee J H, Takahashi T (2003) A dispatching method for automated guided vehicles by using a bidding concept. OR Spectrum 25: 25–44
9. Van der Meer R (2000) Operational control of internal transport. Ph.D. Thesis, Erasmus University, Rotterdam
10. Vis I F A, de Koster R, Roodbergen K J (1999) Determination of the number of AGVs in a semi-automated container terminal. Management Report, No. 32-1999 of Erasmus University
11. Yang C H, Kim Y H, Choi S H, Bae J W, Lee J E (2000) A study on the system design and operations of the automated container terminal. Korea Maritime Institute
12. Zeng L, Wang H P B, Jin S (1991) Conflict detection of automated guided vehicles: A petri net approach. International Journal of Production Research 29: 865–879

Generating scenarios for simulation and optimization of container terminal logistics*

Sönke Hartmann

OR Consulting, Bornstraße 6, 20146 Hamburg, Germany, and
Institut für Betriebswirtschaftslehre, Lehrstuhl für Produktion und Logistik,
Christian-Albrechts-Universität zu Kiel, 24098 Kiel, Germany
(e-mail: soenke.hartmann@epost.de)

Abstract. This paper introduces an approach for generating scenarios of sea port container terminals. The scenarios can be used as input data for simulation models. Furthermore, they can be employed as test data for algorithms to solve optimization problems in container terminal logistics such as berth planning and crane scheduling. A scenario consists of arrivals of deep sea vessels, feeder ships, trains, and trucks together with lists of containers to be loaded and unloaded. Moreover, container attributes such as size, empty, reefer, weight, and destination are included. The generator is based on a large number of parameters that allow the user to produce realistic scenarios of any size. The purpose of this paper is to outline the parameters that are important to produce realistic scenarios of high practical relevance and to propose an algorithm that computes scenarios on the basis of these parameters. The generator discussed here has been developed within the simulation project at the HHLA Container-Terminal Altenwerder in Hamburg, Germany. Nevertheless, its structure is general enough to be applied to any other terminal as well.

Keywords: Container logistics – Container terminal – Scenario generator – Simulation

1 Introduction

Since the 1960s, the container has gained an enormous importance in worldwide trade and transportation of goods. This is due to both increasing containerization (which means that the number of goods transported in containers has steadily grown) and increasing world trade. As a consequence, new container terminals

* This research project has been carried out for the HHLA Container-Terminal Altenwerder in Hamburg, Germany, when the author was with LOGAS Gesellschaft für logistische Anwendungssysteme, Hamburg, Germany.

are being built and existing ones are extended in order to cope with the growing number of containers. In addition, the container terminals face the challenge of turning around not only more but also larger ships in the shortest possible time. To meet these objectives, container terminals employ innovative (and often automated) equipment and optimize their logistic processes. Given the increasing importance of efficient container terminal logistics, it is no surprise that practioners as well as researchers continuously develop new optimization approaches as well as simulation models.

Optimization problems include the allocation of berths to arriving vessels (see Guan and Cheung [11], Imai et al. [14,15], Lim [21]) as well as scheduling the loading and unloading operations of quai cranes (see Daganzo [5], Peterkofsky and Daganzo [24]). A berth assignment approach which simultaneously considers quai crane capacities has been developed by Park and Kim [23]. Internal transportation of containers, particularly between the quai and the stack, has been studied for straddle carriers (see Böse et al. [4], Kim and Kim [18], Steenken et al. [26]) and for automated guided vehicles (see Bae and Kim [2], Evers and Koppers [7], Grunow et al. [10]). The problem of allocating and scheduling stacking cranes has been addressed by Zhang et al. [30]. In addition, a general approach for scheduling equipment such as straddle carriers, automated guided vehicles, and stacking cranes as well as manpower has been proposed by Hartmann [13]. Finally, strategies for locating containers in the yard have been discussed for two different cases, namely import containers (see de Castilho and Daganzo [6], Kim and Kim [16]) and export containers (see Kim et al. [17], Taleb-Ibrahimi et al. [27]). An approach covering both cases has been proposed by Zhang et al. [29]. A more comprehensive overview of literature on optimization issues is given by Meersmans and Dekker [22]. While these papers consider sea port container terminals, an optimization approach for an inland container terminal (i.e., without vessels or ships) has been presented by Alicke [1].

Simulation models are developed to evaluate the dynamic processes on container terminals. This allows to generate and analyze statistics such as average productivity, average waiting time (e.g., of a quai crane waiting for a straddle carrier), and avarage number of shuffle moves in the stack. This way, potential bottlenecks can be identified. Depending on the application, detailed simulation models usually cover both the physical resources (particularly the equipment such as cranes and vehicles) and the components for control and strategies (hence, simulation models provide a testing environment for optimization algorithms). Simulation projects can be carried out when building a new terminal (see Schütt and Hartmann [25]) or when analyzing or modifying an existing one (see Legato and Mazza [20], Yun and Choi [28]). In either case, plans concerning the logistic processes (such as capacity extensions, alternative stacking strategies or new scheduling algorithms) can be tested by means of simulation before they are actually implemented. Moreover, simulation models can be employed as a decision support system for the terminal management (see Gambardella et al. [8,9]). For example, one could use the actual data of the next hours to simulate the next shift in order to evaluate impact of decisions (e.g., on the assignment of resources) in advance.

This paper introduces an approach to generate scenarios for container terminals. The goal of the generator is to produce realistic scenarios that provide all required input data for detailed simulation models of container terminals. Such data is usually necessary for simulations of container terminals to be built or extended. Of course, for such projects, appropriate real data is not available when the simulation is carried out, thus artificially generated data is needed. On the basis of adjustable parameters, the generator computes arrivals for deep-sea vessels, feeder ships, trains, and trucks including lists of containers to be unloaded and loaded. By specifying the parameters appropriately, one can generate realistic scenarios for terminals of different sizes. It is possible to adjust the distributions among the modes of transport, the ship sizes, the arrival distributions over time, and the distributions of various container properties. The generator has been developed within the simulation project of the new HHLA Container-Terminal Altenwerder in Hamburg, Germany (for details on this terminal see Baker [3]). Nevertheless, it is general enough to cover almost any seaport container terminal.

The paper is organized as follows. After a description of the general concept, it outlines the parameters that have been identified as being important to produce realistic scenarios for container terminal logistics. Next, we define a procedure to construct scenarios on the basis of these parameters. The procedure is then evaluated by comparing the generated scenarios with real-world statistics. Finally, we give a detailed discussion of applications of the proposed generator and close the paper with a few concluding remarks.

2 General concept

In this paper, a scenario for a container terminal is defined as the data concerning all ship, train, and truck arrivals within a specified horizon including the information related to the containers being delivered or picked up. That is, for each day within the horizon, the scenario includes the arrival time of any ship, train, and truck together with the number of containers delivered and picked up. For each of these containers, detailed information on size (20' oder 40'), type (reefer, oversized etc.), and destination are given. Before we describe the parameters and algorithms of the generator in detail, this section summarizes some aspects of the overall concept.

2.1 Modes of transportation

The generator distinguishes four different types of transport modes which are depicted in Figure 1. On the seaside we have large vessels for world-wide service as well as (typically smaller) feeder ships for regional service. The landside is made up by trains and trucks. With these four predefined transport modes, it is possible to adjust realistic parameters of the individual transport modes (usually, a vessel carries substantially more containers than a train). Equally important, the distribution of containers among the four transport modes can be specified (for example, one can determine how many of the containers delivered by truck will be picked up by vessel and by feeder ship). Hence, the generator allows for container

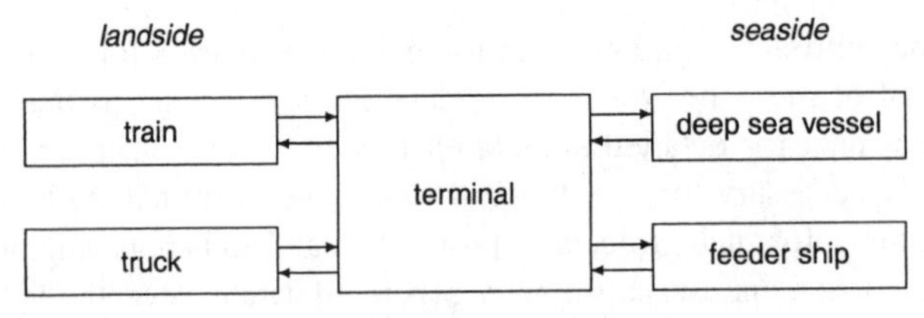

Fig. 1. Modes of transport

flows between any of the four modes. Considering the port of Hamburg (and hence also the Container-Terminal Altenwerder), the container flow between large vessels and feeder ships is of particular importance. Therefore, we distinguish these two categories instead of having just a single seaside mode "ship." Nevertheless, it is possible to adjust the parameters in a way that only one type of ship is considered.

Note that our definition of transport modes is more general than approaches in the literature (see, e.g., Zhang et al. [30]) which often consider only one seaside and one landside mode and container flows either from landside to seaside (often referred to as "export") or from seaside to landside ("import").

2.2 Interdependencies

The interdependencies within a real-world container terminal are considerably complex. This includes the distribution of containers among the transport modes on delivery and pick-up as mentioned above. At the same time, the distribution of arrival times and sizes of the means of transportation must be considered, as must be the dwell time distribution of containers (the dwell time is the time a container spends on the terminal until it is picked up). In other words, all containers must be assigned to a transport mode for delivery and to a transport mode for pick-up such that the distributions on transport modes, transport mode sizes, transport mode arrival times, and container dwell times are matched simultaneously. For example, this means that there has to be a concentration of container deliveries for a specific vessel in the days before the vessel arrives as well as a concentration of container pick-up in the days after the vessel's arrival.

With regard to these interdependencies, the goal of the generator concept is twofold: First, it should provide parameters that allow to control the distributions mentioned above. Second, it should provide an algorithm that observes these distributions simultaneously in order to reflect the interdependencies.

2.3 File concept

The generator works with several files. An overview of the file concept is sketched out in Figure 2. The user specifies the settings of the input parameters in a text file. A detailed description of the parameters is given in Section 3. After reading the parameters and computing a scenario, the generator writes two types of output files which are summarized in the following paragraphs.

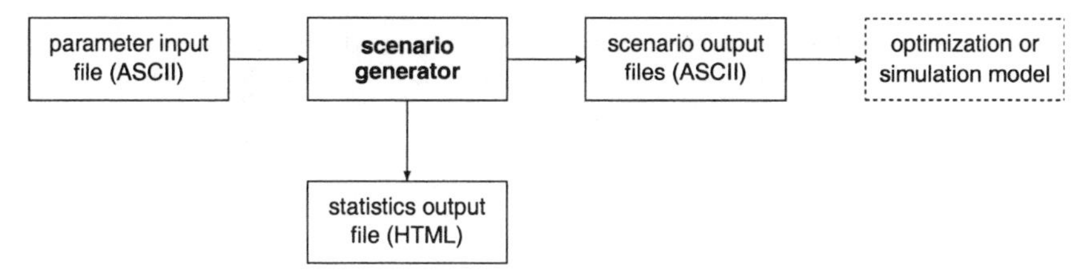

Fig. 2. File concept of the scenario generator

The scenario itself is written into text files which can then serve as input for simulation models or optimization algorithms (for each day in the horizon and each of the four transport modes, a separate file is produced in order to avoid huge output files). The scenario data contains all information that is needed to perform a detailed simulation. The structure of the output files can be summarized as follows. The means of transportation are listed in the order of arrival. Each vessel, feeder ship, train, and truck is described by a line that contains the transport mode, a unique ID, an arrival day and time, and the numbers of containers to be loaded and unloaded. For vessels and feeder ships, also the length and the number of bays are given. Then the attributes of all containers to be loaded and unloaded follow, that is, a unique container ID, size, destination, weight, information on the type (empty, reefer, dangerous content, oversized), as well as pick-up information (which transport mode will pick it up on what day).

For illustration, we consider an example output file for deep sea vessels. Table 1 displays a few lines of such a file. Two types of lines are distinguished. Lines describing a transport mode arrival start with "T:" while lines describing a container start with "C:". After the line for the transport mode arrival, there is a line for each container to be unloaded and, subsequently, a line for each container to be loaded. In what follows, the numbers of Table 1 are underlined in order to ease the explanation. The first line of Table 1 contains the information on the arrival of a vessel. The transport mode is indicated by first number; 1 stands for vessel (2 would stand for feeder, 3 for train, and 4 for truck). The vessel with ID 18 arrives on day 8 of the horizon at 7:40 in the morning. It will unload 824 containers and load 788. The length of the vessel is 280 m, it has 20 bays. The next line specifies the first container to be unloaded. The container with ID 8127 will be picked up by transport mode 2, i.e., a feeder, with ID 147 on day 14. The destination of the container (port, city) is 4. Next, 1 indicates a 40' container (0 would imply a 20' container) with a weight of 28 tons. It is neither a reefer container (0) nor an IMO container (0) nor an OOG container (0). A reefer container would be indicated by 1, and IMO and OOG containers would be indicated by a number > 0 referring to the specific type (Subsect. 3.4 provides an explanation of reefer, IMO, and OOG containers). The last number in the line, 0, indicates that it is not an empty container. The following lines describe containers 8128 and 8129, the remaining containers to be unloaded are skipped in Table 1. Then the containers to be loaded are given, starting with container 32892. The format is the same as for the containers to be unloaded. Note that the pick-up information (transport mode, ID, and day of pick-

Table 1. Example output file of the scenario generator

```
T:  1         18   8  7 40    824    788    280      20
C:      8127   2        147   14  4  1  28  0  0  0 0
C:      8128   2         85    9  3  1  23  0  3  0 0
C:      8129   3        109   12  8  1  36  1  0  0 0
[...]
C:     32892  -1         -1   -1  2  1  41  0  0  0 0
C:     32979  -1         -1   -1  1  0  17  0  0  1 0
C:     33031  -1         -1   -1  1  1  25  0  0  0 0
[...]
```

up) is set to -1 since it makes only sense for containers to be unloaded (of course, all these containers are picked up by vessel 18 arriving on day 8).

The generator produces a separate statistics file (the format of this file is HTML so it can easily be viewed with a browser). This file illustrates the generated scenario by providing an overview, and it offers a possibility to check the scenario with respect to plausibility. The input parameter settings are displayed together with the corresponding aggregated information from the generated scenario (e.g., fraction of 20' containers), which allows for a convenient comparison. For more complex parameters which describe distributions, the correlation between parameters and generated figures is given (e.g., distribution of container dwell times). Some additional information is also provided, including vessel schedules as well as the overall numbers of containers arriving and departing each day, together with the resulting number of containers in the stack.

3 Scenario parameters

The parameters allow the user to specify the scenario to be generated. The goal was to determine parameters for which planning data and/or statistics are typically available in practice because we want to produce scenarios that are as realistic as possible.

The following subsections describe the parameters which are classified into four groups. These are general parameters like the horizon and the number of containers arriving within that horizon, parameters concerning the sizes of the modes of transport, parameters dealing with arrival frequencies, and parameters reflecting container properties like size and weight distribution.

3.1 General parameters

A summary of the general parameters is given in Table 2. In what follows, they are discussed in more detail. The first of the general parameters are related to the horizon of the scenario to be generated. The length T of the horizon (in days) has

Table 2. General parameters

T	horizon (number of days)
T_1	first day, $T_1 \in \{\text{Monday}, \ldots, \text{Sunday}\}$
C_i	number of containers delivered by mode i within the horizon T
c_{ij}	fraction of containers picked up by mode j among all containers arriving by mode i, $i, j \in \{\text{vessel}, \text{feeder}, \text{train}, \text{truck}\}$
τ	maximum dwell time (days)
δ_t	fraction of containers with a dwell time of t days, $t \in \{0, \ldots, \tau\}$

to be specified along with the first day T_1 of the horizon which can be a Monday, Tuesday and so on.

Next, we consider the number of containers which arrive at the terminal within the horizon. We define parameters for the numbers C_{vessel}, C_{feeder}, C_{train}, and C_{truck} of containers that arrive at the terminal by deep sea vessel, feeder ship, train, and truck, respectively, during the horizon of T days. Clearly, this parameter set allows to produce scenarios for container terminals of any size.

For example, a medium sized terminal handles about one million containers per year. Typically, more than 50 % of the containers arrive via the seaside. For the port of Hamburg, typical figures are 45 % arriving by vessel, 15 % by feeder ship, 15 % by train, and 25 % by truck.

The following parameters take care of the distribution of the containers among the four modes of transport when picked up at the terminal. Let us consider two modes of transport $i, j \in \{\text{vessel}, \text{feeder}, \text{train}, \text{truck}\}$ (for the sake of readability, we refer to the transport modes by symbolic constants). We denote as c_{ij} the fraction of those containers that are picked up by mode j among those that arrive by mode i. In other words, of the C_i containers arriving by mode i, $C_i \cdot c_{ij}$ containers are picked up by mode j. Note that this definition leads to the condition $\sum_j c_{ij} = 1$ which must be fulfilled for each mode i. The number of containers picked up by train, for example, can be computed as

$$C_{\text{vessel}} \cdot c_{\text{vessel,train}} + C_{\text{feeder}} \cdot c_{\text{feeder,train}} + C_{\text{train}} \cdot c_{\text{train,train}} + C_{\text{truck}} \cdot c_{\text{truck,train}}.$$

If none of the containers arriving by train will be picked up by train, one simply has to set $c_{\text{train,train}} = 0$. By specifying all C_i and c_{ij}, the user has completely defined the number of containers and their distribution among the four modes of transport on arrival and pick-up.

To give the reader an idea of realistic figures (again for the port of Hamburg), 30 % of the containers arriving by vessel are picked up by feeders, 30 % by train, and 40 % by truck. Between 75 % and 95 % of the containers arriving by the other three transport modes are picked up by vessel. It is interesting to note that some (although only a very few) containers arriving by a landside mode are also picked up by a landside mode (which can be captured by the general parameter structure).

The last set of the general parameters determines the dwell time, that is, the number of days a container spends on the terminal before it is picked up. First, the maximum dwell time τ (in days) has to be specified. Then for each possible dwell time $t = 0, \ldots, \tau$ the fraction d_t of containers with that dwell time has to be given. For example, $d_1 = 0.25$ means that 25 % of the containers are picked up after one day in the stack. Additionally setting $d_2 = 0.15$ and $d_3 = 0.1$, one has specified that approximately 50 % of the containers are picked up in the first three days after arrival (which is a realistic figure). Also note that the average dwell time (which can be derived from the distribution) has a significant impact on the level to which the terminal stack capacity is filled. The average dwell time varies between terminals, but realistic values range between 3 and 6 days.

3.2 Parameters for means of transportation

The second group of parameters determines the modes of transport. An overview is given in Table 3. Let us consider the deep sea vessels first. The user can define several classes of vessels, that is, different vessel sizes. After specifying the number $n_{\text{vessel}}^{\text{class}}$ of classes, each class $k \in \{1, \ldots, n_{\text{vessel}}^{\text{class}}\}$ is described by the following parameters. Each class is associated with a range of the number of containers that a vessel of this class will unload. (Note that a range is given only for containers to be unloaded. It will become clear below that the number of containers to be loaded is determined by the generator without a parameter range.) Next, ranges for the length and the number of bays of the vessel class have to be given. Finally, the fraction (i.e., relative occurance) of vessels of each class has to be specified. In reality, the number of containers unloaded by a deep sea vessel varies within a broad range

Table 3. Parameters for means of transportation

n_i^{class}	number of classes of mode of transport $i \in \{\text{vessel}, \text{feeder}, \text{train}\}$
s_{ik}	fraction of class k for mode $i \in \{\text{vessel}, \text{feeder}, \text{train}\}$
$\underline{S}_{ik}$	minimum number of delivered containers in class k of mode $i \in \{\text{vessel}, \text{feeder}, \text{train}\}$
$\overline{S}_{ik}$	maximum number of delivered containers in class k of mode $i \in \{\text{vessel}, \text{feeder}, \text{train}\}$
$\underline{L}_{ik}$	minimum ship length in class k of mode $i \in \{\text{vessel}, \text{feeder}\}$
$\overline{L}_{ik}$	maximum ship length in class k of mode $i \in \{\text{vessel}, \text{feeder}\}$
$\underline{B}_{ik}$	minimum number of bays in class k of mode $i \in \{\text{vessel}, \text{feeder}\}$
$\overline{B}_{ik}$	maximum number of bays in class k of mode $i \in \{\text{vessel}, \text{feeder}\}$
F	factor for the maximal relation of containers unloaded and loaded by a vessel, feeder, or train

Table 4. Parameters for arrival frequencies

D_{id}	fraction of mode $i \in \{\text{vessel}, \text{feeder}, \text{train}, \text{truck}\}$ arriving on day $d \in \{\text{Monday}, \ldots, \text{Sunday}\}$
n_i^{timeslot}	number of time slots of arrivals of mode $i \in \{\text{vessel}, \text{feeder}, \text{train}, \text{truck}\}$
h_{ik}	fraction of arrivals in timeslot k for mode $i \in \{\text{vessel}, \text{feeder}, \text{train}, \text{truck}\}$
$\underline{H}_{ik}$	first hour of time slot k for mode $i \in \{\text{vessel}, \text{feeder}, \text{train}, \text{truck}\}$
$\overline{H}_{ik}$	last hour of time slot k for mode $i \in \{\text{vessel}, \text{feeder}, \text{train}, \text{truck}\}$

(often between 50 and 1500). This makes the definition of different vessel classes indispensable.

So far, we have described the parameters for deep sea vessels. The parameters for feeder ships are analogous. For trains, the same parameters apply with the exception that length and, of course, number of bays are not considered. Finally, for trucks, different classes are not considered. For the sake of simplicity, we assume that one truck either delivers a single container or picks up a single container (in the real world application this generator was originally designed for, no data on the distribution of trucks carrying different numbers of containers were available).

The last parameter F limits the relation of the number of containers unloaded and loaded by the same vessel, feeder, or train. If the number of containers to be unloaded and loaded are denoted as U und L, respectively, this parameter implies that the condition $L \leq F \cdot U$ must be fulfilled. The idea behind this is that the classes defined above only determine the number of containers to be unloaded. In order to keep the number of containers to be loaded realistic, it can be limited by this parameter. Note that the parameter F is the same for vessels, feeder ships, and trains (since trucks carry only a single container by definition, it does not apply to trucks). The application of parameter F will become more clear in Subsection 4.4.

3.3 Parameters for arrival frequencies

The third group of parameters determines the distribution of arrivals over time (for a summary see Table 4). For each of the four transport modes, the arrival frequencies can be influenced separately. This is necessary because the four modes usually have different arrival distributions over time.

First, for each day $d \in \{\text{Monday}, \ldots, \text{Sunday}\}$ and each transport mode i, the fraction D_{id} arriving on that day has to be given. For example, $D_{\text{feeder},\text{Monday}} = 0.23$ means that 23 % of the feeders arrive on Mondays. Over the seven days of the week, these fractions must sum up to 1, that is, we require $\sum_{d=\text{Monday}}^{\text{Sunday}} D_{id} = 1$ for each mode i. Next, one day can be partitioned into time slots which allow to capture the arrival rates within a day. Note that the arrival characteristics of a transport mode within a day are assumed to be the same on each day. A time slot is given by the start and end hour (ranging between 0 and 24) and by the fraction arriving within that time slot. Hence, a time slot can reflect a single hour or blocks of several hours.

Considering the notation of Table 4, we show how to reflect the arrival behavior of trucks as an example. Due to German law, trucks are not allowed to travel on Sundays which leads to a fraction of $D_{\text{truck,Sunday}} = 0$ for container terminals in Germany. Within a day, trucks arrive rather in the daytime than during the night, and their arrivals peak at early afternoon. In a rather rough approach, we could define $n_{\text{truck}}^{\text{timeslot}} = 4$ time slots of 6 hours each (e.g., the second time slot would cover the hours between $\underline{H}_{\text{truck},2} = 6$ and $\overline{H}_{\text{truck},2} = 12$). The fractions of the four time slots could be $h_{\text{truck},1} = 0.1$, $h_{\text{truck},2} = 0.3$, $h_{\text{truck},3} = 0.4$, and $h_{\text{truck},4} = 0.2$.

3.4 Parameters for container properties

The properties of a container (such as size, weight, and destination port) are important for vessel stowage as well as for determining a location for the container in the yard (cf. Kim et al. [17]). Therefore, they are relevant for detailed simulations. In what follows, we summarize the parameters that allow to specify the distribution of various container attributes. An overview is given in Table 5.

Table 5. Parameters for container properties

$p^{20'}$	fraction of 20' containers ($1 - p^{20'}$ is the fraction of 40' containers)
p^{empty}	fraction of empty containers
p^{reefer}	fraction of reefer containers
p^{IMO}	fraction of IMO containers, i.e., containers with dangerous goods
n^{IMO}	number of different IMO container types
p_k^{IMO}	fraction of containers of IMO type k among all IMO containers
p^{OOG}	fraction of OOG containers, i.e., oversized containers
n^{OOG}	number of different OOG container types
p_k^{OOG}	fraction of containers of OOG type k among all OOG containers
$n^{\text{weight},20'}$	number of weight groups for 20' containers
$p_k^{\text{weight},20'}$	fraction of 20' containers with weight group k
$\underline{w}_k^{20'}$	minimum weight of weight group k for 20' containers
$\overline{w}_k^{20'}$	maximum weight of weight group k for 20' containers
$w_{\text{empty}}^{20'}$	weight of empty 20' container
	(weight parameters for 40' containers are analogous)
n^{dest}	number of container destinations
p_{ik}^{dest}	fraction of containers with destination k among all containers picked up by mode $i \in \{\text{vessel, feeder, train, truck}\}$

The first parameter defines the container size. We distinguish 20' and 40' containers. The user has to enter the fraction $p^{20'}$ of 20' containers; $1 - p^{20'}$ is the fraction of 40' containers. A realistic value for $p^{20'}$ is between 0.4 and 0.5. Next, the fraction p^{empty} of empty containers has to be given.

The following parameters are related to special container types for which specific constraints for yard locations must be observed. Reefer containers require a connection (usually to electricity) to keep their content cool or frozen. These connections are available only in designated areas of the yard. Other containers may contain dangerous goods such as specific chemicals. These containers are often called IMO containers after the International Maritime Association (IMO). Here, complex restrictions have to be considered. For example, a minimum distance between the yard locations of two containers with certain different chemicals must be observed in order to avoid the danger of a chemical reaction. Finally, OOG (out of gauge) containers are oversized containers which require more space than normal containers. For each of these three categories, the fraction has to be given (for example, p^{OOG} denotes the fraction of OOG containers among all containers).

The IMO and OOG containers may be further classified into different groups. In both cases, the number of groups has to be specified, along with the fraction of each group. For illustration, consider the following example of $n^{\text{OOG}} = 2$ different OOG container groups. Let us assume that 30 % of the OOG containers are open top containers for which the load exceeds the container height such that no other container may be put on top of them. Moreover, 70 % of the OOG containers are open on one side such that the width of the container is exceeded by the load. These two OOG groups can then be captured by defining the fractions $p_1^{\text{OOG}} = 0.3$ and $p_2^{\text{OOG}} = 0.7$. If, however, separate groups of OOG containers are not needed, one would simply define a single group by setting $n^{\text{OOG}} = 1$ and $p_1^{\text{OOG}} = 1$. Considering the IMO containers, the parameters allow to apply the standard categorization defined by the International Maritime Association.

The next set of parameters specifies the container weight. Starting with 20' containers, the number of weight groups has to be specified. Then for each weight group the fraction of that group together with a lower and an upper weight limit are entered (the latter two can be interpreted as, e.g., metric tons). Subsequently, weight groups for 40' containers have to be defined. Finally, the weights of empty 20' and empty 40' containers are required.

The last set of parameters for container attributes is concerned with the destination of a container. First, the user specifies the number n^{dest} of destinations. Considering container terminals in Hamburg as an example, the destinations would subsume destination ports for large vessels such as Hong Kong and Singapore, destination ports for feeder ships such as Gothenborg, and destinations for trains such as Munich. As the destination of a container may be unknown (especially if a container will be picked up by a truck), it is useful to define an additional destination representing an unknown destination. Obviously, the destination is related to the means of transportation that picks up the container. Therefore, for each of the four transport modes, a separate distribution among the destinations can be given. For mode of transport $i \in \{\text{vessel}, \text{feeder}, \text{train}, \text{truck}\}$ and destination $k \in \{1, \ldots, n^{\text{dest}}\}$, p_{ik}^{dest} represents the fraction of containers with destination k among all containers picked up by transport mode i. Clearly, this implies $\sum_{k=1}^{n^{\text{dest}}} p_{ik}^{\text{dest}} = 1$ for each mode $i \in \{\text{vessel}, \text{feeder}, \text{train}, \text{truck}\}$. For example, let us assume that destination $k = 3$ refers to an Asian port like Hong Kong. If 10 % of the container picked up by large vessels go to Hong Kong but none of the containers picked up by feeder ships, one will set $p_{\text{vessel},3}^{\text{dest}} = 0.1$ and $p_{\text{feeder},3}^{\text{dest}} = 0$. It should be noted that, even for detailed simulations, it might not be necessary to model "real" destinations and ports exactly. Often, a number of abstract destinations will be sufficient.

4 Algorithm for generating a scenario

In this section, we summarize the algorithm that constructs a scenario. The goal of the algorithm is to compute a scenario in a way that the parameter settings are observed as closely as possible. Before we describe the algorithm which consists of four successive stages, we have a brief look at procedures that will be used in several parts of the algorithm.

4.1 Preliminaries

In several parts of the generator, we have to find an assignment that should match a distribution given as parameters as closely as possible. For example, the user can define classes of vessels (as well as feeders and trains). These classes determine the size of the vessels. Of course, when we compute vessels, the distribution of the generated vessel sizes should match the parameter settings. Similarly, the distributions of the arrival days and times should match the distributions selected by the user.

Let us give an abstract formulation of this problem setting. We assume that we have objects (e.g., vessels or containers) each of which must be assigned one out of m properties (e.g., arrival time slots for vessels or sizes for containers). The goal is to construct an assignment such that property $j \in \{1, \ldots, m\}$ occurs with a fraction f_j (these fractions correspond to the distributions given by the parameters). In the generator, we employ the following two alternative procedures to assign properties to objects.

Randomized assignment. For each object i, the property is assigned randomly where the probability to select property $j \in \{1, \ldots, m\}$ is given by fraction f_j. This is, of course, the straightforward approach. However, if the number of objects is relatively small, this method might generate distributions that do not match the fractions. Therefore, in those cases, we employ the following modified method.

Optimized assignment. In order to match the given fractions more accurately, we define the following simple greedy procedure. The first object $i = 1$ is assigned a property $j \in \{1, \ldots, m\}$ randomly. Then we successively consider objects $i > 1$. The i-th object, $i > 1$, is assigned a property based on the previously generated assignments of objects $1, \ldots, i-1$. Let z_j denote the number of times property j has been selected so far. Object i is assigned property j^* with $f_{j^*} - \frac{z_{j^*}}{i} = \max\{f_j - \frac{z_j}{i} \mid j = 1, \ldots, m\}$ (ties are broken arbitrarily). Subsequently, we update $z_{j^*} := z_{j^*} + 1$. This way, we always select the property with the largest difference between the selected fraction and the currently generated one. The assignment step is repeated until a stopping criterion is fulfilled, that is, until each object has been assigned a property.

4.2 Generating transport modes and containers to be unloaded

The first stage of the generator produces arriving vessels, feeders, trains, and trucks along with the related containers to be unloaded. Let us begin the description of the related procedure for vessels.

We have C_{vessel} containers that arrive by vessel. This number will now be split into individual vessels. Here, the distribition of the vessel classes as given by the parameters has to be observed. We apply the optimized assignment procedure described in Subsection 4.1. In each step of that procedure, a new vessel class $k \in \{1, \ldots, n_{\text{vessel}}^{\text{class}}\}$ is selected. Based on the information related to class k, a specific vessel is determined. The number S of containers to be unloaded by that vessel is drawn randomly from $\{\underline{S}_{ik}, \ldots, \overline{S}_{ik}\}$. Similarly, the length and the number of bays are selected randomly from the respective parameter intervals related to class k. This way, we obtain a new vessel in each step of the greedy procedure. In order to reflect the remaining number of containers, we set $C_{\text{vessel}} := C_{\text{vessel}} - S$ whenever we have determined a new vessel. The above step is repeated until no containers are left ($C_{\text{vessel}} = 0$) or until the number of remaining containers is smaller than the number that would be required for the smallest vessel type (in the latter case, the remaining containers are randomly added to the vessels that have already been defined).

The generation of feeders and trains is analogous with the exception that length and number of bays are not considered for trains. The construction of trucks, however, is different. Since we assume that a truck carries only a single container, we simply define a truck for each of the C_{truck} containers arriving by truck.

At the end of this stage, we have vessels, feeders, trains, and trucks together with the respective numbers of containers to be unloaded.

4.3 Generating arrival day and time

In the second stage of the generator, an arrival day and time is computed for each vessel, feeder ship, train, and truck that was generated in the previous stage. We describe the computation of the deep sea vessels in more detail. The procedure for the remaining three transport modes is analogous.

The first step deals with the day of arrival. So far, we know the fraction of vessels that should arrive on each of the seven days of the week. This information is given by the parameters $D_{\text{vessel},d}$ for day $d \in \{\text{Monday}, \ldots, \text{Sunday}\}$. We now have to transform these parameters such that they reflect the fractions of the days of the horizon. For each day $t = 1, \ldots, T$ of the horizon, we set $D'_{\text{vessel},t} := D_{\text{vessel},d}$ if the t-th day of the horizon corresponds to day $d \in \{\text{Monday}, \ldots, \text{Sunday}\}$ with respect to the first day T_1. Then we calculate the fraction of vessels arriving on day

$t = 1, \ldots, T$ as

$$D''_{\text{vessel},t} = \frac{D'_{\text{vessel},t}}{\sum_{q=1}^{T} D'_{\text{vessel},q}}.$$

With this distribution, we are now ready to compute an arrival day for each vessel. Again, we apply the optimized assignment procedure of Subsection 4.1.

Having fixed the arrival day for each vessel, the second step computes a time on that day. Recall that the parameter settings include a specification of time slots $1, \ldots, k, \ldots, n^{\text{timeslot}}_{\text{vessel}}$ along with a distribution $h_{\text{vessel},k}$. Using the optimized assignment procedure of Subsection 4.1, we determine a time slot for each vessel based on the given distribution. Subsequently, we consider the exact time of arrival. If a vessel is assigned time slot k, we draw the hour of arrival from $\{\underline{H}_{\text{vessel},k}, \ldots, \overline{H}_{\text{vessel},k} - 1\}$. Finally, we draw the minutes past the hour from the interval $\{0, \ldots, 59\}$.

At the end of this stage, we have vessels, feeders, trains, and trucks together with the respective number of containers to be unloaded as well as the arrival day and time.

4.4 Generating container pick-up

So far, we have generated lists of arriving means of transportation together with arrival date and time as well as the numbers of containers delivered. While we already know which container is delivered by which vessel, feeder ship, train, or truck, we now determine for each container which vessel, feeder ship, train, or truck will pick it up. For all containers, the following three steps are executed. Let us consider a container for which we want to compute the pick-up information. The assignment of pick-up properties to the container under consideration is done using the randomized assignment procedure of Subsection 4.1.

In the first step, we determine the transport mode that will pick up the container. If the container is delivered by transport mode i, then we use c_{ij} as the probability that it will be picked up by mode j ($i, j \in \{\text{vessel}, \text{feeder}, \text{train}, \text{truck}\}$), see Table 2. Let $j^* \in \{\text{vessel}, \text{feeder}, \text{train}, \text{truck}\}$ denote the transport mode selected to pick up the container.

The second step is to determine the day on which the container will leave the terminal. With t_1 we denote the day on which the container arrives at the terminal. Now we determine the dwell time of the container using the dwell time distribution. Let Δ^* be the selected dwell time. The day on which the container is picked up is then determined as $t_2 = t_1 + \Delta^*$.

Finally, the third step is to find a means of transportation of mode j^* that arrives on day t_2. Among the means of transportation that fulfill these requirements, some might not be eligible for pick-up because they have already reached their maximal load. Given a means of transportation with U containers to be unloaded that has already been assigned L containers to load, it can only receive more containers to pick up if we have $L + 1 \leq F \cdot U$ (for the definition of parameter F see again Table 3). Let M be the set of the eligible means of transportation of mode j^* arriving on day t_2. When we select one of them to pick up the container under consideration, we want larger means of transportation (e.g., larger vessels) to have a higher probability to be selected. To do so, we reflect the size of the means of transportation $m \in M$ by the number U_m of containers to be unloaded (recall that the number of containers to be unloaded has already been fixed). More precisely, the probability to chose $\mu \in M$ is defined as $p(\mu) = \frac{U_\mu}{\sum_{m \in M} U_m}$. Three special cases that may occur in the third step should briefly be mentioned:

- If no eligible means of transportation exist on that day (either because none of mode j^* arrive or because those that arrive have already reached their maximal load), then we start again with the first step.
- If day t_2 is not in the horizon (i.e., $t_2 > T$), then a means of transportation is not determined.
- If the transport mode j^* is truck, we define a new truck to pick up the container given that truck arrival on that day is possible (recall that we assumed trucks to deliver or pick up a single container, thus the selection mechanism of existing means of transportation as described above cannot be applied).

Observe that at the end of this stage we know for each vessel, feeder ship, train, and truck which containers it unloads and which containers it loads. Of course, we also know for each container which vessel, feeder ship, train, or truck delivers it and which one picks it up. Hence, we have completed the relationship between means of transportation and individual containers. We have taken into account the parameters concerning the classes (or sizes) of the means of transportation, the arrival time distributions over the days and within a day, and the dwell time distribution. In particular, the generation of container pick-up as described above ensures that there is a concentration of container deliveries for a specific vessel in the days before the vessel arrives (and, analogously, a concentration of container pick-up in the days after the vessel's arrival).

4.5 Generating container properties

At this point, we have individual containers, each of which is associated with individual means of transportation for arrival and pick-up. The last stage of the generator consists of the assignment of properties to the containers. The procedure is quite straightforward. It makes use of the parameters of Section 3.4 and the randomized assignment of Subsection 4.1.

Using the fractions $p^{20'}$, p^{empty}, p^{reefer}, p^{IMO}, and p^{OOG} as probabilities, the related properties are fixed for each container. Next, for the containers that have been assigned the IMO attribute, a specific IMO type is selected. This is done using the fractions of the different IMO types as probabilities. Analogously, OOG containers are assigned an OOG type.

The destination of a container is determined in a similar way. However, the choice of the destination depends on the transport mode that picks up the container. That is, if a container is picked up by mode $i \in \{\text{vessel}, \text{feeder}, \text{train}, \text{truck}\}$, then the probability to select destination $k \in \{1, \ldots, n^{\text{dest}}\}$ is given by p_{ik}^{dest}.

Finally, a container is assigned a weight. Here, the weight parameters related to the container size (20' or 40') are applied. If the container has been defined as empty, the related parameter for the weight of empty containers is used. Otherwise, a weight group k is drawn according to the related fractions which are employed as probabilities. The actual weight w is randomly chosen from the interval determined by the minimal and maximal weight of selected weight group k, that is, we select $w \in \{\underline{w}_k^{20'}, \ldots, \overline{w}_k^{20'}\}$ or $w \in \{\underline{w}_k^{40'}, \ldots, \overline{w}_k^{40'}\}$, respectively.

5 Experimental validation

5.1 Evaluation of generated scenarios

In this section, we show that the proposed algorithm is appropriate for constructing scenarios that match the selected parameters. We report on experiments in which we generated scenarios based on the following setting. We have a horizon of 4 weeks and a total of 76,600 containers arriving within the horizon (this corresponds to a medium sized terminal with one million containers per year). The remaining parameter selections were done using statistics of the HHLA container terminal Burchardkai in Hamburg, Germany. For this setting, ten scenarios were generated.

The size of the output on disk is approximately 8.6 MB for such a scenario (i.e., all transport mode and container property information). On a Pentium 4 based computer with 1.6 GHz, the computation took on the average 9.9 seconds for one scenario (we coded the generator in *C* and used the lcc compiler under Windows XP). Hence, the algorithm is reasonably fast.

In order to validate the computed scenarios, we compared the statistics computed from the generated scenarios with the real-world statistics that were used as parameters. For the sake of brevity, we restrict the presentation of the results to the most important statistics concerning transport modes, arrival frequencies, and container dwell time. Table 6 summarizes the respective correlations between scenario characteristics and real-world statistics. With the exception of the dwell time distribution, the correlations are measured separately for the transport modes (recall that the respective parameters are given separately for the transport modes as well). The correlations are given as average values over the ten scenarios generated. As we can see, all correlations are clearly above 0.8, and most are even above 0.95. Only some of the correlations related to vessels are below 0.95 which is due to the fact that the number of vessels arriving within the horizon is relatively small.

Table 6. Average correlations of generated data and control parameters

Correlation type	Overall	Vessel	Feeder	Train	Truck
Distribution of unloaded containers among transport modes	–	1.000	1.000	1.000	1.000
Distribution of transport mode size classes	–	0.994	1.000	0.986	–
Distribution of arrivals among weekdays	–	0.842	0.996	0.995	0.958
Distribution of arrivals among time slots within a day	–	0.886	0.994	1.000	1.000
Distribution of container dwell times	0.996	–	–	–	–

Table 7. Average correlations of generated data and control parameters w.r.t. assignment method

Correlation type	Assignment	Vessel	Feeder	Train	Truck
Distribution of arrivals among time slots within a day	randomized	0.350	0.650	0.998	1.000
Distribution of arrivals among time slots within a day	optimized	0.886	0.994	1.000	1.000

The results indicate that the algorithm of the generator is capable of constructing scenarios in which the selected parameters are observed considerably well. Hence, if real-world statistics are used as input parameters, the generator produces realistic scenarios. These findings were confirmed in several other experiments based on different parameter settings.

5.2 Effect of assignment method

This subsection briefly analyzes the impact of the two different procedures for assigning properties to objects described in Subsection 4.1. Recall that we have defined a simple randomized method and an optimized method for assignment. Considering the assignment of time slots, Table 7 compares the resulting correlations if these two methods are used. For vessels and feeders, the optimized approach leads to time slot distributions that match the parameters by far better. The difference for trains is much smaller, but this is due to our parameter selection (due to the real-world statistics, we have only three time slots for trains whereas we have 24 time slots for the other three modes). Finally, since we have a huge number of trucks, the randomized method would be sufficient there.

Summarizing, the results confirm our decision to employ the optimized assignment approach for time slot selection. Similar results were obtained for the other steps of the generator.

6 Applications

Parameter-based generation of data is an important field in operations research (see, e.g., Hall and Posner [12], Kolisch et al. [19]). In scientific research, generated data is needed to carry out experiments with algorithms or modeling approaches. Parameters are employed to generate data systematically according to a specific experimental design. In practice, decisions on algorithms or strategies in control systems are made only if they have been tested before. While real-world data is often preferred for such tests, artificially generated data is used if real-world data is unavailable. This is the case for new container terminals to be built, for existing ones that will be expanded, and for analyzing future developments such as more ship arrivals or larger vessels. For applications in practice, it is of particular importance that the artificially generated data is realistic (otherwise, the results would not be an appropriate basis for decision support). The following subsections outline several applications of the generator proposed in this paper.

6.1 Stack simulation at the Container-Terminal Altenwerder

The generator described in this paper has been developed within the simulation project of the HHLA Container-Terminal Altenwerder in Hamburg, Germany. It was applied of thewithin a simulation study of the strategies for selecting yard blocks and slots for arriving containers. The study was carried out in the planning phase of the terminal, thus real-world data was not available. The purpose of the study was to test, improve, and parameterize the stacking strategies. A simulation-based approach was necessary to analyze the dynamic container arrivals and departures in a realistic online environment.

When a container arrives at the terminal, a position in the stack has to be determined. This is done using strategies which are associated with the following goals. First and most important, the number of shuffle moves has to be minimized. A shuffle move occurs if two containers stand on top of each other and the lower one has to be picked up first. In such a case, the upper one has to be moved to another position in the stack, which reduces the stacking crane (or straddle carrier) capacities for productive moves. While shuffle moves cannot be totally avoided, stacking strategies select positions such that the number of shuffle moves is kept to a minimum. Second, the distance over which a stacking crane (or straddle carrier) transports a container through a yard block should be minimized in order to use the equipment capacities efficiently.

In order to achieve the goals mentioned above, the stacking strategies make the selection on the basis of the information associated with a container. Considering the shuffle move minimization, two containers with the same properties (in particular, same vessel for pick-up and same weight class) can be stacked on top of each other (in such a case, the upper one can be picked first). It can also be a good idea to put a container on top of another if the expected departure time of the upper container is earlier than that of the lower one. The second goal implies that, e.g., a position on the seaside should be preferred for a container which arrives with a feeder and will be picked up by a vessel. Such considerations are usually employed within a priority based evaluation method for all available positions in the stack, and the best stack is selected (for reasons of confidence, details of the stacking strategies at the Container-Terminal Altenwerder cannot be given here).

The generator proposed in this paper was employed to produce the input data for the simulation of the stacking strategies. The parameters of the generator were adjusted such that they captured the specific planning assumptions of the Container-Terminal Altenwerder (e.g., the expected number of containers turned over per year). The remaining parameter settings were done on the basis of statistics from the HHLA Container-Terminal Burchardkai. In addition to the transport mode arrivals over time, the container properties were of particular importance because they are required by the stacking strategies. Each scenario (and thus also each simulation run) covered a horizon of several weeks (note that the impact of stacking strategies could not be observed within a shorter horizon).

The simulation model was was developed using the emPlant software package. It was designed as follows. In a first step, a list of events is constructed from the scenario under consideration. For example, for a truck arrival with a request of a specific container, a related pick-up event is created for the truck's arrival time. Considering the arrival of a vessel, the arrival times of the containers to be unloaded are calculated on the basis of an assumed quai crane productivity. That is, for the containers arriving by a vessel, one event related to a single container is created every n seconds starting with the vessel's arrival time (the loading process as well as feeder and train arrivals are treated analogously). In the second step, the simulation moves through the event list, that is, a container related to an arrival event is put in the stack, and a container associated with a pick-up event is removed from the stack. The equipment (quai cranes, automated guided vehicles, stacking cranes) is not modeled explicitly since the transportation process was not in the focus of this study (in fact, modeling the transportation process would have slowed down the simulation runs). The stacking strategies and the stack itself are modeled on a detailed level. An arrival event triggers the stacking strategy computation and, subsequently, an update of the stack data with respect to the selected position. A pick-up event triggers again an update of the stack data with respect to the position from which the requested container is removed. In reality, an arriving container may be associated with missing or incorrect information on its pick-up time and transport mode. Since this is important when analyzing stacking strategies, the simulation model contains parameters to distort the input scenarios. Several statistics were incorporated into the simulation model in order to evaluate different strategies, in particular number of shuffle moves (overall and for different container types), number of free groundslots, stacking crane distance per container, stacking height for different container types, and level of stack utilization over time. The model also contains a visualization with a view of individual containers as well as container types in the yard blocks.

For each generated scenario, a large number of simulation runs were carried out in order to test alternative stacking strategies and parameter settings for strategies (again, for reasons of confidence, the simulation results cannot be given here). The originally considered strategies did not perform well, thus they were modified on the basis of the simulation. Hence, the simulation led to improved stacking strategies before the terminal started operation.

6.2 Further applications

The main purpose of the generator presented in this paper is to produce scenarios that can be used as input data for simulation models. Simulation approaches are used by researchers and practitioners to study the dynamic processes in container terminal logistics. Typically, a simulation model covers the terminal resources (particularly equipment such as quai cranes, straddle carriers, and stacking cranes), the stack, and the control strategies (e.g., equipment scheduling and stack reservation). The subject of a simulation study determines the level of detail to which a component is modeled. Generally speaking, arrivals of transport modes together with arriving and requested containers are the events that cause the execution of processes on the terminal (e.g., transportation tasks, stack reservation). Therefore, an event-based simulation model requires input data concerning transport mode arrivals (and hence container arrivals and requests) over time. This is the data that is produced by the generator proposed here. Given that realistic input data is provided, simulation models can be employed to support decisions on required resource capacities, to detect possible bottlenecks, and to select and adjust strategies, objective functions, and algorithms (e.g., for equipment scheduling).

In addition to simulation studies, the proposed generator can also be applied in experiments with optimization approaches that are not embedded into an online environment. The following well-known optimization problems require input data that can be produced by our generator.

In the berth allocation problem (see Guan and Cheung [11], Imai et al. [14,15], Lim [21]), a list of ship arrivals is given. These ships have to be assigned a berth at the quai (which, of course, has a limited length). The objective is to minimize the waiting time for a free berth and hence the total time in port for the ships. Our generator produces ship arrivals over time which are needed as input data for this optimization problem. In addition to the arrival date and time of the ships, it also generates the number of containers to be loaded and unloaded. On the basis on a productivity rate, this number of containers can easily be converted into the time needed at the quai. Adding the quai length, one has obtained all problem data for berth allocation.

The quai crane scheduling problem (see Daganzo [5], Peterkofsky and Daganzo [24]) deals with the assignment of a limited number of quai cranes to vessels which arrive over time. A quai crane is allowed to move from ship to ship. Hence, a quai crane schedule determines for each quai crane at what time it is working on which vessel. Again, the objective is to minimize the time in port. Similarly to the berth planning problem, the generator produces the input data for the quai crane scheduling problem.

Both problems have been combined to a more realistic problem in which the berthing time of vessels depends on quai crane capacities (see Park and Kim [23]). Also for this integrated problem, the generator can be employed to produce test data. Note that, whereas the full scenario data was required for the simulation study described in Subsection 6.1, only parts of the scenarios are necessary for these optimization problems. Both the landside information (i.e., arrivals of trains and trucks) and the container properties can be skipped here.

7 Conclusions

In this paper, we presented an approach to generate scenarios for container terminals. A scenario contains data on arrivals of vessels, feeder ships, trains, and trucks together with lists of the containers to be delivered and picked up and the container attributes (size, weight etc.). The generation of a scenario is controlled by means of various parameters. The goal was to develop an easy-to-use generator that produces realistic scenarios. Therefore, we selected parameters that allow to use statistics and planning information that are typically available in practice.

By employing real-world information for the parameters, the artificially generated scenarios become realistic.

After a description of the parameters and the algorithms of the generator, we have demonstrated by experiments based on real-world statistics that the algorithm produces scenarios that match the parameter settings appropriately well. Subsequently, we have sketched out various applications in research and practice. While the main purpose of the generator is to produce input data for simulation models of container terminals, it can also be used to generate test data for well-known optimization problems such as berth planning and quai crane scheduling. In partcular, we have described a practical application of the generator within the simulation project at the HHLA Container-Terminal Altenwerder in Hamburg, Germany. In this project, the strategies for determining a position in the stack for arriving containers were examined. This application has shown that the generator is well suited for projects in practice.

Although the generator is already considerably general, extensions of the generator might be promising tasks in future research. A possible extension of the generator could be to compute the positions of the individual containers on the vessels and feeder ships. These positions are relevant when stowage plans are part of the problem considered. This is of particular importance if the process of loading vessels is examined in full detail. Often, each position on the vessel is associated with a specific container. This imposes a partial order on the containers to be loaded. Alternatively, clusters of cells may be associated with container attributes rather than individual containers. In the latter case, there are more degrees of freedom concerning the loading order, and specific strategies are needed to exploit this setting. Hence, the design of an extended generator (and particularly that of the additional parameters) would depend on the application of the stowage plans.

References

1. Alicke K (2002) Modeling and optimization of the intermodel terminal Mega Hub. OR Spectrum 24: 1–17
2. Bae J W, Kim K H (2000) A pooled dispatching strategy for automated guided vehicles in port container terminals. International Journal of Management Science 6: 47–70
3. Baker C (1999) Altenwerder – the details. Port Development International 1999 (07/08): 24–25
4. Böse J, Reiners T, Steenken D, Voß S (2000) Vehicle dispatching at seaport container terminals using evolutionary algorithms. In: Sprague R H (ed) Proceedings of the 33rd Annual Hawaii International Conference on System Sciences, pp 377–388. IEEE, Piscataway
5. Daganzo C F (1989) The crane scheduling problem. Transportation Research B 23: 159–175
6. de Castilho B, Daganzo C F (1993) Handling strategies for import containers at marine terminals. Transportation Research B 27: 151–166
7. Evers J J M, Koppers S A J (1996) Automated guided vehicle traffic control at a container terminal. Transportation Research A 30: 21–34
8. Gambardella J M, Bontempi G, Taillard E, Romanengo D, Raso G, Piermari P (1996) Simulation and forecasting of an intermodal container terminal. In: Bruzzone A G,

Kerckhoffs E J H (eds) Simulation in industry – Proceedings of the 8th European Simulation Symposium, pp 626–630. SCS, Ghent, Belgium
9. Gambardella L M, Rizzoli A E, Zaffalon M (1998) Simulation and planning of an intermodal container terminal. Simulation 21: 107–116
10. Grunow M, Günther H O, Lehmann M (2004) Dispatching multi-load AGVs in highly automated seaport container terminals. OR Spectrum 26: 211–235
11. Guan Y, Cheung R K (2004) The berth allocation problem: Models and solution methods. OR Spectrum 26: 75–92
12. Hall N G, Posner M E (2001) Generating experimental data for computational testing with machine scheduling applications. Operations Research 49: 854–865
13. Hartmann S (2004) A general framework for scheduling equipment and manpower at container terminals. OR Spectrum 26: 51–74
14. Imai A, Nagaiwa K, Tat C W (1997) Efficient planning of berth allocation for container terminals in Asia. Journal of Advanced Transportation 31: 75–94
15. Imai A, Nishimura E, Papadimitriou S (2001) The dynamic berth allocation problem for a container port. Transportation Research B 35: 401–417
16. Kim K H, Kim H B (1999) Segregating space allocation models for container inventories in port container terminals. International Journal of Production Economics 59: 415–423
17. Kim K H, Park Y M, Ryu K R (2000) Deriving decision rules to locate export containers in container yards. European Journal of Operational Research 124: 89–101
18. Kim K Y, Kim K H (1999) A routing algorithm for a single straddle carrier to load export containers onto a container ship. International Journal of Production Economics 59: 425–433
19. Kolisch R, Sprecher A, Drexl A (1995) Characterization and generation of a general class of resource-constrained project scheduling problems. Management Science 41: 1693–1703
20. Legato P, Mazza R M (2001) Berth planning and resources optimisation at a container terminal via discrete event simulation. European Journal of Operational Research 133: 537–547
21. Lim A (1998) The berth planning problem. Operations Research Letters 22: 105–110
22. Meersmans P J M, Dekker R (2001). Operations research supports container handling. Technical Report EI 2001-22, Econometric Institute, Erasmus University Rotterdam
23. Park Y M, Kim K H (2003) A scheduling method for berth and quai cranes. OR Spectrum 25: 1–23
24. Peterkofsky R I, Daganzo C F (1990) A branch and bound solution method for the crane scheduling problem. Transportation Research B 24: 159–172
25. Schütt H, Hartmann S (2000) Simulation in Planung, Realisierung und Betrieb am Beispiel des Container-Terminals Altenwerder. In: Möller D P F (ed) Frontiers in Simulation — Simulationstechnik, 14. Symposium in Hamburg, pp 425–430. SCS, Ghent, Belgium
26. Steenken D, Henning A, Freigang S, Voß S (1993) Routing of straddle carriers at a container terminal with the special aspect of internal moves. OR Spectrum 15: 167–172
27. Taleb-Ibrahimi M, de Castilho B, Daganzo C F (1993) Storage space vs. handling work in container terminals. Transportation Research B 27: 13–32
28. Yun W Y, Choi Y S (1999) A simulation model for container-terminal operation analysis using an object-oriented approach. International Journal of Production Economics 59: 221–230
29. Zhang C, Liu J, Wan Y W, Murty K G, Linn R J (2001) Storage space allocation in container terminals. Technical report, Hong Kong University of Science and Technology
30. Zhang C, Wan Y W, Liu J, Linn R J (2002) Dynamic crane deployment in container storage yards. Transportation Research B 36: 537–555

Fleet sizing and vehicle routing for container transportation in a static environment*

Pyung Hoi Koo[1], **Woon Seek Lee**[1], **and Dong Won Jang**[2]

[1] Department of Industrial Engineering, Pukyong National University, San100 Yongdang, Namgu Busan, 608-739 Korea (e-mail: phkoo@pknu.ac.kr)

[2] Home Delivery Planning Team, Chunil Cargo Transportation, Joungsanri 848-8 Mulgeum, Yangsan Gyungnam, 628-810, Korea

Abstract. Busan is one of the busiest seaports in the world where millions of containers are handled every year. The space of the container terminal at the port is so limited that several small container yards are scattered in the city. Containers are frequently transported between the container terminal and container yards, which may cause tremendous traffic problems. The competitiveness of the container terminal may seriously be aggravated due to the increase in logistics costs. Thus, there exist growing needs for developing an efficient fleet management tool to resolve this situation. This paper proposes a new fleet management procedure based on a heuristic tabu search algorithm in a container transportation system. The proposed procedure is aimed at simultaneously finding the minimum fleet size required and travel route for each vehicle while satisfying all the transportation requirements within the planning horizon. The transportation system under consideration is static in that all the transportation requirements are predetermined at the beginning of the planning horizon. The proposed procedure consists of two phases: In phase one, an optimization model is constructed to obtain a fleet planning with minimum vehicle travel time and to provide a lower bound on the fleet size. In phase two, a tabu search based procedure is presented to construct a vehicle routing with the least number of vehicles. The performance of the procedure is evaluated and compared with two existing methods through computational experiments.

Keywords: Container transportation – Vehicle routing – Fleet sizing – Tabu search

---

* This work was supported by Korea Research Foundation Grant. KRF-2001-003-E00080.

Correspondence to: P. H. Koo

1 Introduction

Busan is one of the busiest seaports in the world, which handles about 10 million twenty-foot equivalent units (TEUs) of containers, more than 90 % of the total container volumes exported from and imported to Korea. The container terminal area in Busan is so limited that all the containers cannot be stored in on-dock container yards. Hence, a significant amount of the containers are stored and handled in off-the-dock container yards (ODCYs) located near the port container terminal. The containers are moved by container trucks between the container terminal yard and the ODCYs. The container transportation within the city causes tremendous traffic problems in the port city and increases the logistics cost which may aggravate the competitiveness of the container terminal.

Figure 1 shows a simple container transportation environment under consideration. Import containers are unloaded from a container vessel entering the port, and placed at a marshalling area in the container terminal. Then the containers are moved to on-dock container yards, ODCYs, rail container yards, inland container depots, and local coastal port yards. The flow of the export containers would be reversed. Each container to be delivered has its own destination. For example, in Figure 1, seven containers are to be moved from ODCY3 to the seaport container terminal and 21 containers from seaport container terminal to ODCY3. A container is delivered by a single vehicle and a vehicle carries only a single container at a time. Each container will not be split during the travel and, thus is considered a transportation unit load.

This paper deals with a static transportation problem in which all the transportation jobs are ready to be picked up at the beginning of a planning horizon. It is assumed that the number of containers to be moved between two locations is determined at the beginning of the planning horizon, and travel times between locations as well as loading and unloading times are deterministic and known in advance. At the beginning of the planning horizon (e.g., one shift), several identical vehicles are ready at a location. This transportation environment is referred to as

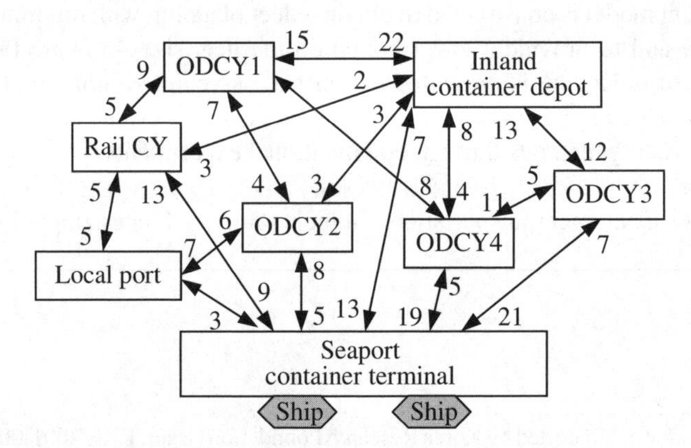

Fig. 1. An example of container transportation

a tractor-trailer transportation system (Bodin et al., 1983) or a static dial-a-ride problem with multiple vehicles of single capacity.

It is desirable to satisfy all the transportation requirements within the planning horizon with the minimum number of vehicles or fleet size. The fleet size can be found if the total vehicle travel time in the planning horizon is known. The lower bound on the required fleet size is the total vehicle travel time divided by the length of the planning horizon or time period available for a vehicle within the planning horizon. The total vehicle travel time consists of empty travel time, loading time, loaded travel time, and unloading time. Among these, the loaded travel time can be found using a from-to chart indicating transportation requirements between locations and travel time matrix. The loading and unloading times can be estimated from the number of loadings and unloadings performed in the planning horizon. The fleet size can thus be determined by the empty travel time. However, estimation of the empty travel time is a complex task since it requires information about vehicle routing.

This paper presents a two-phase fleet sizing and vehicle routing procedure. The objective of the procedure is to provide a multiple vehicle routing to complete all the transportation requirements with the minimum fleet size. Phase one uses an optimization model to produce a lower bound on the required fleet size, and phase two applies a tabu search based heuristic to generate vehicle routing along with an appropriate fleet size.

2 Previous research works

Many existing research works on freight transportation deal with how to determine the sequence of vehicle's visit to the request locations, which is closely related to the well-known traveling salesman problem (TSP), Multiple TSP, and general vehicle routing problems. Readers are referred to Laporte and Osman (1995), Crainic and Laporte (1998) and Chao (2002) among others. The vehicle routing problem in container transportation is slightly different from these research works in that a location may be visited as many times as the number of containers to be delivered, and a container should be delivered to a specific destination once loaded. Thus, it is similar to the single-capacity pickup-delivery transportation problem with sequence dependency. The transportation capacity constraint that a vehicle can move only a single container at a time makes the current problem more difficult to solve than the TSP which is known to be NP-hard. This implies that the optimal solution is computationally infeasible to obtain as problem sizes increase.

Most research works on container transportation systems focus on the design and operation of logistics within seaport container terminals. For vehicle operation problems in container terminals, Kim and Bae (1999) address assignment problems of container-delivery tasks to vehicles during ship operation in automated container terminals. Their work is later extended to the case of multiple quay cranes in Bae and Kim (2000). Grunow et al. (2004) present a priority-rule based dispatching procedure for a container terminal where automated guided vehicles (AGVs) with multiple-load capability are used as container transporters. They conclude through numerical experiments that the use of dual load capabilities of the vehicles significantly improves the performance of the transportation system with respect to the total lateness and empty vehicle travel times. Kozan and Preston (1999) present a genetic algorithm based scheduling procedure for multimodal seaport container terminals to determine the optimal storage strategies and container handling schedules. They examine the effect of the number of containers, handling equipment, storage capacities and policies, and container terminal layout through simulation experiments. Kim and Kim (1999) present a routing procedure for a straddle carrier in port container terminals. An integer programming model is formulated and a heuristic is presented to solve the real world problem in an efficient way. Bish et al. (2001) develop a vehicle-scheduling-location heuristic to assign each container to a yard location and dispatch vehicles to the containers so as to minimize the time spent to download all the containers from the ship. Queueing network models (Legato and Mazza 2001) and simulation models (Gambardella et al. 1998; Yun and Choi 1999; Shabayek and Yeung 2002) are also applied to design and analyze the container terminal operations. A variety of decision problems at container terminals are classified and extensively reviewed in Vis and de Koster (2003).

Determining the fleet size is the most fundamental decision in a transportation system whose capacity is directly related to the number of available vehicles. Determining the optimal number of vehicles for a particular system requires a tradeoff between the investment costs of the vehicles and the potential penalties associated with not meeting all the demands. Beaujon and Turnquist (1991) present a nonlinear mathematical model to optimize the fleet size and vehicle allocation in a multi-period transportation planning environment. The model is transformed to a minimum cost network flow problem with a nonlinear objective function that can be solved by using yet another proposed solution procedure based on the Frank-Wolfe algorithm. Du and Hall (1997) address fleet sizing and empty vehicle redistribution for a one-to-many (or hub-and-spoke) transportation structure. Terminals are classified into surplus and shortage terminals based on the balance of the incoming and outgoing transportation requirements. A proper fleet size is determined based on the inventory control theory. It is assumed that operating costs are incurred for an excessive number of vehicles while shortage costs are charged for an insufficient number of vehicles. Vis et al. (2001) present a model and an algorithm to determine the necessary number of AGVs at an automated container terminal. A network flow based model and a polynomial time algorithm are developed to solve the problem in which containers are available for transport at known time instants. Another research arena of fleet sizing for a single-capacity transportation system is the use of AGV systems in automated manufacturing systems. Maxwell and Muckstadt (1982) propose a mathematical model to determine the minimum number of AGVs for a given number of transportation requests during a time window. Each location is associated with a net flow of vehicles which is defined as the difference between the numbers of incoming and outgoing deliveries. Flow balances of locations have to be achieved by empty vehicle movements. The model gives the lower bound on the number of vehicles needed in the system. Rajotia et al. (1998) improve the model of Maxwell and Muckstadt by imposing one more constraint that only a small portion of transportation requests from a location can be served by vehicles being idle at the same location due to the randomness of the vehicle requests. For fleet sizing in a dynamic transportation environment, Kobza et al. (1998) present a model based on a discrete Markov chain and Koo and Suh (2002) present a queueing theory based model to estimate the vehicle waiting time and determine the fleet size in a dynamic transportation environment.

Most existing fleet sizing procedures for static transportation environments ignore vehicle routing in determining the fleet size. For fleet sizing and vehicle routing problems for container transportation, Ko et al. (2000) present a fleet sizing algorithm using an insertion algorithm. Given a planning horizon, they assign transportation orders to a vehicle one by one. When the completion time of a vehicle is larger than a predefined planning horizon, an additional vehicle is introduced and the procedure is repeated.

3 Two-phase fleet sizing and vehicle routing procedure

Figure 2 shows the overall procedure for fleet sizing and vehicle routing proposed in this paper. The procedure consists of two phases. In phase one, given the containers to be transported between locations and the travel times between locations, an optimization model is developed to generate a fleet planning with the minimum empty vehicle travel time. Since the model does not consider routing for each vehicle, actual empty travel times would be larger than those obtained from the optimization model. The minimum fleet size resulting from this optimization model may be regarded as a lower bound on the number of vehicles required. Given the fleet size, a tabu search based algorithm is developed to obtain the vehicle routing in phase two. Finally, the makespan (equivalent to the time taken until all the transportation jobs are finished) of the current solution is compared with the predetermined makespan limit. If not satisfactory, the procedure increases the fleet size by one and continues until the makespan constraint is satisfied. The two-phase procedure is described in more detail in the following sections.

3.1 Phase one: optimization model to obtain the lower bound on fleet size

This section presents an optimization model to obtain the lower bound on the fleet size required. The fleet size depends on the total vehicle travel time required, which consists of empty travel time, loading time, loaded travel time, and unloading time. As discussed in the previous section, loading time, loaded travel time, and unloading time may easily be obtained when the transportation requirements and travel time between the locations are known. However, empty travel time is dependent on how to select the container to be delivered next when a vehicle becomes free. The lower bound on the fleet size, denoted by N_{min}, can be obtained by dividing the total vehicle travel time by the length of planning horizon or the available time of a vehicle (e.g., a shift). In order to reduce the fleet size required, the empty vehicle travel time must be minimized. The optimization model proposed by Maxwell and Muckstadt (1982) is applied to obtain the lower bound on the fleet size. Following notations will be used in the model:

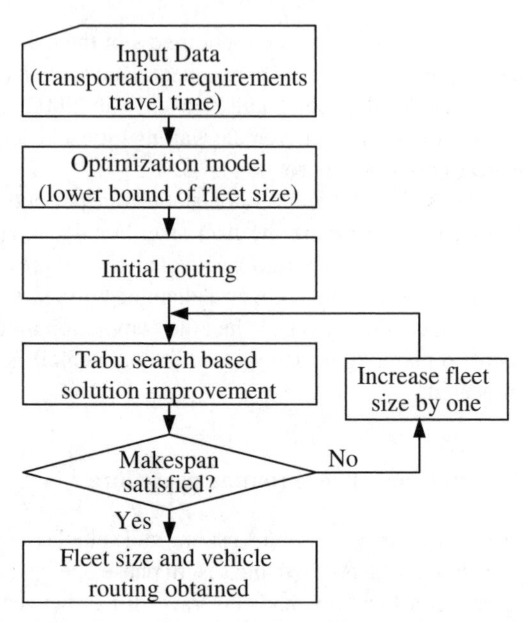

Fig. 2. Two-phase fleet sizing and vehicle routing procedure

x_{ij} the number of empty vehicle trips that should be made from location i to location j

v_{ij} the number of containers to be delivered from location i to location j (or equivalently, the number of loaded vehicle trips that should be made from location i to location j)

t^a_{ij} loaded vehicle travel time from location i to location j, which represents the time spent to load a container on a vehicle, move it from location i to location j, and unload it at location j

t^b_{ij} empty vehicle travel time from location i to location j

Let us take location i for identifying the vehicle trip frequency during a shift. It can be observed that $\sum_j v_{ij}$is the number of containers to be picked up at location i during the shift. This means that $\sum_j v_{ij}$empty vehicles are needed at location i to move the containers. Similarly, $\sum_i v_{ij}$is the number of containers to be delivered to location j, and $\sum_i v_{ij}$vehicles will become empty after they unload the containers. For locations which do not allow overnight parking for vehicles, the total vehicle flow into the location within the shift is equal to the total vehicle flow out. The net flow for location i, denoted by $nf(i)$, is the difference between the total number of containers to be delivered in and the total number of containers to be picked up from there, that is, $nf(i) = \sum_j v_{ji} - \sum_j v_{ij}$. Since there may be requirements on the number of vehicles available at the beginning of a shift or required at the end of the shift, the net flow for location i must be adjusted to satisfy these requirements. For example, if f_i vehicles are available at the beginning of the shift and g_i vehicles are required at the end of the shift, the net flow for location i is as follows:

$$nf(i) = \left(\sum\nolimits_j v_{ji} + f_i\right) - \left(\sum\nolimits_j v_{ij} + g_i\right)$$

In the above equation, the first term on the right hand side indicates the number of vehicles available at location i during a shift while the second term indicates the number of vehicles required at location i during the shift. Hence, the net flow represents the number of empty vehicle trips into or out of the location. The locations with positive net flows would have empty vehicle trips available to be assigned to other locations with negative net flows. Following is the optimization model to find the number of empty vehicle trips from location i to location j:

$$\text{Min} \sum_i \sum_j x_{ij} t^b_{ij} \tag{1}$$

Subject to

$$\sum_j x_{ij} = nf(i), \text{ if the net flow for location } i \text{ is non-negative} \tag{2}$$

$$\sum_k x_{ki} = -nf(i), \text{ if the net flow for location } i \text{ is negative} \tag{3}$$

$$x_{ij} \text{ is a non-negative integer} \tag{4}$$

The objective is to minimize the total empty vehicle travel time. The total vehicle travel time, denoted by z, is then obtained by adding the total empty vehicle travel time and the total loaded travel time, that is, $z = \sum_i \sum_j v_{ij} t^a_{ij} + \sum_i \sum_j x_{ij} t^b_{ij}$. If h hours are available during the planning horizon per vehicle, at least $\lceil z/h \rceil$ vehicles are required to satisfy the transportation requirements, where $\lceil x \rceil$ is the smallest integer greater than or equal to x.

The optimization model yields the minimum number of vehicles required. However the model may not directly be applied to real world situations for the following reasons: First of all, it does not consider individual transportation requirements during the planning horizon. In addition, if the vehicle parking location at the beginning of the planning horizon is the same as the parking location at the end, the net flow of this location will be zero and there will thus be no vehicles starting from the parking location. Consequently, no vehicle trips will be made during the planning horizon, and the actual empty vehicle travel time will be underestimated. The fleet size obtained from the above optimization model will only be used as the lower bound on the fleet size in phase two, where vehicle routing and fleet sizing are solved simultaneously.

3.2 Phase two: tabu search based fleet sizing and vehicle routing

This section provides a vehicle routing and fleet sizing heuristic based on a tabu search (TS) algorithm. TS is a general improvement heuristic first presented by Glover (1989). It explores the solution space repeatedly moving from a solution to its best neighbor. The search process has the mechanisms that allow the objective function to deteriorate in a controlled manner and escape from local optima. Starting from an initial solution, an admissible move leads to the next solution with the minimum cost. If this solution is a local minimum, a non-improving perturbation may be accepted. To prevent cycling in the course of the search, the reverses of a certain number of moves that have recently been performed are forbidden and recorded in a constantly updated tabu list. TS has been successfully implemented in a variety of combinatorial problems such as production scheduling (Franca et al. 1996), vehicle routing problem (Nanry and Barnes 2000, Breedam 2001, and Osman and Wassan 2002), and traveling salesman problem (Gendreau et al. 1999). See Glover (1997) and Osman and Laporte (1996) for an extensive literature review and detailed descriptions on tabu search.

Vehicles pick up containers at a yard and deliver them to another yard. They have a series of transportation jobs to be performed. The main problem is an assignment of containers to vehicles in which all the transportation requirements may be scheduled on identical vehicles with the objective of minimizing makespan. Empty vehicle travels are incurred for each transportation demand, and depend on the sequence of transportation jobs. In our TS implementation for solving fleet sizing and vehicle routing, the fleet size generated in phase one is used as the initial solution, which is then improved through the TS based improvement procedure. The procedure uses the fleet size as the primary decision criterion and makespan as the secondary decision criterion to plan the vehicle routing with the minimum number of vehicles. Note that, with the same number of vehicles, the shorter the makespan is, the more efficiently vehicles are utilized. A shorter makespan can be realized by shorter empty vehicle travel times and transportation load balance among the vehicles. A neighborhood solution is obtained by removing a transportation job from the busiest vehicle (that is, the vehicle with the longest completion time) and inserting it in a vehicle tour with the shortest completion time. As such, the makespan may further be reduced. The transportation jobs should be completed within the predetermined time limit (e.g., 480 minutes for one shift).

The initial vehicle routing follows a greedy solution procedure often used in container transportation business. When a vehicle completes a transportation job, it selects a job which is the nearest to the vehicle. The procedure inherently yields a solution with fairly short empty travel time and makespan.

Initial vehicle routing

Step 0: The lower bound on fleet size N_{min}, container transportation requirements, and travel times between locations t_{ij} are determined.

Step 1: Choose n containers at random and assign them to each vehicle.

Step 2: Select a vehicle with the least $C(V_i)$, where $C(V_i)$ is the time taken for vehicle V_i to leave a depot, perform transportation jobs for all the containers assigned, and return to the depot.

Step 3: Select one from unassigned containers that yields the least empty vehicle travel time when it is appended to the route of the selected vehicle. Append it to the last job of the selected vehicle. Tie breaker is the longest-loaded-vehicle-time-first rule. This rule is selected because the longest processing time (LPT) rule is known to perform well in parallel machine scheduling with the objective of minimizing makespan.

Step 4: Repeat Step 3 until all the transportation jobs are assigned.

Now we have an initial solution where each vehicle has its own route. The next step uses the concept of TS to improve the current solution.

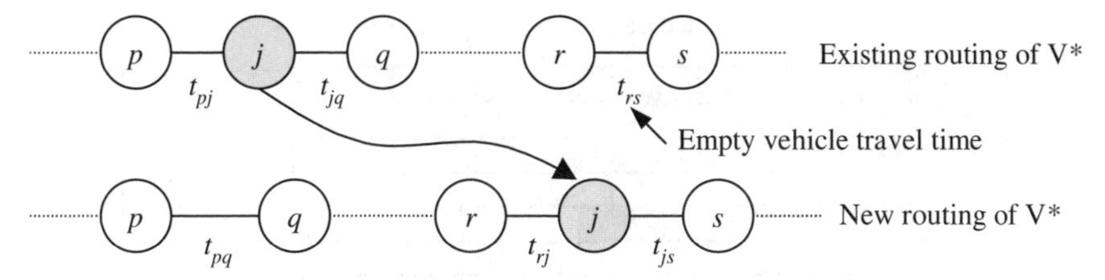

Fig. 3. Internal insertion process on V^*

TS Solution improvement

Step 0: Set the iteration counter c=0.

Step 1: Select the busiest vehicle V^* (that is, the vehicle with the largest $C(V_i)$) in the current solution.

Step 2: *Internal insertion.* This step polishes the route of V^* by moving individual transportation jobs forward or backward in the same route. For each transportation job assigned to V^*, a sequence change operation is performed (i.e., a job is deleted from the sequence of V^* and inserted in a different position of the sequence of the same vehicle). For example, in Figure 3, transportation job $j$to be moved right after job p and right before job q is removed from the sequence and inserted back to a position between job r and job s. Then the completion time is reduced by $(t_{pj}+t_{jq}+t_{rs})-(t_{pq}+t_{rj}+t_{js})$. The relocation of the job sequence resulting in the largest reduction in completion time is selected and the current solution is changed. The insertion process in this step is called internal insertion process. If we have at least one vehicle with larger $C(V_i)$ than $C(V^*)$, then go to Step 1. Otherwise go to Step 3.

Step 3: *External insertion.* This step attempts to reduce the makespan by moving a transportation job of V^* to another vehicle route with the least completion time. Suppose jobs i, j, and k are consecutive jobs on the route of V^*. Calculate $s_j = (t_{ij} + t_{jk} - t_{ik})$ for each transportation job j, where s_j is the empty vehicle travel time reduced by the removal of job j in V^*. Since less vehicle travel times are preferred, select job j^* to be removed from V^*, where $j* = \max(s_j)$, unless the move is in the tabu list. The selected job j^* is inserted in the tabu list. Now identify a vehicle that has the smallest $C(V_i)$, and insert j^* in this vehicle tour in a way that the completion time of the selected vehicle increases least. The insertion process in Step 3 is called external insertion process. Figure 4 shows the external insertion operation, where shaded and white areas indicate loaded and empty vehicle travels, respectively. One of the transportation requirements of vehicle #3 (V^*) is removed from the current route and inserted in the route of vehicle #1 which has the smallest completion time. The figure shows that the makespan is reduced after the external insertion process.

Step 4: Update the incumbent solution if the makespan of the current solution is less than that of all the solutions so far. Update iteration counter ($c = c$+1) and tabu list. If c has not reached the predetermined iteration limit or the

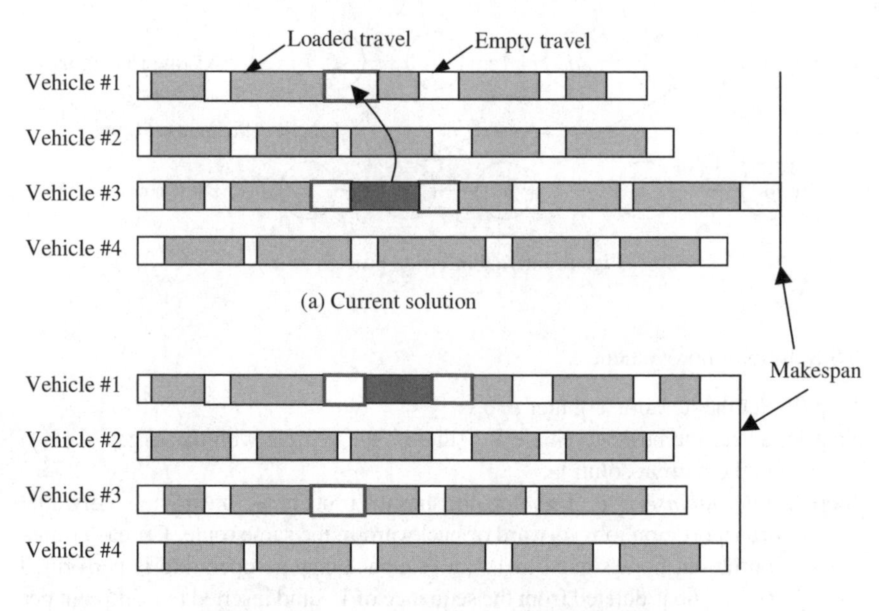

Fig. 4a,b. External insertion operation

maximum number of iterations without improvement has not been reached, go to Step 1.

As seen in Figure 2, the makespan of the incumbent solution is then checked. If the makespan of the incumbent solution exceeds the predefined planning horizon, the fleet size is increased by one and phase two should be repeated. If the makespan lies within the planning horizon, the incumbent solution is final.

4 Computational experiment

A sample test problem is adopted from Ko et al. (2000) where the container transportation environment is almost the same as in this paper. Tables 1 and 2 show from-to matrices for container transportation requirements and vehicle travel times, respectively.

The AMPL modeling tool (Fourer et al. 1993) and CPLEX solver are used to solve the optimization model. Since the optimization problem is similar to the classical transportation problem, solutions are found quite fast. The optimization model produces empty vehicle travel frequencies between locations as shown in the following:

Loaded vehicle travel time: 4,620 minutes
Empty vehicle travel time: 2,550 minutes
Total vehicle travel time: 7,170 minutes
Empty vehicle travel frequency

- $E \rightarrow A$: 6 vehicles
- $E \rightarrow B$: 28 vehicles
- $E \rightarrow C$: 17 vehicles
- $H \rightarrow A$: 33 vehicles
- $I \rightarrow A$: 1 vehicle

Minimum fleet size: $\lceil 7,170/480 \rceil = \lceil 14.9 \rceil = 15$

Table 1. Container transportation requirements

		To								
		A	B	C	D	E	F	G	H	I
From	A	–				15			47	2
	B		–			28				
	C			–		22			5	2
	D				–					
	E	3		10		–				1
	F						–			
	G							–		
	H	21		2					–	
	I								4	–

Table 2. Vehicle travel time matrix

		To								
		A	B	C	D	E	F	G	H	I
From	A	–	50	30	35	40	35	30	30	30
	B	50	–	30	35	40	35	30	35	35
	C	30	30	–	5	10	25	30	35	35
	D	35	35	5	–	5	20	25	30	30
	E	40	40	10	5	–	15	20	25	25
	F	35	35	25	20	15	–	10	15	15
	G	30	30	30	25	20	10	–	5	5
	H	30	35	35	30	25	15	5	–	5
	I	30	35	35	30	25	15	5	5	–

Suppose each vehicle can operate for 480 minutes per shift. At least 15 vehicles (the smallest integer greater than or equal to 7,170/480) are required to satisfy all the transportation requirements within the shift. If a vehicle finishes all the transportation jobs, it may not need to travel empty to somewhere else. However the optimization model counts this unnecessary empty vehicle travel, which results in an increase in vehicle travel time. In order to tackle this problem, two nodes, J and K, are introduced where J is the location at which all the vehicles are parked at the beginning of the planning horizon while K is the location for the vehicles to be parked at the end of the time period. The new locations could be real sites such as depots or artificial locations for preventing additional empty vehicle travels. If they are artificial locations, the travel time between the existing locations and the artificial sites is set to zero. In our experiments, it is assumed that 15 vehicles are parked at an artificial location J at the beginning and they are returned to an artificial location K after they finish their transportation jobs. The optimization model again produces empty vehicle travel frequencies as follows. Now, it can be seen that the lower bound on the fleet size is 14.

Loaded vehicle travel time: 4,620 minutes
Empty vehicle travel time: 1,950 minutes
Total vehicle travel time: 6,570 minutes
Empty vehicle travel frequency
- E $\rightarrow$ B: 19 vehicles
- E $\rightarrow$ C: 17 vehicles
- E $\rightarrow$ K: 15 vehicles
- H $\rightarrow$ A: 33 vehicles
- I $\rightarrow$ A: 1 vehicle
- J $\rightarrow$ A: 6 vehicles
- J $\rightarrow$ B: 9 vehicles

Minimum fleet size: $\lceil 6,570/480 \rceil = \lceil 13.7 \rceil = 14$

Based on the lower bound on the fleet size, the tabu search based algorithm yields a vehicle routing. The tabu tenure (i.e., the time period for which a move is prohibited) is set to three after some preliminary experiments. That is, the reverse move is prohibited for three periods after a move is performed. If the tabu tenure is too small, the probability of cycling increases. On the other hand, if it is too large, there is a possibility that the search space is too restricted, which may degrade the performance of the algorithm. The maximum number of iterations is set to 500. With 14 vehicles, the algorithm produces a solution with 500 minutes of the makespan and 6,965 minutes of total travel time. The total travel time is larger than the result obtained from the optimization model in phase one by 395 minutes. Since the makespan exceeds the predetermined time limit of 480 minutes, we increase the fleet size by one and run the experiment again. As a result, with 15 vehicles, total vehicle travel time and makespan are 6,740 and 460 minutes, respectively. The container transportation requirements can thus be met with 15 vehicles.

The heuristic is coded in the Visual Basic programming language. Figure 5 shows a screen capture of the experimental result when 15 vehicles are used. For example, the final solution of the tabu search based procedure generates the route of the first vehicle as follows:

Depot – $2 \rightarrow 5 - 2 \rightarrow 5 - 3 \rightarrow 5 \rightarrow 9 \rightarrow 8 - 1 \rightarrow 8 - 1 \rightarrow 8 - 1 \rightarrow 8 \rightarrow 1 \rightarrow 5 - 3 \rightarrow 5 - 3 \rightarrow 5$ – Depot.

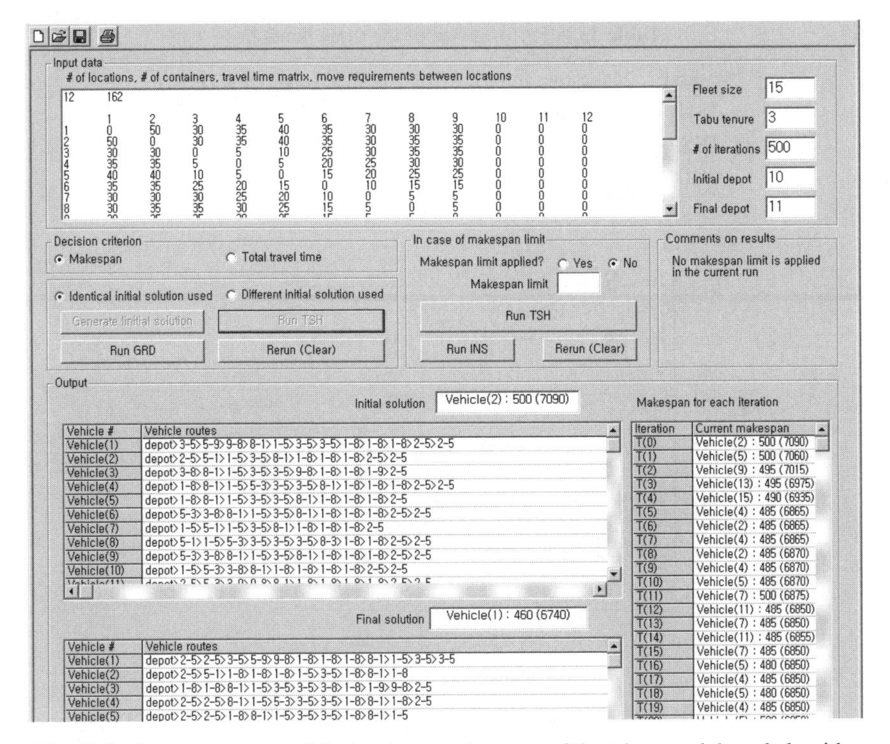

Fig. 5a,b. A screen capture of the implemented system of the tabu search based algorithm

Here, '→' indicates the loaded vehicle travel between two locations while '–' denotes empty vehicle travel. It can be observed that the first vehicle operates 460 minutes during the planning horizon, 300 minutes for loaded travels and 160 minutes for empty travels.

The performance of the proposed two-phase heuristic (TSH) is compared with two existing methods, the insertion algorithm based heuristic (INS) of Ko et al. (2000) and a greedy procedure (GRD). In INS, given a planning horizon, transportation jobs are assigned to a vehicle on by one. INS first selects a vehicle for scheduling. A transportation job is then randomly selected and assigned to the vehicle. Among all the transportation jobs which are not assigned, a job which increases the vehicle travel time the least is selected and inserted in the route of the vehicle. The assignment procedure is repeated until the total travel time of the vehicle exceeds the planning horizon, when a new vehicle is introduced and the assignment procedure is repeated. GRD makes a myopic decision to select transportation jobs. When a vehicle becomes free, it selects the nearest job. The first stage of the proposed procedure uses GRD in order to obtain an initial solution.

Table 3 shows the makespan and total vehicle travel time when 15 vehicles are used. TSH completes all the transportation jobs within 460 minutes while INS and GRD requires 495 and 500 minutes, respectively. It is observed that TSH also produces the vehicle routing with the least total vehicle travel time.

Table 3a,b. Performance of the three heuristics

Method	Makespan (min.)	Total vehicle travel time (min.)
TSH	460	6740
INS	495	6860
GRD	500	7000

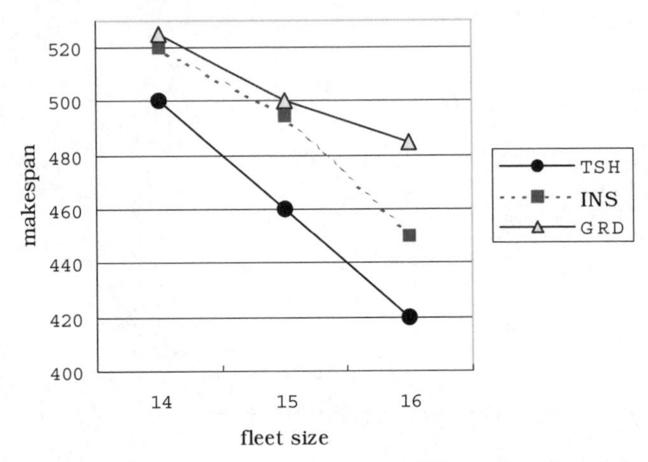

Fig. 6a,b. Change of makespan over different fleet sizes

Figure 6 compares the makespans of the three heuristics against the fleet size. As expected, with more vehicles, the makespan decreases for all the heuristics. If the makespan is restricted to only 480 minutes per shift, TSH requires 15 vehicles while INS and GRD requires more vehicles, 16 and 17, respectively. It can also be observed that, with 14 vehicles, TSH requires 20 minutes of overtime while INS and GRD needs 40 and 45 minutes, respectively.

In this paper, the travel time between two locations is assumed to be fixed. However, this assumption is often invalid in practice since the containers usually travel through congested streets. As pointed out in Park et al. (2000), a lognormal or triangular distribution may often be adequate to model the travel time between locations. The performance of the proposed procedure with stochastic travel times is investigated by assuming a triangular distribution with variations ranging from 5 % to 25 %. When the mean travel time between two locations is 50 minutes with a time variation of 20%, the travel time follows a triangular distribution with mode 50, minimum 40, and maximum 60. The experiments are repeated 10 times for each method, and the results are summarized in Table 4. As the variation increases, the makespans of TSH and INS slightly increase. The makespan of GRD seems insensitive to the travel time variations. Overall, the proposed procedure performs well even when the travel time variations are large.

The performance of the proposed procedure under various conditions is also investigated by running experiments with 40 problem instances. The average makespans of the three heuristics, TSH, INS and GRD are summarized in Table 5. TSH

Table 4a,b. Makespan comparison under probablistic travel times

Method	Travel time variation					
	0%	5%	10%	15%	20%	25%
TSH	500.0	503.8	507.7	511.5	515.5	519.7
INS	540.0	543.7	547.5	551.2	555.0	558.8
GRD	580.0	577.1	577.0	577.4	578.3	577.4

Table 5a,b. Comparison of three methods with 40 different cases

Method	Fleet size	Makespan	Total vehicle travel time
TSH	18.2	464.0	8405.3
INS	21.3	479.1	9607.4
GRD	18.2	534.1	9105.8

requires 18.2 vehicles on average to satisfy the transportation requirements within 480 minutes while INS requires 21.3 vehicles. GRD is experimented with the same fleet size as in TSH, and it is found that its makespan is 534.1 minutes, 70.1 minutes larger than that of TSH.

5 Conclusions

This paper proposes a new heuristic procedure for fleet sizing and vehicle routing in a static container transportation environment. The heuristic consists of two phases. The first phase determines the lower bound on the fleet size by using an optimization model and the second phase constructs a vehicle routing by applying the concept of tabu search. The proposed procedure has been compared with two existing methods through computational experiments. It has been observed that the new procedure consistently provides good quality solutions in terms of makespan and total vehicle travel time.

A heuristic tabu search algorithm is applied in the second phase to solve the container transportation problem discussed above. It may also be meaningful to investigate the application of other meta-heuristics such as simulated annealing and genetic algorithm to this problem. Finally, a container transportation problem is investigated in a static environment in which all transportation requirements are ready to be picked up at the beginning of the planning horizon. However, this may not usually be the case in the real world since containers may arrive in the middle of the planning horizon. Then, the fleet sizing and vehicle routing problems should be addressed from different perspectives. These subjects may provide a good opportunity for further studies.

References

Bae J W, Kim K H (2000) A pooled dispatching strategy for automated guided vehicles in port container terminals. International Journal of Management Science 6: 47–67

Beaujon G J, Turnquist M A (1991) A model for fleet sizing and vehicle allocation. Transportation Science 25: 19–45

Bish E K, Leong T Y, Li C L, Ng J W C (2001) Analysis of a new vehicle scheduling and location problem. Simchi-Levi D. Naval Research Logistics 48: 363–386

Bodin, L D, Golden, B L, Assad A A, Ball M O (1983) Routing and scheduling of vehicles and crews: the state of the art. Computers and Operation Research 10: 63–211

Breedam A V (2001) Comparing descent heuristics and metaheuristics for the vehicle routing problem. Computers and Operations Research 28: 289–315

Chao IM (2002) A tabu search method for truck and trailer routing problem. Computers and Operations Research 29: 33-51

Crainic T G, Laporte G (1998) Fleet management and logistics. Kluwer, Amsterdam

Du Y, Hall R (1997) Fleet sizing and empty equipment redistribution for center-terminal transportation networks. Management Science 43: 145–157

Fourer R, Gay D M, Kernighan B W (1993) AMPL a modeling language for mathematical programming. Boyd and Fraser, Massachusetts

Franca P M, Gendreau M, Laporte G, Muller F M (1996) A tabu search heuristic for the multiprocessor scheduling problem with sequence dependent setup times. International Journal of Production Economics 43: 79–89

Gambardella L M, Rizzoli A E, Zaffalon M (1998) Simulation and planning of an intermodal container terminal. Simulation 71: 107–116

Gendreau M, Laporte G, Vigo D (1999) Heuristics for the traveling salesman problem with pickup and delivery. Computers and Operations Research 26: 699–714

Glover F (1989) Tabu search, part I. ORSA Journal on Computing 1: 190–206

Glover F (1997) Tabu search. Kluwer, Boston

Grunow M, Gunther H O, Lehmann M (2004) Dispatching multi-load AGVs in highly automated seaport container terminals. OR Spectrum 26: 211–235

Kim K H, Bae J W (1999) A dispatching method for automated guided vehicles to minimize delays of containership operations. International Journal of Management Science 5 (1): 1–26

Kim K Y, Kim K H (1999) A routing algorithm for a single straddle carrier to load export containers onto a containership. International Journal of Production Economics 59: 425–433

Ko C S, Chung K H, Shin J Y (2000) Determination of vehicle fleet size for container shuttle service. Korean Management Science Review 17: 87–95

Kobza J E, Shen Y C, Reasor R J (1998) A stochastic model of empty-vehicle travel time and load request service time in light-traffic material handling systems. IIE Transactions 30: 133–142

Koo P H, Suh J D (2002) Fleet sizing under dynamic vehicle dispatching. Journal of Korean Institute of Industrial Engineering 28: 256–263

Kozan E, Preston P (1999) Genetic algorithms to schedule container transfers at multimodal terminals. International Transactions in Operational Research 6: 311–329

Laporte G, Osman H (1995) Routing problems: a bibliography. Annals of Operations Research 61: 227–262

Legato P, Mazza R M (2001) Berth planning and resources optimization at a container terminal via discrete event simulation. European Journal of Operational Research 133: 537–547

Maxwell W L, Muckstadt J A (1982) Design of automated guided vehicle systems. IIE Transactions 14: 114–124

Nanry W P, Barnes J W (2000) Solving the pickup and delivery problem with time windows using reactive tabu search. Transportation Research Part B 34: 107–121

Osman I H, Laporte G (1996) Metaheuristics: a bibliography. Annals of Operations Research 63: 513–628

Osman I H, Wassan N A (2002) A reactive tabu search metaheuristic for the vehicle routing problem with backhauls. Journal of Scheduling 5: 263–285

Park C H, Jun G S, Koh S Y, Kim D N, Kim Y C, Suh S D, Seol J H, Yun H M, Lee S M, Jang H B, Choi G J, Choe J S (2000) Introduction to transportation engineering. Yongji Monhwasa, Seoul

Rajotia S, Shanker K, Batra J L (1998) Determination of optimal AGV fleet size for an FMS. International Journal of Production Research 36: 1177–1198

Shabayek A A, Yeung W W (2002) A simulation model for the Kawi Chung container terminals in Hong Kong. European Journal of Operational Research 140: 1–11

Vis, I F A, de Koster R, Roodbergen K J (2001) Determination of the number of automated guided vehicles required at a semi-automated container terminal. Journal of the Operational Research Society 52: 409–417

Vis, I F A, de Koster R (2003) Transshipmment of containers at a container terminal: An overview. European Journal of Operational Research 147: 1–16

Yun W Y, Choi Y S (1999) A simulation model for container-terminal operation analysis using an object-oriented approach. International Journal of Production Economics 59: 221–230

The berth allocation problem: models and solution methods

Yongpei Guan[1] **and Raymond K. Cheung**[2]

[1] School of Industrial and Systems Engineering, George Institute of Technology, Atlanta, GA 30332, USA (e-mail: guanyp@isye.gatech.edu)
[2] Department of Industrial Engineering and Engineering Management, The Hong Kong University of Science and Technology, Clear Water Bay, Kowloon, Hong Kong (e-mail: rcheung@ust.hk)

Abstract. In this paper, we consider the problem of allocating space at berth for vessels with the objective of minimizing total weighted flow time. Two mathematical formulations are considered where one is used to develop a tree search procedure while the other is used to develop a lower bound that can speed up the tree search procedure. Furthermore, a composite heuristic combining the tree search procedure and pair-wise exchange heuristic is proposed for large size problems. Finally, computational experiments are reported to evaluate the efficiency of the methods.

Keywords: Berth allocation – Tree search procedure – Heuristics

We consider the problem of allocating berth space for vessels in container terminals, which we refer to as the *berth allocation problem*. The research is motivated by such a problem in Hong Kong. Being the world's busiest container ports in most of the last decade, Hong Kong had a container throughput of over 18 million twenty-foot equivalent units (TEU) in 2002. As the berth space is very limited in Hong Kong and thousands of containers have to be handled everyday, an effective berth allocation is critical to the efficient management of container traffic flow. A typical berth at the container terminals can accommodate multiple vessels at the same time. When no berth space is available, a vessel needs to wait for mooring. For simplicity, we call the sum of the waiting time and the processing time of a vessel as its *flow* time. Our objective is to allocate berth space to vessels and to schedule the vessels such that the total weighted flow time is minimized, where the weights reflect the relative importance of vessels.

---

* The authors would like to thank the helpful comments of two anonymous referees and the editors. The research was supported in part by Grant HKUST6039/01E of the Research Grant Council of Hong Kong

Correspondence to: R.K. Cheung

Several berth allocation models have appeared in the literature. They differ in the assumptions being made, such as whether vessel waiting is allowed, whether multiple vessels mooring on a berth is possible, whether vessel arrival times are considered, and whether the processing times are proportional to vessel size. Models that assume no vessel waiting are considered in Chen and Hsieh [1], Brown et al. [2] and Lim [12]. Vessels do not need to wait when berth space is abundant or parallel mooring is possible. Chen and Hsieh [1] develop an integer programming model in the form of generalized network with side constraints to minimize the total berthing cost. The costs in the model reflect the distances between the vessel mooring locations and the container locations in a terminal. Brown et al. [2] study the berthing of submarines at a naval base where parallel mooring and shifting of submarines along the berth are allowed. The problem is formulated as an integer program that minimizes the total berthing cost. For commercial ports, however, parallel mooring and shifting of vessels along berths are rare. On the other hand, Lim [12] considers the problem of minimizing the total berth length used for a given set of vessels. The problem is modeled as a network and the decision variables are the directions of arcs, which indicate the relative positions of vessels along the berth.

Models that permit vessel waiting are considered in [3,4,7–11,13]. Imai et al. [7] assume that only one vessel is moored to a berth at a time and develop a model for scheduling the currently waiting vessels to moor so as to minimize the total flow time. Imai et al. [8] extend the model by considering both the currently waiting vessels and the incoming vessels. A Lagrangian relaxation based procedure is developed. Allowing multiple vessels to be serviced at a berth simultaneously, Daganzo [3] discusses several principle-based heuristics for scheduling shore cranes so as to minimize the vessel waiting time, whereas Peterkofsky and Daganzo [13] develop a branch-and-bound method to solve such a problem. Lai and Shih [10] use simulation models to evaluate different berthing policies. The numerical results indicate that different policies may be used for different vessel arrival patterns. Kim and Moon [9] develop a mixed-integer programming model for the berth scheduling problem. The objective is to minimize the costs resulting from berthing vessels at the non-ideal locations and the delay costs for the vessel departures. When the arrival times of vessels are not considered, Li et al. [11] study the berth allocation problem where the processing time of a vessel is agreeable with the vessel size. That is, a larger vessel requires a longer processing time. The paper proposes a multiple-job-on-one-processor scheduling model with the objective of minimizing the latest vessel departure time (or makespan in scheduling terminology). A heuristic is developed of which the worst case analysis is performed. Finally, Guan et al. [4] consider a similar model but with the objective of minimizing the total weighted flow time. The problem is formulated as a multi-processor-for-one-job scheduling model and a new heuristic is proposed and analyzed.

In this paper, we study a more general berth allocation model that allows multiple vessel mooring per berth, considers vessel arrival times, and has the objective of minimizing the total weighted flow time. First, we consider two mathematical formulations for the model. The first one is similar to the model of [9] but with a slightly different objective. Our formulation has a different objective and the formulation is used to develop a tree search procedure for obtaining an exact solution of the problem. The other formulation is used to develop a tight lower bound that can speed up the tree search procedure. Next, we develop a composite heuristic in which the tree search procedure developed can be combined with the heuristic of Guan et al. [4] and a simple pair-wise exchange procedure. Finally, we conduct numerical experiments for evaluating the efficiency and the effectiveness of the methods.

The remainder of this paper is organized as follows: Section 1 presents the two mathematical formulations. Section 2 describes a tree search procedure while Section 3 introduces the heuristics. Section 4 reports the numerical results. Finally, Section 5 summarizes our research and outlines some potential research directions.

1 Formulations

In our model, we discretize the continuous time and space into integer units. The sections of the berth are indexed by $1, 2, \dots, S$, the time units over the planning horizon are indexed by $1, 2, \dots, T$, and the vessels are indexed by $1, 2, \dots, N$. We use the notion of $[a, b]$ to represent the set of integers between and including a and b. For each vessel i, define

p_i = Processing time for vessel i,
s_i = Size of vessel i, measured in the number of berth sections and $s_i \leq S$,
a_i = Arrival time of vessel i,
w_i = Weight assigned for vessel i.

We make a number of assumptions in our model:

1. Vessel arrivals can be grouped into batches. Our model is for tactical planning where container shipping companies provide approximate vessels' arrival times in the format of time ranges, such as the afternoon of a particular day. We group the vessels with a similar arrival time range as a batch and set the arrival times of this group of vessels as the beginning of the time range.
2. Once a vessel is moored, it will remain in its location until all the required container processing is done. In practice, any interruption of the container processing during mooring is costly. Thus, this assumption is valid in practice.
3. Vessel processing times and vessel sizes are agreeable, meaning that if $s_i \geq s_j$, then $p_i \geq p_j$, and vice versa. This assumption reflects that a larger vessel requires a longer processing time.
4. A berth section handles at most one vessel at a time. Multiple mooring of vessels at a section happens at specialized terminals (such as submarine base or barge terminal) but is not common for large ocean-going terminals.

The berth allocation problem can be represented by a time-space diagram where the horizontal axis and the vertical axis represent the time units and berth sections respectively. A vessel can be viewed as a rectangle whose length is the processing time and whose height is the vessel size. We call vessel i mooring at berth section ℓ and time t, then the vessel occupies the consecutive berth sections between ℓ and $\ell + s_i - 1$ and from time units t to $t + p_i - 1$. Let

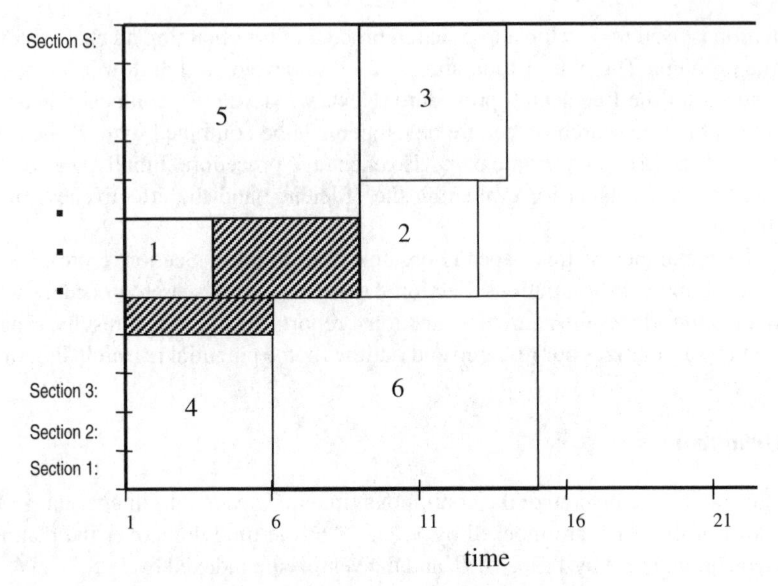

Fig. 1. Time-space representation

u_i = The mooring time of vessel i,
v_i = The starting berth section occupied by vessel i,
c_i = The departure time of vessel i.

We say that vessel rectangle i is immediately *left* of vessel rectangle j if $c_i = u_j$ (that is, vessel j is moored right after vessel i is finished) and vessel rectangle i is immediately *above* vessel rectangle j if $v_i = v_j + s_j$. Figure 1 shows the time-space diagram of 6 vessels.

1.1 Relative position formulation

One way to represent the problem is to consider the relative position of the vessel rectangles in the time-space diagram. Let

$$\sigma_{ij} = \begin{cases} 1, & \text{if vessel rectangle } i \text{ is completely on the left of vessel rectangle } j \\ & \text{and the two rectangles are not overlapped,} \\ 0, & \text{otherwise;} \end{cases}$$

$$\delta_{ij} = \begin{cases} 1, & \text{if vessel rectangle } i \text{ is completely below vessel rectangle } j \text{ and} \\ & \text{the two rectangles are not overlapped,} \\ 0, & \text{otherwise.} \end{cases}$$

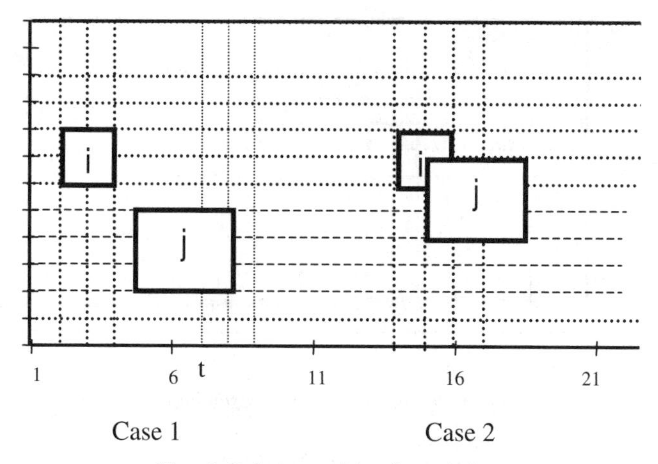

Fig. 2. Relative position formulation

The mathematical formulation of the berth allocation can be written as:

$$\min \sum_{i=1}^{N} w_i(c_i - a_i) \tag{1}$$

subject to

$$u_j - u_i - p_i - (\sigma_{ij} - 1)T \geq 0 \quad \forall i, j \tag{2}$$

$$v_j - v_i - s_i - (\delta_{ij} - 1)S \geq 0 \quad \forall i, j \tag{3}$$

$$\sigma_{ij} + \sigma_{ji} + \delta_{ij} + \delta_{ji} \geq 1 \quad \forall i, j \tag{4}$$

$$\sigma_{ij} + \sigma_{ji} \leq 1 \quad \forall i, j \tag{5}$$

$$\delta_{ij} + \delta_{ji} \leq 1 \quad \forall i, j \tag{6}$$

$$p_i + u_i = c_i \quad \forall i \tag{7}$$

$$u_i \in [a_i, T - p_i + 1], v_i \in [1, S - s_i + 1] \quad \forall i \tag{8}$$

$$\sigma_{ij} \in \{0, 1\}, \delta_{ij} \in \{0, 1\} \quad \forall i, j \tag{9}$$

We refer to the formulation defined by (2)–(8) as the Relative Position Formulation (RPF). Constraints (2)–(3) enforce the definitions of σ_{ij} and δ_{ij}. To see that, suppose that vessel rectangle i is on the left of vessel rectangle j. By definition, $\sigma_{ij} = 1$ and (2) becomes $u_j \geq u_i + p_i$. This reflects the requirement that vessel j cannot be started until vessel i is finished. Conversely, if vessel rectangle i is not on the left of vessel rectangle j, then the requirement is not needed. In this case, $\sigma_{ij} = 0$ and (2) becomes redundant since $u_j + T \geq u_i + p_i$ is always satisfied. Constraints (4)–(6) ensure that no vessel rectangles overlap. For example, in case 1 of Figure 2, rectangle i is at the left-and-above position of rectangle j and thus $\sigma_{ij} = 1$, $\sigma_{ji} = 0$, $\delta_{ij} = 0$ and $\delta_{ji} = 1$. In case 2 of Figure 2, the rectangles are overlapped and the values of all σ_{ij}, σ_{ji}, δ_{ij}, and δ_{ji} are all 0. In this situation, constraint (4) is violated. Constraint (7) defines the departure time as the vessel mooring time plus the processing time.

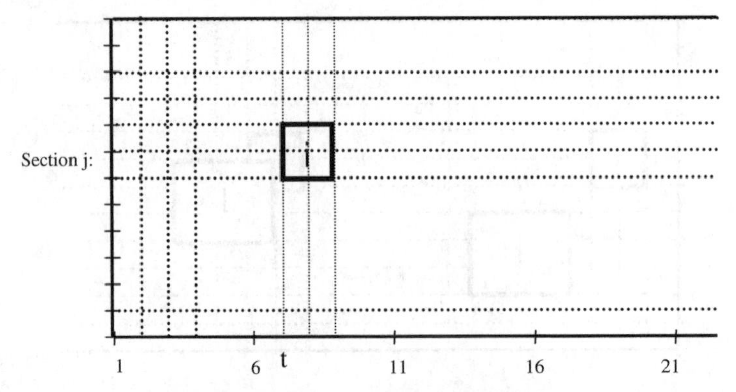

Fig. 3. Position assignment formulation

One feature of RPF is that, as stated in the following proposition, there is an optimal solution in which each vessel rectangle j is immediately right of another vessel rectangle or starting at period a_j, and is immediately above another vessel rectangle or starts at berth section 1.

Proposition 1 *There exists an optimal solution in which for a vessel j, (a) either $u_j = a_j$ or $u_j = u_i + p_i$ for some i; and (b) either $v_j = 1$ or $v_j = v_i + s_i$ for some i.*

Proof. The proof is similar to the proofs of Properties 1–3 in [9] when the penalty costs of not berthing at the optimal locations in their model are not considered and the delay costs for the vessels are set to our weights. □

1.2 Position assignment formulation

An alternative formulation is to consider the space *covered* by the vessel rectangles, where we refer to as the Position Assignment Formulation (PAF). Suppose the space-time area is partitioned into $S \cdot T$ unit blocks (each block has the height of one berth section and the width of one time unit). Let block (m, n) be the block for berth section m, $m \in [1, S]$ and time unit n, $n \in [1, T]$. For instance, in Figure 3, the vessel having a size of 2 and a processing time of 2 units is moored at section j at time t, the blocks (j, t), $(j, t+1)$, $(j+1, t)$, and $(j+1, t+1)$ are covered by this vessel. Let

$$y_{ijt} = \begin{cases} 1, & \text{if the left-bottom corner of vessel rectangle } i \text{ is located} \\ & \text{at section } j \text{ at unit } t, \\ 0, & \text{otherwise;} \end{cases}$$

$$x_{imn} = \begin{cases} 1, & \text{if block } (m, n) \text{ is covered by vessel rectangle } i, \\ 0, & \text{otherwise.} \end{cases}$$

The mathematical formulation can be given as

$$\min \sum_{i=1}^{N} w_i(c_i - a_i) \tag{10}$$

subject to

$$\sum_{j=1}^{S-s_i+1} \sum_{t=a_i}^{T-p_i+1} y_{ijt} = 1 \qquad \forall i \tag{11}$$

$$\sum_{m=j}^{j+s_i-1} \sum_{n=t}^{t+p_i-1} x_{imn} - p_i s_i - (y_{ijt}-1)M \geq 0 \qquad \forall i,j, \text{ and } t \geq a_i \tag{12}$$

$$\sum_{i=1}^{N} x_{imn} \leq 1 \qquad \forall m,n \tag{13}$$

$$p_i + \sum_{j=1}^{S-s_i+1} \sum_{t=a_i}^{T-p_i+1} t y_{ijt} = c_i \qquad \forall i \tag{14}$$

$$x_{imn} \in \{0,1\} \qquad \forall i,m,n \tag{15}$$

$$y_{ijt} \in \{0,1\} \qquad \forall i,j,t \tag{16}$$

In this formulation, the objective function is the same as the one of RPF. Constraint (11) shows that each vessel must be moored exactly once. Notice that vessel i must start within sections 1 to $S - s_i + 1$ at a time period between a_i and $T - p_i + 1$. Constraint (12) ensures that the blocks occupied by a particular vessel must be consecutive. Constraint (13) says that a block can only be occupied by at most one vessel rectangle. Constraint (14) relates the processing time, the start time, and the completion time of a vessel. To see it, assume that vessel i starts at time t' at section j', that is, $y_{ij't'} = 1$. By (11), we know that $\sum_{j=1}^{S-s_i+1} \sum_{t=a_i}^{T-p_i+1} t y_{ijt} = t'$. When $j \neq j'$ or $t \neq t'$, we set $y_{ijt} = 0$. Thus, the left hand side of (14) gives us the sum of the mooring time and the processing time of vessel i. By definition, this time is the departure time of vessel i.

1.2.1 A relaxed problem

Suppose that constraint (13) is relaxed and let μ_{mn} be the corresponding Lagrangian multipliers for a given pair of m and n. Let μ be the vector of μ_{mn}. The relaxed problem can be written as

$$L(\mu) = \min \sum_{i=1}^{N} w_i(c_i - a_i) + \sum_{m=1}^{S} \sum_{n=1}^{T} \mu_{mn} \left(\sum_{i=1}^{N} x_{imn} - 1 \right)$$

subject to (11), (12), (14), (15), and (16).

We now show that the value $L(\mu)$ for a given μ is easy to obtain. Substituting (14) into the objective function and rearranging the terms, the relaxed problem becomes

$$L(\mu) = \min \sum_{i=1}^{N} \sum_{j=1}^{S-s_i+1} \sum_{t=a_i}^{T-p_i+1} tw_i y_{ijt} + \sum_{m=1}^{S} \sum_{n=1}^{T} \mu_{mn} \sum_{i=1}^{N} x_{imn} + \sum_{i=1}^{N} w_i p_i - \sum_{i=1}^{N} w_i a_i - \sum_{m=1}^{S} \sum_{n=1}^{T} \mu_{mn}$$

subject to (11), (12), (15), and (16).

The last three terms are constants and can be ignored from the optimization point of view. Thus, solving the relaxed problem is equivalent to solving

$$\min \sum_{i=1}^{N} \sum_{j=1}^{S-s_i+1} \sum_{t=a_i}^{T-p_i+1} tw_i y_{ijt} + \sum_{m=1}^{S} \sum_{n=1}^{T} \mu_{mn} \sum_{i=1}^{N} x_{imn} \tag{17}$$

subject to (11), (12) (15), and (16).

Let us consider the second term in the objective function of (17). Since x_{imn} can only be 0 or 1, the sum $\sum_{m=j}^{j+s_i-1} \sum_{n=t}^{t+p_i-1} x_{imn}$ is at most $p_i s_i$. On the other hand, if $y_{ijt} = 1$, then (12) implies

$$\sum_{m=j}^{j+s_i-1} \sum_{n=t}^{t+p_i-1} x_{imn} \geq p_i s_i \quad \forall i, j, t \geq a_i.$$

Therefore, when $y_{ijt} = 1$, we have $x_{imn} = 1$ for $m \in [j, j+s_i+1]$ and $n \in [t, t+p_i-1]$. Because our problem is a minimization problem and the multipliers $\mu_{mn} \geq 0$, we know that x_{imn} must be 0 for $m \notin [j, j+s_i+1]$ or $n \notin [t, t+p_i-1]$ in the optimal solution. Thus, for vessel i, if $y_{ijt} = 1$, we have

$$\sum_{m=1}^{S} \sum_{n=1}^{T} \mu_{mn} x_{imn} = \sum_{m=j}^{j+s_i-1} \sum_{n=t}^{t+p_i-1} \mu_{mn}.$$

According to constraint (11), for each i, there is only one $y_{ijt} = 1$ for $j \in [1, S - s_i + 1]$ and $t \in [a_i, T - p_i + 1]$. This implies that the relaxed problem (17) becomes

$$\min \sum_{i=1}^{N} \sum_{j=1}^{S-s_i+1} \sum_{t=a_i}^{T-p_i+1} y_{ijt} \left(tw_i + \sum_{m=j}^{j+s_i-1} \sum_{n=t}^{t+p_i-1} \mu_{mn} \right)$$

subject to (11) and (16) (18)

which is separable in i. Therefore, the solution of the relaxed problem can be obtained, for each i, by setting $y_{ij't'} = 1$ where the pair (j', t') is obtained as

$$(j', t') = \arg \min_{(j, t \geq a_i)} \left(tw_i + \sum_{m=j}^{j+s_i-1} \sum_{n=t}^{t+p_i-1} \mu_{mn} \right). \tag{19}$$

1.2.2 Lower bounds

For fixed values of μ_{mn}, the value $L(\mu)$ gives a lower bound to which we refer to as the Lagrangian Relaxation (LR) bound. To get a better lower bound, we can use the classical subgradient method to determine μ_{mn} such as the one in [5]. In addition to the LR bound, we can obtain lower bounds from the LP relaxations of the two formulations. However, it is well known in the literature that the LR bounds are tighter than the LP bounds.

2 A tree search procedure

In this section, we describe a tree search procedure that utilizes the properties of both RPF and PAF. In the procedure, we add vessel rectangles to the time-space diagram one at a time. If k vessel rectangles have been inserted to the time-space diagram, we call it a k-partial solution. The tree in our procedure has N levels. A node in the tree at level k corresponds to a k-partial solution. The child node of this node is a $(k+1)$-partial solution which is the k-partial solution of its parent node combined with an additional vessel rectangle. Notice that each additional vessel rectangle can be assigned at different positions in the unoccupied space of the diagram. Therefore, a node at level k can have more than $N-k$ child nodes and the corresponding tree has a huge number of nodes. In the following parts, we will describe several approaches to reduce the number of tree nodes, which include efficient branching scheme, tight lower bound for each tree node and duplication reduction.

According to Proposition 1, we can branch a search tree in the following way. The k inserted vessel rectangles form a stair on each tree node. The steps of the stair help indicate potential feasible positions for inserting other vessel rectangles. When we insert a vessel rectangle on one step, the assignment time of the vessel rectangle is the maximum of the step start time and the vessel arrival time. Consider the small example shown in Figure 4a. Suppose that we have inserted vessel rectangle 4 (with $a_4 = 1$, $s_4 = 4$ and $p_4 = 5$) to the left-most bottom position of the time-space diagram, vessel rectangle 1 (with $a_1 = 10$, $s_1 = 2$ and $p_1 = 3$) on the right of vessel rectangle 4 at time 10 (since $a_1 = 10 > 6 = c_4$), and vessel rectangle 3 on the top of vessel rectangle 4. The three insertions result in the 3-partial solution which occupies the left bottom region of the time-space diagram. The boundary of this region forms a stair, as the bold line shown in Figure 4a. This stair has three steps: step 1 starts at time 13 and berth section 1; step 2 starts at time 12 and section 3; and step 3 starts at time 0 and section 10. Suppose that we next insert vessel rectangle 2 (with $a_2 = 11$, $s_2 = 2$ and $p_2 = 3$). According to Proposition 1, vessel rectangle 2 can be put on either step 1, step 2 or step 3. Assume that it is put on step 2. It will start at time $\max\{a_2, \bar{u}_2\} = \bar{u}_2$, where $\bar{u}_2$ represents the start time of the second step. After this inserting operation, another stair with three steps is formed as shown by the dashed bold line in Figure 4a, where the next vessel rectangles can be put on the steps.

To reduce the branching effort, lower and upper bounds are used to provide cuts in the tree searching process. For a given k-partial solution, the LR bound for this

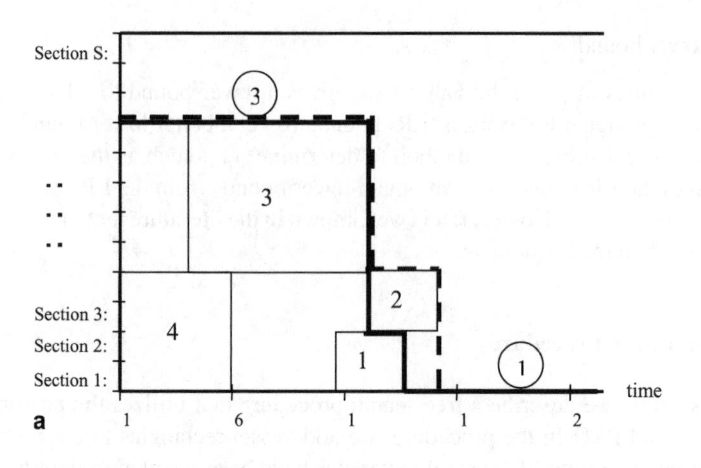

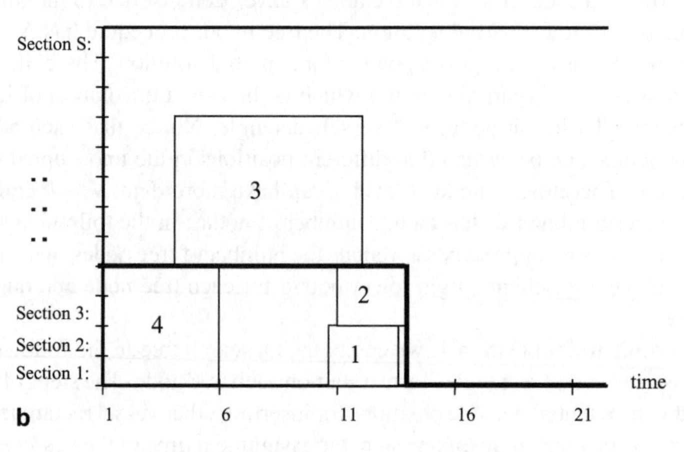

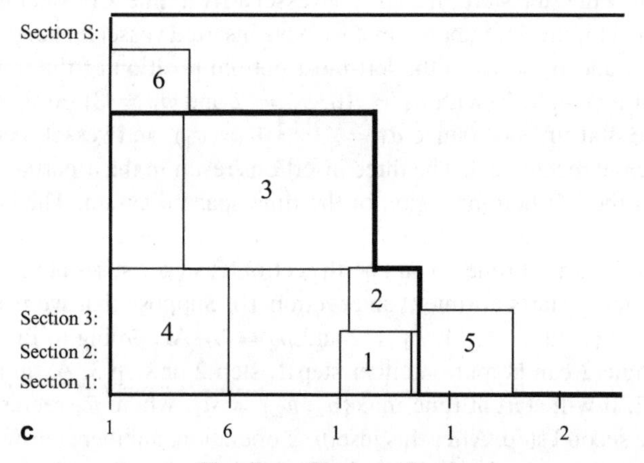

Fig. 4a–c. Tree search procedure

k-partial solution is calculated in such a way that we apply Lagrangian relaxation method to the remaining $N - k$ vessel rectangles on the unoccupied time-space diagram. To get an upper bound, we simply insert the $N - k$ un-inserted vessel rectangles according to the left-most bottom positions each time to obtain a feasible solution. The objective value of this solution represents an upper bound. This upper bound is updated whenever a better feasible solution is observed. When the lower bound of this tree node is larger than the current upper bound, the sub-tree from this node will be pruned.

In our approach, we consider duplication reduction operations on each tree node. For instance, as shown in Figure 4c, consider the six vessel pattern which can be obtained either by the insertion sequence 4-1-2-3-5-6 or the sequence 4-1-2-3-6-5. The two sequences produce duplicated patterns and one will be removed in our tree search approach. When inserting a vessel rectangle on a step of the current stair, we check if the same step exists in the parent node's solution. If so, we insert this vessel rectangle at the same berth section and time unit on the respective steps of the stairs in the partial solution of both the current node and the parent node. We prune the branch from the current node if starting berth section of this vessel rectangle has a smaller value than that of the last inserted vessel rectangle at the current node. For example, if the current node represents the partial solution of 4-1-2-3-6 and its parent node is for 4-1-2-3, and vessel rectangle 5 is to be inserted. This rectangle can be inserted to the position of berth section 1 and time 14. This position is on the current step which exists in both the stairs of the current partial solution 4-1-2-3-6 and that of its parent node 4-1-2-3. Since the starting section number of vessel rectangle 5 is smaller than the starting section number of vessel rectangle 6, the branch 4-1-2-3-6-5 is deleted. Depth-first search strategy is used in our tree search procedure, which helps reduce the memory requirement as a child node differs from its parent node by one vessel rectangle. Therefore, in general, this left-most bottom strategy together with depth-first search approach considerably reduce the need of the complete enumeration.

3 Heuristics

We now propose a heuristic that first groups the vessels into batches according to their arrival times (see assumption 1). For each batch, we use the heuristic of [4] to generate an initial solution. Next, we exchange the vessel rectangles in two consecutive batches if the exchange can reduce the value of the objective function. Then, the exact tree search procedure is applied for each of the resulting batches. Notice that in practice, the number of vessels in each batch is small enough that the tree search procedure can be applied. We review the key steps of the heuristic in [4] in Section 3.1, describe the pair-wise exchange procedure in Section 3.2 and list the steps of the overall procedure in Section 3.3.

3.1 A heuristic for common vessel arrival

When vessels have a common arrival time, the heuristic H of [4] can be applied. In the heuristic, the vessel rectangles are assumed in ascending order of vessel size.

We start by selecting the first group of (smallest) vessel rectangles whose total size is just smaller than S, and assign them to the berth sections. Next, we select the next group and assign them to the berth sections and so on. For an odd group, the vessel rectangles are assigned to the sections according to the increasing order of berth section numbers such that section S will be occupied by the largest vessel rectangle in this group. For an even group, the vessel rectangles are assigned to the sections according to the decreasing order of the section numbers such that section 1 will be occupied by the largest vessel rectangle in this group. Associated with this method is a lower bound (R1) with no more than 100% error determined as follows. First, a vessel rectangle with processing time p_i and size s_i is divided into s_i *sub-rectangles*, each has the height of 1 berth section and the processing time p_i. Then, apply the shortest-processing-time-first (SPT) rule to the sub-rectangles of the relaxed problem. The rule is optimal for the relaxed problem and thus produces a lower bound for the original problem.

3.2 A pair-wise exchange heuristic

Assume that there are K batches. Let B_k be the set of vessels with arrival times $A_k, k \in [1, K]$ such that $A_1 < A_2 < A_3 < \ldots < A_K$. For two consecutive batches, say B_k and B_{k+1}, the pair-wise exchange heuristic considers a pair of vessel rectangles, one from B_k and one from B_{k+1}. If exchanging the positions of the two vessel rectangles can reduce the objective value, such an exchange will be executed. During the exchange process, other *vessel rectangles'* positions can be moved left, right, up or down to make room for the exchange. After an exchange, we can move other vessel rectangles to the left such that each vessel can be started as early as possible. We illustrate the heuristic using Figure 5. Vessel rectangles $1, 2, 3, 4, 5$ are in B_1 with batch arrival time A_1 and vessel rectangles I, II, III are in B_2 with batch arrival time A_2. Consider the pair vessel rectangle 4 in B_1 and vessel rectangle I in B_2. Initial solution provided by Heuristic H is shown in Figure 5a. Vessel rectangle 4 is right after vessel rectangle 3 and on the top of vessel rectangle 5. Vessel rectangle I is below vessel rectangle II and after vessel rectangle 5. During the exchange operation, we move vessel rectangle II to the right to make it possible to let vessel rectangle I assigned at time A_2, on top of vessel rectangle 5 and before vessel rectangle II. Then, vessel rectangle 4 is put below vessel rectangle II and after vessel rectangle I. Finally, we move vessel rectangle III to the left to reduce the objective value, as shown in Figure 5b.

3.3 A heuristic for batch arrivals

After obtaining an initial solution by Heuristic H for each batch and then applying the pair-wise exchange heuristic, we can use the tree search procedure of Section 2 for each updated batch to further improve the objective value. The process can be repeated until no improvement is found. We call this heuristic for batch arrival as Heuristic HB which has the following steps:

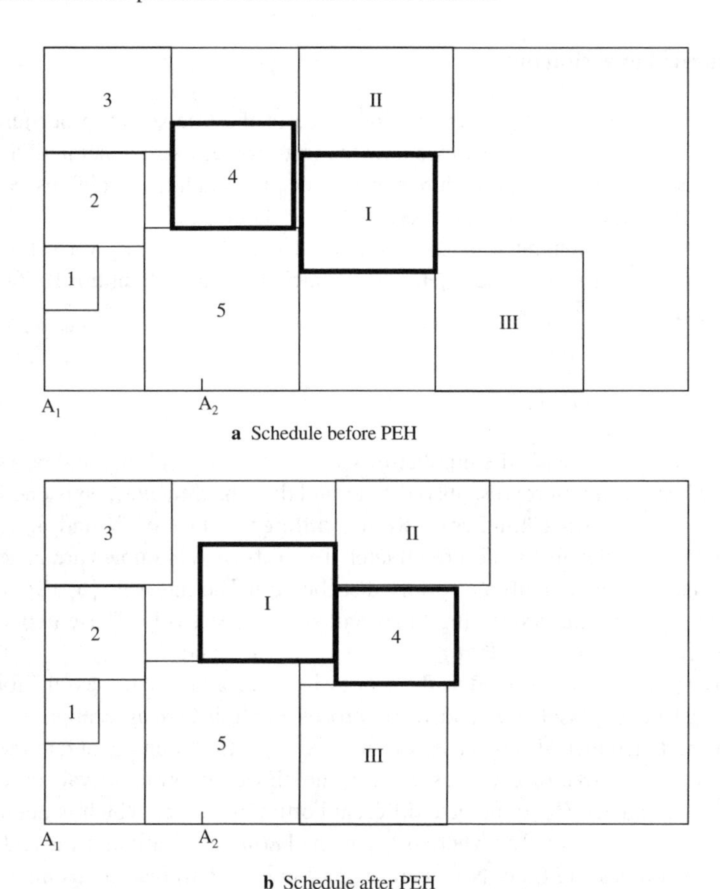

a Schedule before PEH

b Schedule after PEH

Fig. 5a,b. An example for the pair-wise exchange heuristics

Heuristic HB

Step 0: Obtain initial assignment of vessels for each batch B_i, $\forall i \in [1, K]$ by Heuristic H.

Step 1: Sequence B_i, $\forall i \in [1, K]$ according to batch arrival time A_i. Set $i = 1$.

Step 2: For $i < K$, do:

- Step 2.1: For each pair of vessel rectangles with one in B_i and one in B_{i+1}, if the pair-wise exchange heuristic can reduce the objective value, then declare this pair as a candidate. If there is no candidate, go to Step 2.3.
- Step 2.2: Select the candidate pair with the maximum reduction and implement the exchange. Go to Step 2.1.
- Step 2.3: Apply the tree search procedure in B_i and move vessel rectangles in $B_\ell, \ell \in [i+1, K]$ to the left as much as possible. Set $i \leftarrow i+1$ and go to Step 2.

Step 3: If the objective value can be reduced in Step 2, then set $i = 1$ and go to Step 2. Otherwise, terminate.

4 Numerical experiments

In this section, we investigate the performance of the tree search procedure and Heuristic HB. Section 4.1 describes how the test problems are generated. Section 4.2 reports the tree search procedure performance for small test problems. Section 4.3 shows the performance of the heuristic HB for large test problems. All solution approaches are implemented in C++ using the CPLEX 6.0 library for solving the related linear programs. Experiments were carried out on a Pentium II 300MHz processor PC.

4.1 Random test problems

In our experiment, we set the number of sections S to 10 and the time horizon T to 168, where each unit represents one hour and thus the time horizon is one week. For the common arrival time case, we use different values of N and s_{max} (the maximum vessel length) as described later. Two sets of N numbers are generated independent and identically uniformly distributed in the intervals [6, 12] and [1, s_{max}] to represent the processing times and sizes respectively. These two sets of N numbers are sorted such that $p_1 \leq p_2 \leq \ldots \leq p_N$ and $s_1 \leq s_2 \leq \ldots \leq s_N$. Each pair p_i and s_i is assigned as the processing time and size of vessel i for all i in [1, N]. Hence, vessel sizes and vessel processing times are agreeable.

For the batch arrival time case, we first fix s_{max} to 10 and generate the processing time and size of each vessel as in the above common arrival time case. We use a parameter R_a to indicate different berth utilizations which is defined as $\sum_i p_i s_i / (S \cdot (a_{max} + 12))$, where a_{max} is the latest vessel arrival time and 12 is the longest processing time. Note here a_{max} can be determined by using a given R_a and the generated p_i and s_i. Next, we generate vessel arrival times a_i which are independent and identically uniformly distributed in the interval [0, a_{max}]. Then, we set the number of batches K to be the rounded-up integer of $a_{max}/12$ and set batch arrival times A_k to be 0, 12, 24, ... etc. Finally, we group vessel i to batch k if $a_i \geq A_k$ and $a_i < A_{k+1}$ and update vessel arrival time a_i to be A_k.

4.2 Exact solution methods

There are three exact solution methods for the berth allocation problem. They are CPLEX solutions for RPF and PAF and the developed tree search procedure as described in Section 2. When implementing the tree search procedure, a larger value of $k_{\max}$ (the number of iterations in the subgradient method) will produce a tighter LR bound and thus the tree search requires a smaller number of nodes. However, it takes a longer CPU time. We did some preliminary experiments and found that the overall CPU time was smaller when the value of $k_{\max}$ was between 5 and 10. Thus, we set $k_{\max} = 10$ for problems with less than 10 vessels and $k_{\max} = 5$ otherwise. In this experiment, we fix $s_{max} = S$. As the formulations are integer programs, CPLEX employs the branch-and-bound method that involves a large number of tree nodes. Direct application of CPLEX to PAF cannot provide

Table 1. Computational results for the tree search procedure and CPLEX

N	RPF (by CPLEX)		Tree search	
	#nodes	CPU(sec)	#nodes	CPU(sec)
4	83	0.11	34	0.11
10	244.4K	660	1.9K	49
15	–	–	120.4K	55325

an optimal solution for the cases that have more than 4 vessels. In the experiment, N is set to be 4, 10, 15 respectively. We take five samples for each value of N and record the average value, as shown in Table 1. The tree search procedure requires a substantially smaller number of tree nodes than the branch-and-bound method. Consequentially, the tree search procedure can solve problems up to 15 vessels. In practice, the largest berth in Hong Kong's terminals can hold up to four ocean going vessels where the processing time for a vessel typically ranges from 6 to 12 hours. Thus, if we have one batch per day, then the tree search procedure can still be used.

4.3 Performance of the heuristics

First, we would like to evaluate the performance of Heuristic H for the common arrival time case as there is no numerical results in [4]. We set N to be 5, 10, 15, $\ldots$, 80, and s_{max} to be 2, 3, $\ldots$, 7. For each combination, 100 random samples are generated. For each sample, we obtain the objective value Z^H by Heuristic H, lower bound R1 (see Section 3.1) and the Lagrangian relaxation lower bound LR. Average and maximum ratios of Z^H to each lower bound for each combination are shown in Table 2. Computational results show the robust performance of Heuristic H. The worst case sample error is around 25% when $N = 10$ and $s_{max}/S = 0.7$, and the average error is no more than 15%. It is observed that Heuristic H performs better when $s_{\max}$ is small. Computational results also show that LR provides tighter lower bounds than R1 as the ratio $s_{\max}/S$ increases and the latter provides tighter lower bound only when the ratio is very small. In our experiment, Heuristic H took only 0.06 second for the largest problem.

Second, we would like to evaluate the effectiveness of Heuristic HB. We create different combinations by setting $R_a = 0.6, 0.75, 0.9$ and N=12, 18, $\ldots$, 42 respectively. For each combination, we apply Heuristic H for each batch, then use the pair-wise exchange heuristic to swap vessel rectangles between batches, and finally apply the tree search procedure. Let Z^H, Z^P, and Z^T be the objective values resulting from these three procedures respectively. Let Z_l be the LR bound. We take 30 samples for each combination and report the average and maximum gaps for the three procedures, defined as $(Z^H\text{-}Z_l)/Z^H$, $(Z^P\text{-}Z_l)/Z^P$ and $(Z^T\text{-}Z_l)/Z^T$ which are indicated by H, PE, and TS in Table 3 respectively. Experiment results indicate that the pair-wise exchange heuristic and the tree search procedure can reduce the error

Table 2. Computational results for the common arrival time case

S_{max}/S	Z^H/Z_u	N=10		N=20		N=40		N=80	
		Ave	Max	Ave	Max	Ave	Max	Ave	Max
0.2	R1	1.02	1.03	1.03	1.07	1.02	1.03	1.01	1.02
	LR	1.03	1.05	1.05	1.07	1.02	1.04	1.02	1.03
0.3	R1	1.10	1.14	1.06	1.09	1.06	1.08	1.05	1.07
	LR	1.10	1.13	1.05	1.06	1.04	1.05	1.04	1.05
0.4	R1	1.12	1.19	1.14	1.21	1.12	1.17	1.11	1.17
	LR	1.09	1.14	1.09	1.12	1.08	1.10	1.07	1.10
0.5	R1	1.12	1.22	1.11	1.17	1.12	1.14	1.11	1.17
	LR	1.10	1.15	1.08	1.11	1.08	1.09	1.08	1.11
0.6	R1	1.23	1.36	1.19	1.23	1.20	1.26	1.18	1.31
	LR	1.15	1.24	1.10	1.12	1.10	1.14	1.11	1.13
0.7	R1	1.26	1.39	1.24	1.41	1.28	1.32	1.27	1.38
	LR	1.15	1.24	1.10	1.15	1.13	1.15	1.14	1.20

Table 3. Error gaps produced by the steps of heuristics HB for the batch arrival case

		Ra=0.6		Ra=0.75		Ra=0.9	
#vessels		Ave	Max	Ave	Max	Ave	Max
12	H	0.12	0.18	0.28	0.4	0.29	0.4
	PE	0.11	0.15	0.21	0.31	0.15	0.29
	TS	0.05	0.08	0.09	0.19	0.06	0.11
18	H	0.33	0.44	0.37	0.45	0.4	0.5
	PE	0.23	0.32	0.26	0.32	0.24	0.31
	TS	0.05	0.08	0.08	0.15	0.07	0.1
24	H	0.44	0.56	0.45	0.58	0.49	0.56
	PE	0.26	0.32	0.24	0.36	0.27	0.33
	TS	0.11	0.14	0.09	0.14	0.12	0.19
30	H	0.46	0.54	0.5	0.58	0.53	0.62
	PE	0.28	0.38	0.29	0.41	0.24	0.32
	TS	0.14	0.19	0.11	0.13	0.15	0.19
36	H	0.54	0.63	0.55	0.65	0.57	0.65
	PE	0.25	0.32	0.34	0.41	0.23	0.38
	TS	0.16	0.22	0.17	0.22	0.12	0.18
42	H	0.54	0.59	0.59	0.62	0.6	0.66
	PE	0.23	0.32	0.26	0.35	0.29	0.34
	TS	0.17	0.23	0.17	0.2	0.16	0.21

gaps significantly. Average relative errors for batch arrival time case are around 10%–15%. The CPU times for Heuristic H, the pair-wise exchange heuristic and the tree search procedures are 0.05 second, 26.25 seconds and 1671.81 seconds respectively for the largest problem where $R_a = 0.9$ and $N = 42$. This indicates our solution approaches are very efficient in solving the berth allocation problem.

5 Summary

In this paper, we study a berth allocation model that allows multiple vessel mooring per berth, considers vessel arrival times, and has the objective of minimizing the total weighted flow time. In our solution approach, we assume vessel arrivals can be grouped into batches. After describing the *berth allocation problem* and reviewing current researches on this problem, we consider two formulations with which several bounds are derived. Next, a tree search procedure is described which provides an exact solution and it performs better than the direct application of CPLEX. Then, we develop a composite heuristic HB that combines Heuristic H of [4], a pair-wise exchange heuristic and the tree search procedure. The computational experiments show that Heuristic H is efficient to generate good initial solutions while the composite heuristic HB is quite effective.

In the future research, we may consider different objectives, i.e., vessel tardiness and weighted vessel tardiness when vessel due dates are given. In this research, we make the assumption that vessel processing times are constant in the models. In practice, vessel processing time can be adjusted by changing the number of cranes that are operating on the vessel. In this situation, the number of cranes on berth assigned for each vessel can be considered as a decision variable.

References

1. Chen C-Y, Hsieh T-W (1998) A time-space network model for the berth allocation problem. First International Conference of Maritime Engineering and Ports, Genoa
2. Brown G, Cormican K, Lawphogpanich S, Widdis D (1997) Optimizing submarine berthing with a persistence incentive. Naval Research Logistics **44**, 301–318
3. Daganzo C (1989) The crane scheduling problem. Transportation Research B **23B**, 159–175
4. Guan Y, Xiao W-Q, Cheung RK, Li C-L (2002) A multiprocessor task scheduling model for berth allocation: heuristic and worst case analysis. Operations Research Letters **30**, 343–350
5. Held M, Karp RM (1971) The traveling salesman problem and minimum spanning trees: Part II. Mathematical Programming **1**, 6–25
6. Hadjiconstantinous E, Christofides N (1995) An exact algorithm for general, orthogonal, two-dimensional knapsack problems. European Journal of Operational Research **83**, 39–56
7. Imai A, Nagaiwa K, Chan WT (1997) Efficient planning of berth allocation for container terminals in Asia. Journal of Advanced Transportation **31**, 75–94
8. Imai A, Nishimura E, Papadimitriou S (2001) The dynamic berth allocation problem for a container port. Transportation Research B **35**, 401–417

9. Kim KH, Moon KC (2003) Berth scheduling by simulated annealing. Transportation Research B **37(6)**, 541–560
10. Lai KK, K Shih (1992) A study of container berth allocation. Journal of Advanced Transportation **26**, 45–60
11. Li C-L, Cai X, Lee C-Y (1998) Scheduling with multiple-job-on-one-processor pattern. IIE Transactions **30,** 433–445
12. Lim A (1998) The berth planning problem. Operations Research Letters **22**, 105–110
13. Peterkofsky RI, Daganzo CF (1990) A branch and bound solution method for the crane scheduling problem. Transportation Research B **24B**(3), 159–172
14. Sim CW, Lin LS, Soon NH (1992) Berth allocation in the port of Singapore Authority: a total approach. Proceedings of the First Singapore International Conference on Intelligent Systems, 237–241

A scheduling method for Berth and Quay cranes*

Young-Man Park and Kap Hwan Kim

Department of Industrial Engineering, Pusan National University, Changjeon-dong, Kumjeong-ku, Busan 609-735, Korea (e-mail: {ymanpark,kapkim}@pusan.ac.kr)

Abstract. This paper discusses a method for scheduling Berth and Quay cranes, which are critical resources in port container terminals. An integer programming model is formulated by considering various practical constraints. A two-phase solution procedure is suggested for solving the mathematical model. The first phase determines the Berthing position and time of each vessel as well as the number of cranes assigned to each vessel at each time segment. The subgradient optimization technique is applied to obtain a near-optimal solution of the first phase. In the second phase, a detailed schedule for each Quay crane is constructed based on the solution found from the first phase. The dynamic programming technique is applied to solve the problem of the second phase. A numerical experiment was conducted to test the performance of the suggested algorithms.

Key words: Scheduling – Transportation – Optimization – Container Terminal – Berth planning

1 Introduction

The Berth is the most critical resource for determining the capacity of container terminals because the cost of constructing a Berth is very high compared to the investment costs for the other facilities in the terminal. An alternative way of increasing the capacity of the Berth is to improve the productivity of the Berth by utilizing it efficiently. Planners in container terminals usually construct a Berth schedule, which shows the Berthing position and the arrival time of each vessel. To construct a Berth schedule, the calling schedule of vessels, favorable Berthing

* This research has been supported in part by Brain Korea 21 Program (1999–2002).
Correspondence to: Y.-M. Park

locations of vessels, and the number of available cranes must be considered simultaneously.

Lai and Shih [6] studied the problem of assigning one of the discrete segments of a Berth to vessels and suggested several simple rules for the assignment. By considering various practical constraints, Brown et al. [1] formulated an integer-programming model for assigning available sections of a Berth to vessels. They also assumed a Berth to be a collection of discrete Berthing sections. Lim [8] considered a Berth to be a continuous line rather than a collection of discrete segments and discussed how to minimize the sum of the lengths of vessels that are supposed to Berth at the same time by optimally locating the Berthing positions. Li et al. [7] considered the Berth-scheduling problem to be a scheduling problem for a single processor (Berth) that can simultaneously perform multiple jobs (vessels). They suggested various algorithms based on First-Fit-Decreasing (FFD) heuristics and tested the algorithms by a simulation study. Imai et al. [5] also assumed a Berth to be a collection of discrete Berthing sections. They attempted to minimize the waiting time of vessels and provided a mixed-integer programming model for allocating Berthing sections to vessels. They also provided a heuristic procedure based on the Lagrangean relaxation of the original problem. Also, Nishimura et al. [10] suggested genetic algorithms to solve the problem suggested by Imai et al. [5] with a small computational effort.

Daganzo [2] was the first who discussed that the limitation in the length of a Berth must be considered simultaneously during the crane scheduling. However, more emphasis was placed on schedules of Quay cranes than that of the Berth which is the main issue of this study. Regarding the crane-scheduling problem, Daganzo [2] suggested an algorithm for determining the number of cranes to assign to ship-bays of multiple vessels. Peterkofsky and Daganzo [11] also provided an algorithm to determine the departure times of multiple vessels and the number of cranes to assign to each hold of vessels at a specific time segment. They also attempted to minimize the delay costs. The crane-schedules by Daganzo [2] and Peterkofsky and Daganzo [11] – in which Quay cranes are assigned not only to vessels but also to ship-bays of vessels – were more detailed than those in this paper. However, this study assumes that the detailed crane-schedule for each vessel should be constructed based on the rough crane-schedule in this paper – which specifies only the starting and the ending times of the assignment of each Quay crane to a specific vessel – and the amount of works at each ship-bays. Thus, the detailed crane-scheduling problem is another promising issue for further studies.

In the previous studies, the Berth-scheduling and the crane-scheduling problems have been considered to be independent of each other. However, the duration of Berthing of a vessel depends on the number of cranes allocated to the vessel. When the number of cranes allocated to a vessel increases, the duration of Berthing of the vessel is reduced. This is why the Berth-scheduling and the crane-scheduling problems were considered simultaneously in this study.

When the number of cranes assigned to each vessel is fixed, the Berth-scheduling problem can be considered as a special case of the two-dimensional stock-cutting problem. It is known that the two-dimensional stock-cutting problem is an NP-hard problem. Because the problem in this paper considers the crane-scheduling prob-

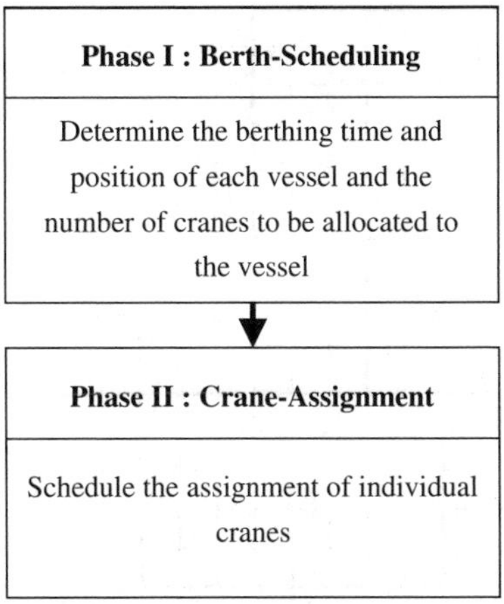

Phase I : Berth-Scheduling

Determine the berthing time and position of each vessel and the number of cranes to be allocated to the vessel

↓

Phase II : Crane-Assignment

Schedule the assignment of individual cranes

Fig. 1. The Berth and crane-scheduling procedure

lem in addition to the Berth-scheduling problem and the Berth-scheduling problem has the same complexity as the two-dimensional stock-cutting problem (Lim [8]), the problem in this paper is also NP-hard.

The following is assumed for the formulation of the problem:

1. Each vessel has a maximum and a minimum number of cranes to be assigned. Sometimes, contract terms between terminal operating companies and shipping companies specify the minimum number of cranes to be assigned to a vessel. The maximum number of cranes that can be simultaneously assigned to a vessel is limited by the length of the vessel.
2. The duration of Berthing of a vessel is inversely proportional to the number of cranes assigned to the vessel. The linearity of Berthing time with the number of cranes may not be true. However, the linearity was assumed for the simplicity of the analysis.
3. For each vessel, a penalty cost is incurred by Berthing earlier or later than the previously committed time. A departure of a vessel later than the previously committed departure time also incurs a penalty cost. Vessels and the terminal operating company continuously communicate with each other for adjusting the Berthing schedule. In the process, a vessel may be requested to arrive at the terminal earlier than her previously committed time. It will require speeding up her voyage to catch up the schedule, which results in an extra fuel consumption. Also, delayed Berthing or departure of a vessel beyond the committed time may lead to a trouble in meeting the schedule at the next port.
4. Every vessel has a most favorable location of Berthing. The most favorable location is the location nearest to the marshaling yard where outbound containers for the corresponding vessel are stacked. The preference of a Berthing location over another may also be due to the depth of water or the strength and direction of waves.

The Berth and crane-scheduling procedure in this study can be summarized as shown in Figure 1.

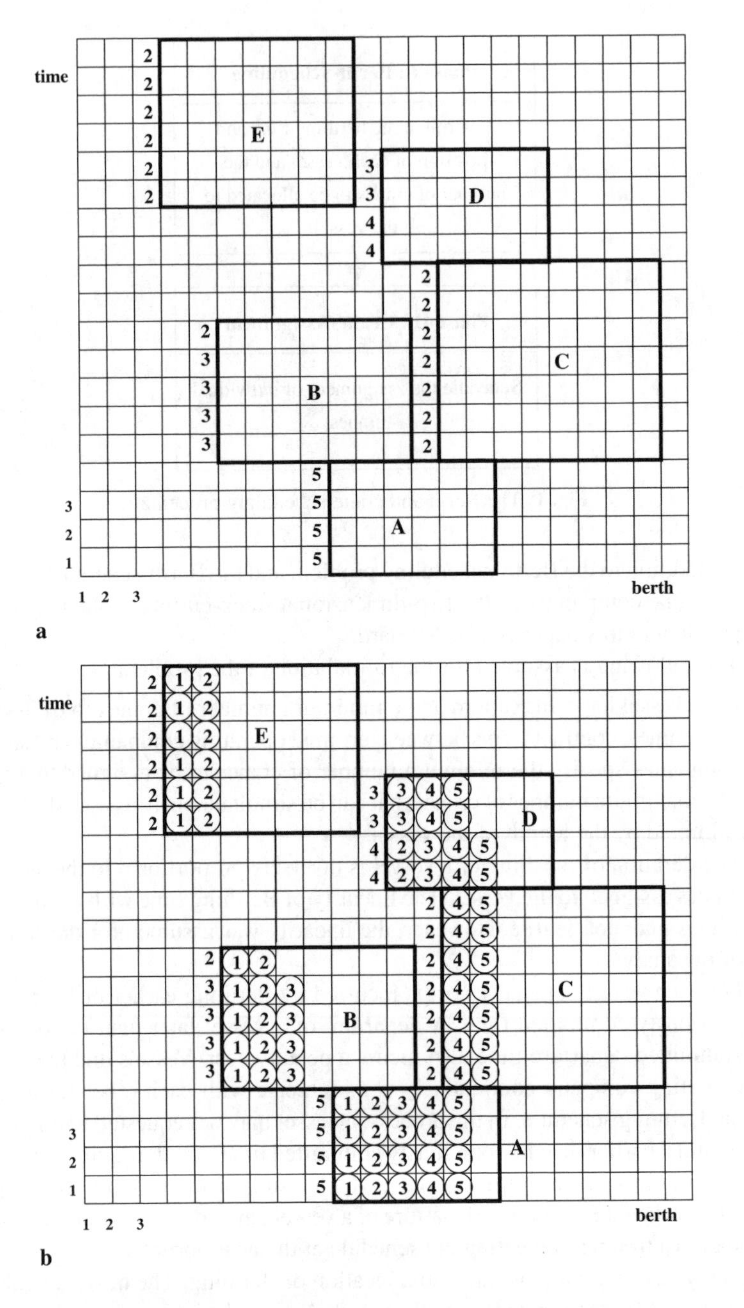

Fig. 2a,b. An illustration of outputs from the two-phase procedure. Sample output of the Berth-scheduling phase. Sample output of the crane-assignment phase

As shown in Figure 1, the Berth and crane-scheduling procedure can be decomposed into two phases. In the first phase, which we call the "Berth-scheduling phase", the Berthing position and the arrival time of vessels and the number of cranes allocated to each vessel are determined. In the second phase, which we call the "crane-assignment phase", which is based on the results of the first phase, the starting and the ending times of the operation by each crane for each vessel are scheduled. In the second phase, the objective is to minimize the number of setups – which include travels by cranes and assistants from one vessel to another, various delays for the preparation to start the operation in a different vessel. Illustrations of output of the Berth and crane-scheduling phase are given in Figure 2. Figure 2a illustrates the output of the first phase. Each rectangle represents the Berthing schedule of a vessel. The positions of the horizontal sides represent the Berthing locations, and the lengths of the horizontal sides correspond to the lengths of the corresponding vessels. Also, the position of a vertical side corresponds to the duration from the starting to the ending time of the operation of a vessel.

The numerical values on the left-hand side of each rectangle denote the number of cranes allocated to each vessel at each time segment. Figure 2b illustrates the output from the second phase based on the same example as the one in Figure 2a. In Figure 2b, specific cranes are assigned to each segment of the operation time for each vessel. In the example shown in Figure 2b, cranes $1\sim 5$ are assigned to vessel A for 4 time segments, cranes $1\sim 3$ are assigned to vessel B, and cranes $4\sim 5$ are assigned to vessel C. However, crane 3, which was assigned to vessel B, stops serving vessel B at the last time segment for vessel B. The remaining part of Figure 2b can be interpreted in the same way.

The following section describes the Berth-scheduling phase. Section 3 explains the crane-assignment phase in detail. Results of a numerical experiment are provided in Section 4. Finally, the concluding remarks are provided in Section 5.

2 The Berth-scheduling phase

This section suggests a mathematical formulation for the Berth-scheduling problem and provides a subgradient optimization method for solving the formulation.

2.1 Problem formulation

The following notations will be used for the formulation of the Berth-scheduling problem:

l : The total number of vessels
m : The total number of different Berthing positions
n : The total number of time segments

e_k : The expected time of arrival (*ETA*) of vessel k. *ETA* is a kind of agreement between carriers and the terminal operator regarding the arrival time of vessels. Thus, a Berthing earlier than the promised Berthing time causes the corresponding vessel to speed up, which in turn causes the extra consumption of fuel, and a Berthing later than the promised Berthing time may incur complaints from carriers

a_k : The total operation time of cranes, which is expressed as the number of crane-time segments, required to unload and load all the containers for vessel k

b_k : The length of vessel k

d_k : The due time for the departure of vessel k

s_k : The least-cost Berthing location of the reference point of vessel k

c_{1k} : The container handling cost per unit distance of vessel k between the Berth and the yard

c_{2k} : The penalty cost of vessel k per unit time of arrival before e_k

c_{3k} : The penalty cost of vessel k per unit time of arrival after e_k

c_{4k} : The penalty cost of vessel k per unit time of delay beyond the due time

l_k : The minimum number of cranes that can be assigned to vessel k

u_k : The maximum number of cranes that can be assigned to vessel k

c : The total number of available cranes ($c > u_k$)

B_k : The Berthing position of vessel k on the Berth axis (decision variable). B_B in Figure 2a is 6

T_k : The Berthing time of vessel k on the time axis (decision variable). T_B in Figure 2a is 5

Y_{jk} : The number of cranes allocated to vessel k at time segment j (decision variable). $l_k \le Y_{jk} \le u_k$, if $T_k \le j < C_k$; 0, otherwise. In Figure 2a, $Y_{5B} = Y_{6B} = Y_{7B} = Y_{8B} = 3$, and $Y_{9B} = 2$

C_k : The completion time of container handling for vessel k that is the maximum j such that $Y_{jk} > 0$ plus one (decision variable). In Figure 2a, $C_B = 10$

X_{ijk} : 1, if the grid square (i, j) is covered by the large rectangle for vessel k; 0, otherwise (decision variable). $X_{ijk} = 1$, if $B_k \le i < B_k + b_k$ and $T_k \le j < C_k$; 0, otherwise

To help readers understand the structure of the mathematical model – which is sufficient to understand the algorithm that follows – without going into too much detailed expressions, notations with more realistic meaning (B_k, T_k, and C_k) were used to express the objective function and the constraints that relate B_k, T_k, and C_k

with X_{ijk} and Y_{jk} are omitted from the model. However, a detailed mathematical model is also provided in Appendix for the completeness.

Note that, once Y_{jk} for $j = 1, \ldots, n$, are given, then C_k is fixed. Also, once B_k, T_k, and C_k are given, then X_{ijk} for all i and j are fixed. Thus, to determine the Berth schedule for vessel k, it is sufficient to determine (B_k, T_k) and Y_{jk} for all j.

Let $Y_k = (Y_{1k}, Y_{2k}, ..., Y_{nk})$, $B = ((B_1, T_1), (B_2, T_2), ..., (B_l, T_l))$, and $Y = (Y_1, Y_2, ..., Y_l)$. Then, the Berth-scheduling problem can be formulated as follows:

$$(P) \underset{B,Y}{Min} \sum_{k=1}^{l} \{c_{1k}|B_k - s_k| + c_{2k}(e_k - T_k)^+ \tag{1}$$

$$+c_{3k}(T_k - e_k)^+ + c_{4k}(C_k - d_k)^+\}$$

$$\text{subject to} \sum_{k=1}^{l} X_{ijk} \leq 1 \text{ for } i = 1, \ldots, m, j = 1, \ldots, n \tag{2}$$

$$\sum_{k=1}^{l} Y_{jk} \leq c \text{ for } j = 1, \ldots, n \tag{3}$$

$$\sum_{j=1}^{n} Y_{jk} \geq a_k \text{ for } k = 1, \ldots, l \tag{4}$$

$$X_{ijk} = 0 \text{ or } 1 \tag{5}$$

$$Y_{jk} = 0, l_k, \ldots, u_k \tag{6}$$

The objective function to minimize the weighted sum of the handling cost of containers, and the cost depends on the distance from the Berthing location of a vessel to the location in the marshaling yard where outbound containers for the corresponding vessel are stacked, the penalty cost incurred by Berthing earlier or later than the expected time of arrival (*ETA*), and the penalty cost incurred by the delay of the departure beyond the promised due time. A grid square must be covered by only one vessel, according to constraint (2). Constraint (3) restricts the maximum number of cranes utilized at a time segment. Constraint (4) implies that the total amount of handling operation for a vessel must be satisfied. Constraints (2) and (3) couple variables of different vessels. By relaxing constraints (2) and (3), a relaxed version of the formulation, can be obtained as shown in the following subsection.

2.2 Lagrangean relaxation and the subgradient optimization procedure

When constraints (2) and (3) of problem (P) is relaxed, a relaxed version of problem (P) is obtained as follows:

$$(R)\ \underset{B,Y}{Min} \sum_{k=1}^{l} \{c_{1k}\,|B_k - s_k| + c_{2k}(e_k - T_k)^+$$
$$+c_{3k}(T_k - e_k)^+ + c_{4k}(C_k - d_k)^+\}$$
$$+\sum_{i=1}^{m}\sum_{j=1}^{n} \pi_{ij}\left(\sum_{k=1}^{l} X_{ijk} - 1\right) + \sum_{j=1}^{n} \pi_j \left(\sum_{k=1}^{l} Y_{jk} - c\right) \tag{7}$$

subject to constraints (4) through (6) and $\pi_{ij},\ \pi_j \geq 0$.

The Lagrangean multiplier π_{ij} and π_j are related to constraints (2) and (3), respectively. The subgradient optimization technique is used to solve the problem. The dual problem to problem (R) can be represented as follows:

$$(D)\ \underset{\pi_{ij},\ \pi_j}{Max}\ L, \tag{8}$$

$$\text{where } L \equiv \left[\begin{array}{l} \underset{B,Y}{Min} \sum_{k=1}^{l} \{c_{1k}\,|B_k - s_k| + c_{2k}(e_k - T_k)^+ \\ +c_{3k}(T_k - e_k)^+ + c_{4k}(C_k - d_k)^+\} \\ +\sum_{i=1}^{m}\sum_{j=1}^{n} \pi_{ij}(\sum_{k=1}^{l} X_{ijk} - 1) + \sum_{j=1}^{n} \pi_j(\sum_{k=1}^{l} Y_{jk} - c) \end{array}\right]$$

subject to constraints (4) through (6) and $\pi_{ij},\ \pi_j \geq 0$.

The subgradient optimization procedure attempts to find π_{ij}^* and π_j^* that maximizes L where L is the minimum relaxed objective function (7) of B and Y for given values of multipliers, (π_{ij})s and (π_j)s. π_{ij}^* and π_j^* will be searched for in a systematic way, while L will be evaluated repeatedly by solving the relaxed problem (R) whenever values of (π_{ij})s and (π_j)s are changed. Suppose that π_{in} for all i and j and π_j for all j are given. Then, by deleting the constant term from (7), the following expression is obtained:

$$\underset{B,Y}{Min} \sum_{k=1}^{l} \{c_{1k}\,|B_k - s_k| + c_{2k}(e_k - T_k)^+ + c_{3k}(T_k - e_k)^+ + c_{4k}(C_k - d_k)^+\}$$
$$+\sum_{i=1}^{m}\sum_{j=1}^{n}\sum_{k=1}^{l} \pi_{ij} X_{ijk} + \sum_{j=1}^{n}\sum_{k=1}^{l} \pi_j Y_{jk} \tag{9}$$

The above expression can be decomposed into sub-problems, or (Q_k)'s, each of which has variables related to only one vessel, as follows:

$$(Q_k)\quad \underset{(B_k, T_k), Y_k}{Min}\quad L_k, \tag{10}$$

where

$$L_k = \left\{c_{1k}\,|B_k - s_k| + c_{2k}(e_k - T_k)^+ + c_{3k}(T_k - e_k)^+ + c_{4k}(C_k - d_k)^+\right\}$$
$$+\sum_{i=1}^{m}\sum_{j=1}^{n} \pi_{ij} X_{ijk} + \sum_{j=1}^{n} \pi_j Y_{jk}$$

subject to constraints (4) through (6).

The following provides the overall procedure for the subgradient optimization procedure (Held [4]; Geoffrion [3]):

Step 1: Let $\pi_j = 0$, and $\pi_{ij} = 0$. Obtain the initial upper bound (Z_{UB}) by a simple procedure as follows.

Step 1.1: Sort vessels in the increasing order of e_k. $k = 0$.

Step 1.2 : $k = k + 1$. If $k > l$, go to step 1.4. Otherwise, set Berthing location and Berthing time of the kth vessel at s_k and e_k, respectively.

Step 1.3: Check if the solution for the kth vessel satisfies constraints (2) and (3) under the condition that the solutions for the preceding vessels are given. If the kth vessel does not violate any constraint, go to step 1.2. Otherwise, move the location of the kth vessel in the positive direction of the time axis until the current vessel reaches a feasible location. Go to step 1.2.

Step 1.4: To obtain the initial upper bound, calculate the penalty cost for every vessel and sum the costs for all the vessels.

Step 2: Using the Lagrangean multipliers (π_j for all j, π_{ij} for all i, and j), solve the decomposed sub-problem (Q_k). The value of equation (7) will become the lower bound of the optimal objective value (Z_{LB}). Calculate the value of G_j , G_{ij}, α_1, and α_2, all of which will be used to update π_j, and π_{ij}, as follows:

$$G_j = \sum_{k=1}^{l} Y_{jk} - c \text{ for all } j, \quad G_{ij} = \sum_{k=1}^{l} X_{ijk} - 1 \text{ for all } i \text{ and } j \tag{11}$$

$$\text{And } \alpha_1 = \lambda(Z_{\text{UB}} - Z_{\text{LB}}) / \sum_{j=1}^{n} G_j^2,$$

$$\alpha_2 = \lambda(Z_{\text{UB}} - Z_{\text{LB}}) / \sum_{i=l}^{m} \sum_{j=1}^{n} G_{ij}^2 \tag{12}$$

Update the maximum lower bound by

$$Z_{MAX} = \max\left(Z_{MAX}, Z_{\text{LB}}\right). \tag{13}$$

Note that λ is a parameter for adjusting the step size of the changing π_j and π_{ij}. The initial value of λ is set to be 2, and it is reduced to (1/2) λ if no improvement is found during N consecutive iterations in the experiment, which will be discussed in detail in the following section. When Z_{MAX} increases again, λ is reset to be 2. When λ becomes less than 0.005, the procedure is stopped.

Step 3: Update π_j, π_{ij} for all i, j, as follows:

$$\pi_j = \max(0, \pi_j + \alpha_1 G_j), \pi_{ij} = \max(0, \pi_{ij} + \alpha_2 G_{ij}) \tag{14}$$

Go to step 2.

Solving decomposed primal problems

The decomposed primal problem for vessel k can be solved optimally, as follows (*Solve-Primal-1*):

For all the combinations (i, j), $i = 1, \dots, m, j = 1, \dots, n$, perform the following procedure:

Step 0: Set the initial value of the minimum objective value to be a large value. Set $i = 0$ and $j = 0$.

Step 1: Move (i, j) to the next grid in the feasible region. If there remains no more (i, j) un explored, stop. Otherwise, locate the reference corner of the rectangle corresponding to vessel k at (i, j) on the Berth-time coordinate. If the sum of the first three terms of L_k in (10) greater than the minimum objective value, go to the beginning of step 1. Otherwise, go to step 2.

Step 2: Assign l_k – which is two in the example of Figure 3 – cranes to vessel k and calculate the required Berthing time for vessel k. Evaluate the objective value (L_k) of the decomposed primal problem. The current solution becomes a candidate solution.

Step 3: Based on the candidate solution obtained in step 2, two different sets of candidate solutions are additionally generated. The first set of candidate solutions is generated by removing the cranes assigned to the last time segment in the candidate solution of step 2, adding one crane to each of the earliest possible time segments with the least number of cranes assigned, and repeating the same procedure until no more cranes can be removed without violating constraints. The first set of candidate solutions are illustrated in Figure 3b1, c1, d1, and e1. The second set of candidate solutions is generated by adding removed cranes to the latest possible time segments with the least number of cranes assigned while the other procedures are the same as those for the first set of candidate solutions, as illustrated in Figure 3b2, c2, d2, and e2.

Step 4: Update the minimum objective value and the current best solution by using the least-cost solution among candidate solutions found in steps 2 and 3. Go to step 1.

Figure 3 illustrates steps 2 and 3 of Solve-Primal-1. Assume that seven time segments are required to complete all the ship operations of vessel k when two cranes are assigned. Further assume that the maximum number of cranes that can work together is 5. We considered 9 different cases as candidate solutions for the crane allocation, as shown in Figure 3. Numerical values in the bottom-right corners are the objective values. The solution shown in Figure 3c1, which has the minimum objective value, will be selected in step 4.

A simple heuristic method (*Solve-Primal-2*) is also suggested for solving the decomposed primal problem. Instead of enumerating all the combinations of (i, j) to find the minimum cost location of the reference point of vessel k every time multipliers are updated, the neighboring area of the previous location is searched until no movement is possible without increasing the objective value. The changes in the cost by movements of the rectangle in four different directions (up, down, left,

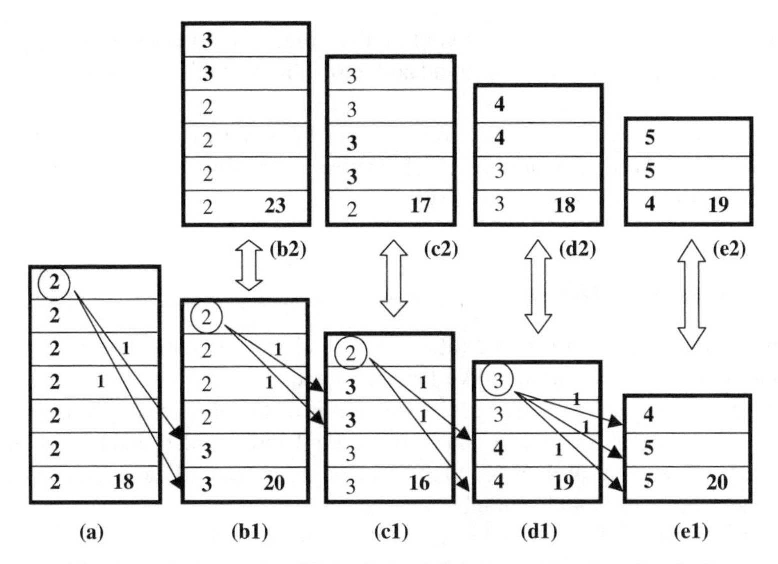

Fig. 3a–e. An example of Solve-Primal-1 for vessel k at location (i, j)

and right) from the location of the rectangle in the previous iteration are evaluated. The rectangle moves in the direction with the largest cost reduction. The process continues until no cost reduction is possible for the current vessel. At each location of the rectangle, the assignment of cranes with the minimum cost is searched, as shown in Figure 3.

A method for making a solution feasible

The iterative procedure for solving the decomposed sub-problems and updating the Lagrangean multipliers may fail to obtain a feasible solution, even when the duality gap – which is the difference between the upper bound and the lower bound of (1) – becomes very small. The infeasibility results from the violation of constraints (2) and (3). The infeasibility implies overlaps of rectangles and an allocation of cranes to time segments exceeding the maximum number of cranes available.

Infeasible solutions can be made feasible, as follows:

Step 1: $k = 0$. $F = I = \emptyset$. Sort vessels in the increasing order of e_k (*ETA*).

Step 2: $k = k+1$. If $k > l$, go to step 3. Otherwise, check if the current solution for the kth vessel satisfies constraints (2) and (3) under the condition that the solutions for the preceding vessels in F are given. If the current solution of the kth vessel violates neither constraints (2) nor (3), add the kth vessel to the set (F) of feasible vessels. Otherwise, add the kth vessel to the set (I) of infeasible vessels. Go to the beginning of this step.

Step 3: Sort vessels in I in the increasing order of e_k. Select the next vessel in I as the current vessel. Find the least-cost feasible solution for the current vessel by using Solve-Primal-1 under the condition that solutions of vessels in F are given. Remove the current vessel from I and add it to F. Repeat this step until I becomes empty.

Note that Solve-Primal-1 in step 3 should be modified so that the least-cost feasible solution for the current vessel satisfies constraints (2) and (3) under the condition that solutions in F are given.

2.3 A numerical example

Figure 4 illustrates the first two iterations of the subgradient optimization technique. Figure 4a shows the results of Solve-Primal-1 under the condition that all the values of the multipliers are zero. The initial solution is infeasible because overlaps exist between rectangles A and B, A and C, A and D, B and D, and C and D, and cranes on the time segment of 4, 5, 6, 7, 8, 9, and 10 are excessively assigned (Assume that the maximum available number of cranes is 5).

After updating the multipliers, their values can be easily obtained, as shown by the numerical values in Figure 4b. The numerical values inside the Berth-time space represent π_{ij}. The numerical values in the rightmost-side represent values of π_j's. By using the revised values of the multipliers, new solutions of the decomposed primal problems can be obtained, as shown in Figure 4b. Then, the values of the multipliers are revised again, as shown in Figure 4c.

The following illustrates the method of obtaining π_4 and $\pi_{10,8}$ in Figure 4c, based on the solutions of the decomposed problems and the values of the multipliers in Figure 4b. It is assumed that $Z_{\text{UB}} = 42$ and $Z_{\text{LB}} = 0$. It can be easily shown that $\sum_{i=1}^{m}\sum_{j=1}^{n} G_{ij}^2 = 1,154$ and $\sum_{j=1}^{n} G_j^2 = 1,238$.

From $\lambda = 2$, we can obtain $\alpha_1 = 0.068$ and $\alpha_2 = 0.073$.
Thus, $\pi_4 = \max(0,\ \pi_4 + \alpha_1 G_4) = \max(0,\ 0.42 + 0.068 \times 6) \cong 0.83$
and $\pi_{10,8} = \max(0,\ \pi_{10,8} + \alpha_2 G_{10,8}) = \max(0,\ 0.12 + 0.073 \times (-1)) \cong 0.05$.

The values of π_1, π_2, π_3, and π_4 are increased because the number of cranes assigned during time segments 1∼4 exceeded the limitation. Also, the π_{ij}'s of the grid squares on which the rectangles are overlapped are increased. Based on the new values of the multipliers, the decomposed primal problems were solved again to obtain the rectangles as shown in Figure 4c.

Figure 5 shows the final iteration of the numerical example. Figure 5a shows the final value of the multipliers and the resulting solutions of the decomposed primal problems. Note that the rectangles of the vessels do not overlap. However, there are excessive allocation of cranes at time segments 5, 6, 13, and 14. Now, the procedure for making the infeasible solution feasible is applied to obtain the solution as shown in Figure 5b.

3 The crane-assignment phase

Using the results of the Berth-scheduling of phase I, this phase constructs the schedule for an individual crane. The objective of the crane-assignment problem is to

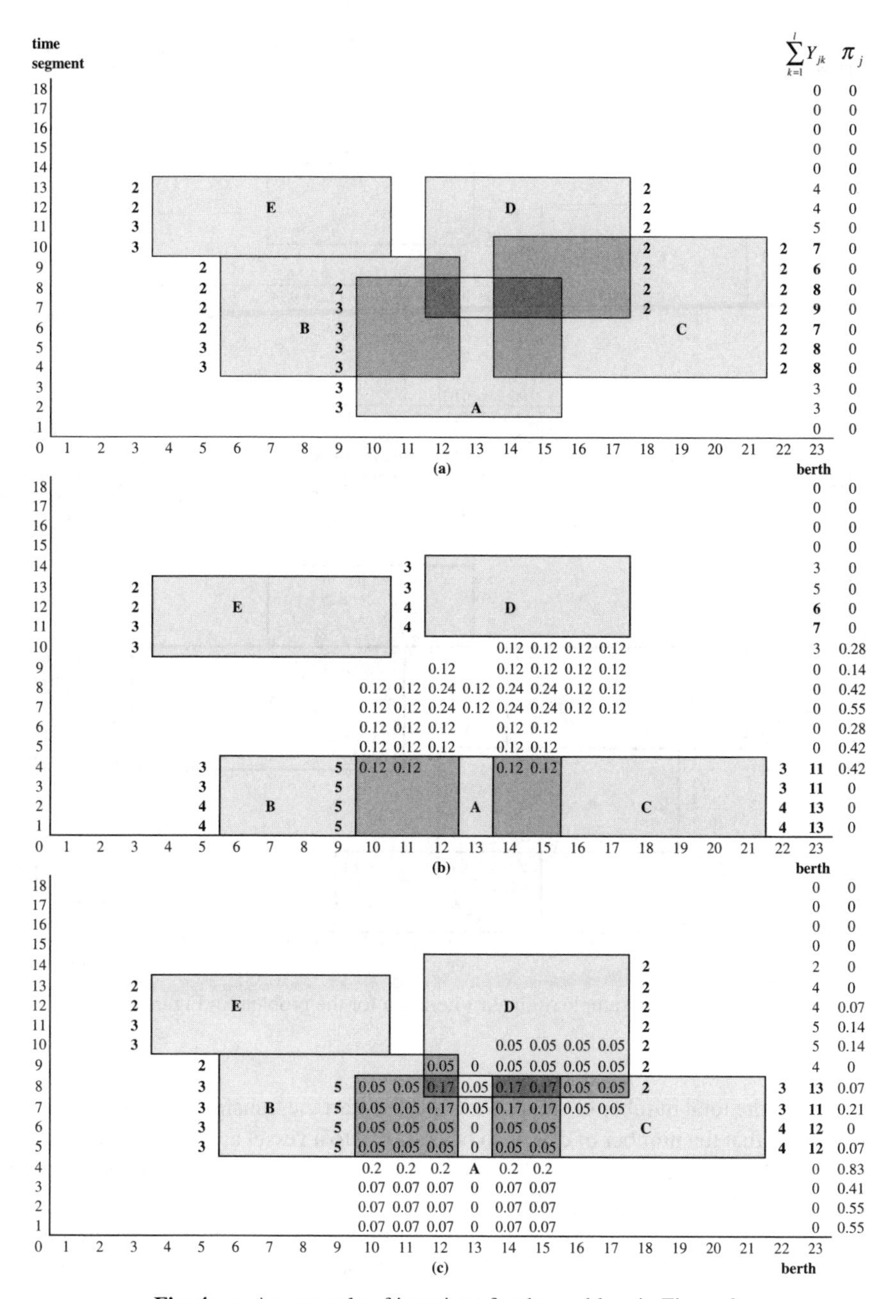

Fig. 4a–c. An example of iterations for the problem in Figure 2

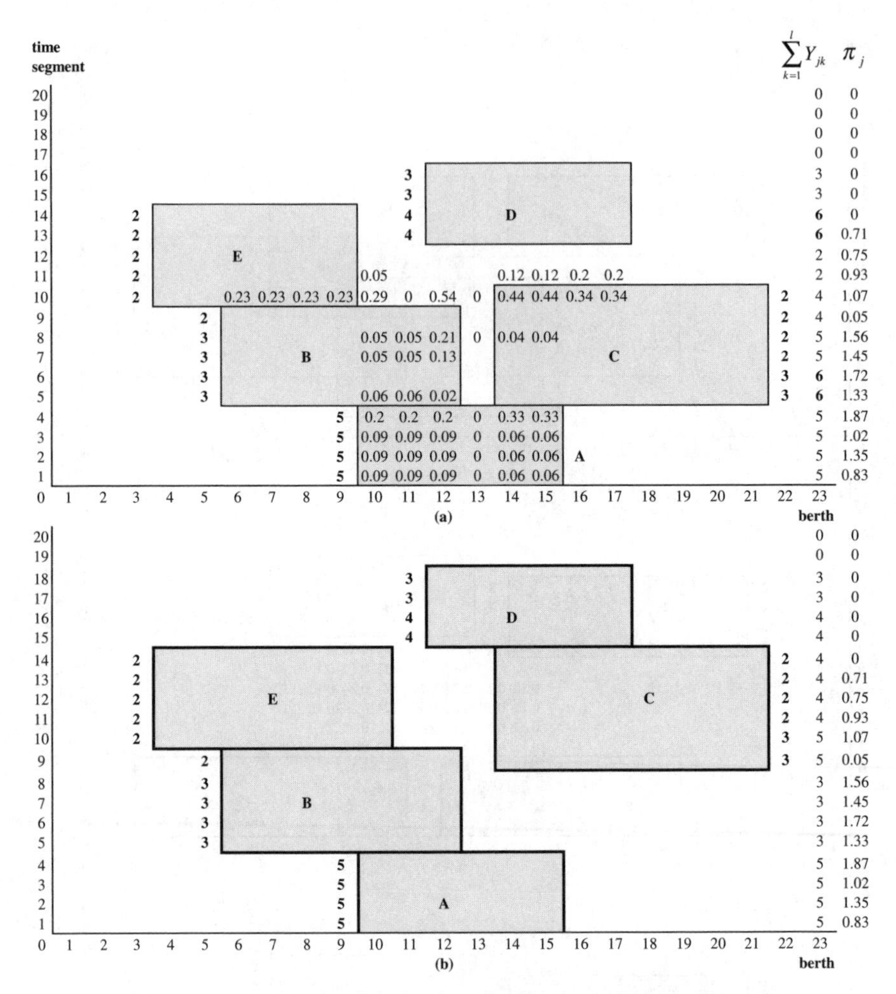

Fig. 5a,b. An example of the last iteration for the problem in Figure 2

minimize the total number of setups for cranes to start the transfer operation of vessels. Note that the number of cranes to be assigned to a vessel at each time segment has already been determined in the Berth-scheduling phase. Figure 6 illustrates a result of the Berth-scheduling phase.

The crane-assignment problem can be formulated by dynamic programming (DP). Two types of events are used to define the stages for DP. One is the arrival of a vessel and the other is the changing of the number of cranes assigned to a vessel. Then, the time when either event occurs is considered as a "stage" of DP. The state of DP can be expressed by the sequence of left-most cranes assigned to each vessel that is Berthing at the stage. For example, in Figure 6, there are 8 stages. In Figure 8, the states of the first, second, and third stages are (6), (1, 6), and (1, 3, 6), respectively. The crane-assignment problem in Figure 6 can be represented by the network shown in Figure 7. From the initial node (S) to the terminal node (T),

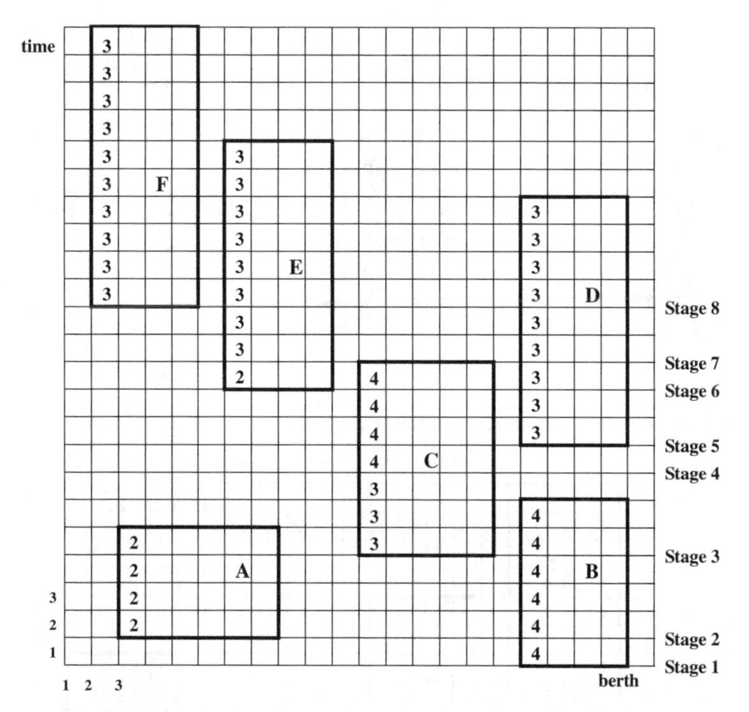

Fig. 6. An illustration of the results of the first phase (the number of vessels = 6, the number of cranes = 9)

there are eight stages and, at each stage, there are the same number of nodes as the number of states of the stage.

The recursive equation of DP for the crane-assignment problem can be written as follows:

$$S_{n,i} = \underset{j}{Min}[S_n(j,i) + S_{n-1,j}], \tag{15}$$

where $S_{n,i}$ is the minimum total number of setups from stage 1 through stage n under the condition that the state of stage n is i. And, $S_n(j,i)$ is the number of setups when the states at stage $(n-1)$ and n are j and i, respectively.

Figure 8 shows the final solution for the example problem shown in Figure 6 and Figure 7. The path, (S) $\rightarrow$ (6) $\rightarrow$ $(1,6)$ $\rightarrow$ $(1,3,6)$ $\rightarrow$ (3) $\rightarrow$ $(3,7)$ $\rightarrow$ $(1,3,7)$ $\rightarrow$ $(4,7)$ $\rightarrow$ $(1,4,7)$ $\rightarrow$ (T), is the optimal solution to the problem shown in Figure 7 and can be interpreted as the solution, as shown in Figure 8. The total number of setups becomes 21 (4+2+4+3+5+3).

4 A numerical experiment

Pusan Eastern Container Terminal (PECT) in Pusan (Korea) was used as the physical model for the numerical experiment. In PECT, there are eleven Quay cranes and a Berth of 1200 m. The discharging and loading operations of a vessel is performed

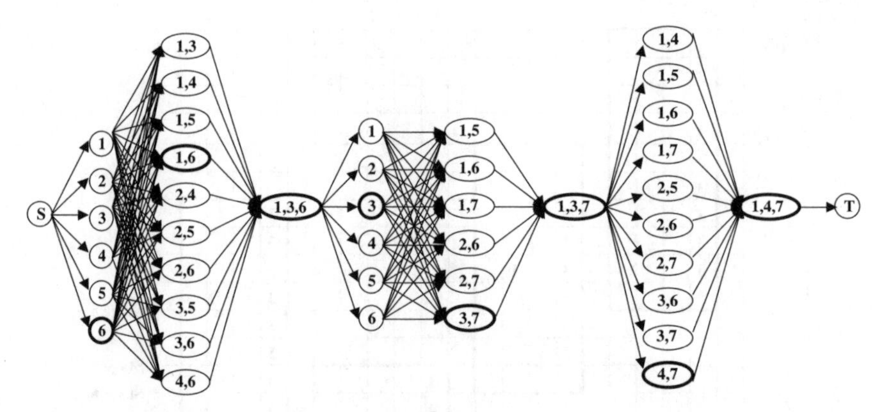

Fig. 7. A network representation of the crane-assignment problem shown in Figure 6

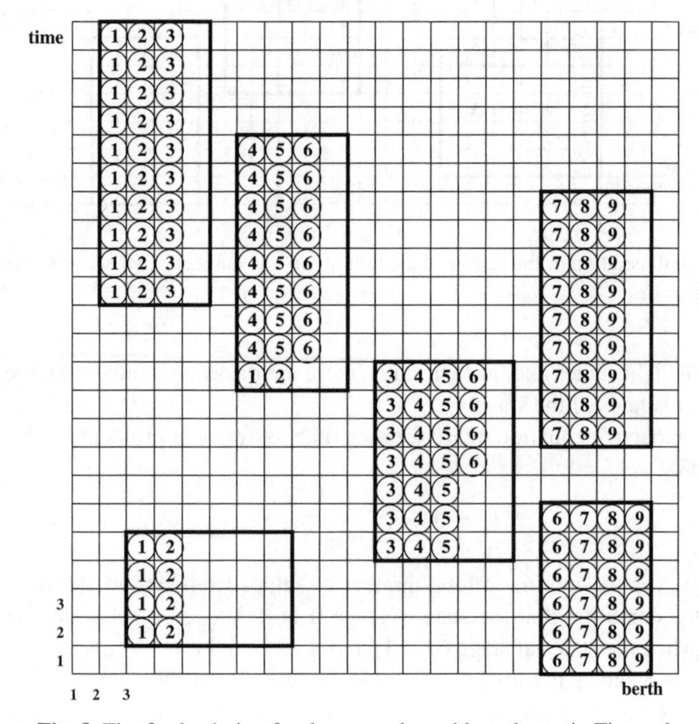

Fig. 8. The final solution for the example problem shown in Figure 6

by 2∼5 cranes. Human planners schedule the Berth and the cranes once per week. The Berth-time space was partitioned into 120×300 grid squares. The real size of a grid square corresponds to 10 m–1 hour in the Berth-time space.

The set of data consisted of 50 randomly generated sets of data for the experiment. In the data set, ten problems with 20 through 40 vessels were generated. In the random data, the arrival times of vessels, the ship-operation times of vessels, the lengths of vessels, and the preference positions for vessels were randomly generated from a uniform distribution of U(1, 170), U(10, 48), U(15, 35), and U(1, 120), respectively. For the numerical experiment, algorithms in this study were programmed by using C++ language and the programs were run on an IBM Pentium III (1 GHz).

4.1 A numerical experiment for the Berth-scheduling algorithm

For all k, the cost coefficient of c_{1k}, c_{2k}, c_{3k}, and c_{4k} was assumed to be \$1000, \$1000, \$1000, and \$2000, respectively. The value of λ was reduced to (1/2) λ if no improvement was found during N consecutive iterations. The value of N was set to be 30.

The performances of two different methods for solving the decomposed primal problems, Solve-Primal-1 and Solve-Primal-2, were compared.

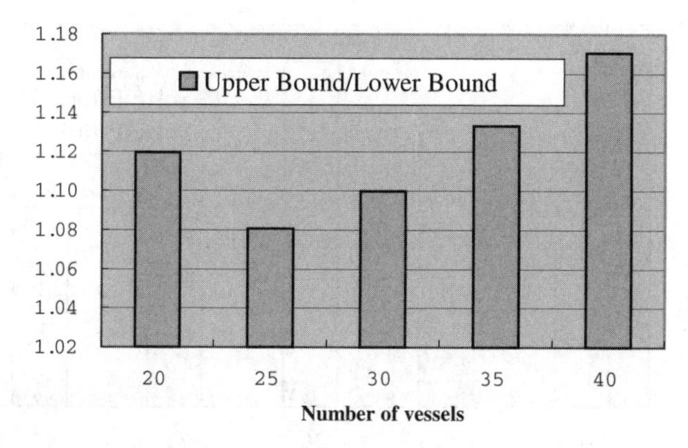

Fig. 9. The ratio of the upper bound to the lower bound for the generated problems (Solve-Primal-1)

Figure 9 shows the ratio of the final upper bound to the final lower bound of the objective value (1) when Solve-Primal-1 is used as the algorithm for the decomposed primal problem. Note that when the size of the problem is 25 vessels, the duality gap is the lowest. However, the gap becomes larger as the size of the problem increases or decreases. Figure 10 shows the CPU time becomes larger as the size of the problem increases.

Note that the final upper bounds are the final feasible objective values that can be obtained by the subgradient optimization procedure. Figure 11 compares the final objective values for when Solve-Primal-1 is used to solve the decomposed primal problems, versus those when Solve-Primal-2 was used. On the average, the final objective values by Solve-Primal-1 were lower than those by Solve-Primal-2. However, because both methods are basically heuristic algorithms, Solve-Primal-2 outperformed Solve-Primal-1 for some large-sized problems. Also, Table 1 shows that the method using Solve-Primal-2 consumed only one-half of the computational time that the method using Solve-Primal-1 did. Thus, the method using Solve-

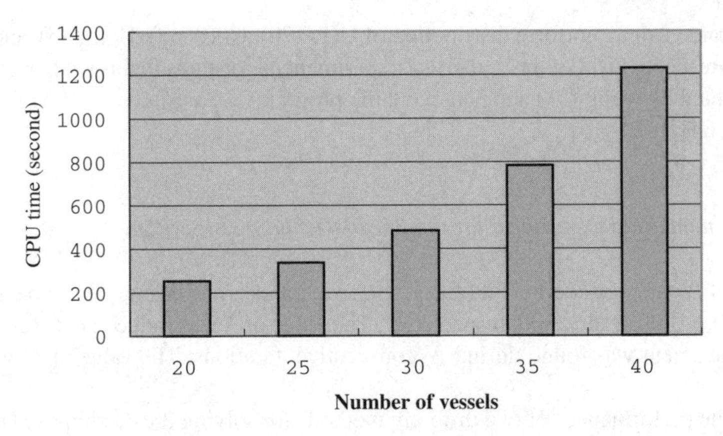

Fig. 10. The CPU time for the generated problems (Solve-Primal-1)

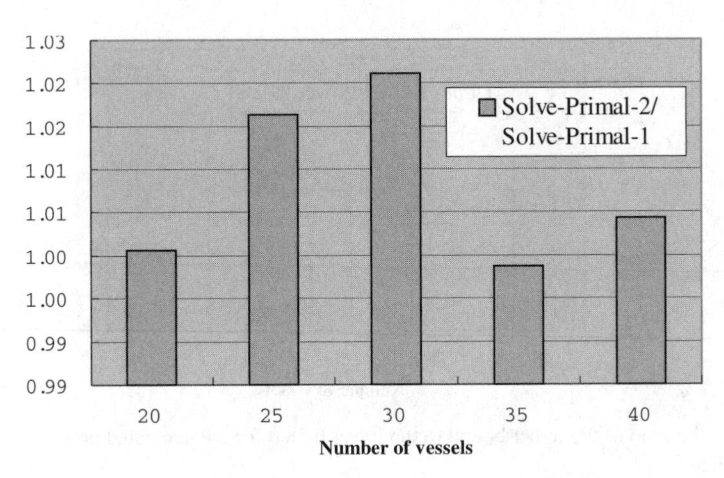

Fig. 11. The ratio of the final upper bound found by Solve-Primal-2 to that found by Solve-Primal-1 for the generated problems

Primal-2 can be recommended as the best solution method for the Berth-scheduling problem.

4.2 A numerical experiment for the crane-assignment algorithm

Fifty test problems, shown in Table 1 for crane-assignments, were solved. The average computation time for the fifty sets was about 5 seconds. It could be concluded that the dynamic programming model can be used in practice.

Table 1. The experiment results of the generated problems

Problem # (No. of vessels)	Method 1 (Solve-Primal-1)			Method 2 (Solve-Primal-2)			Method 2 / Method 1	
	UB*	LB*	CPU time (sec.)	UB	LB	CPU time	UB	CPU time
1 (20)	53	44	285	54	46	121	1.019	0.425
2 (20)	93	81	252	93	81	107	1.000	0.425
3 (20)	161	154	238	161	156	119	1.000	0.500
4 (20)	91	80	341	91	80	191	1.000	0.560
5 (20)	79	75	271	78	75	153	0.987	0.565
6 (20)	31	28	270	31	28	144	1.000	0.533
7 (20)	93	80	216	93	80	120	1.000	0.556
8 (20)	47	41	287	47	40	107	1.000	0.373
9 (20)	65	57	179	65	57	95	1.000	0.531
10 (20)	156	149	190	156	149	93	1.000	0.489
11 (25)	85	80	196	85	80	116	1.000	0.592
12 (25)	127	122	357	126	124	180	0.992	0.504
13 (25)	146	136	377	145	137	213	0.993	0.565
14 (25)	64	60	345	64	60	252	1.000	0.730
15 (25)	86	73	351	86	74	191	1.000	0.544
16 (25)	163	148	337	165	153	153	1.012	0.454
17 (25)	127	118	338	127	119	255	1.000	0.754
18 (25)	142	134	436	147	135	276	1.035	0.633
19 (25)	69	63	268	78	63	154	1.130	0.575
20 (25)	213	204	372	213	204	163	1.000	0.438
21 (30)	109	104	430	109	105	257	1.000	0.598
22 (30)	221	204	426	221	204	220	1.000	0.516
23 (30)	190	180	624	203	188	253	1.068	0.405
24 (30)	77	72	308	77	72	189	1.000	0.614
25 (30)	174	171	446	177	175	209	1.017	0.469
26 (30)	130	119	424	130	119	276	1.000	0.651
27 (30)	103	98	454	110	99	292	1.068	0.643
28 (30)	171	147	445	179	149	226	1.047	0.508
29 (30)	230	195	405	230	198	234	1.000	0.578
30 (30)	94	77	865	95	79	253	1.011	0.292
31 (35)	158	134	753	165	138	290	1.044	0.385
32 (35)	138	133	631	147	132	265	1.065	0.420
33 (35)	146	121	818	136	122	319	0.932	0.390
34 (35)	219	189	949	208	190	387	0.950	0.408
35 (35)	245	221	753	251	222	283	1.024	0.376
36 (35)	169	152	748	170	156	306	1.006	0.409
37 (35)	187	172	1211	203	178	506	1.086	0.418
38 (35)	219	174	703	196	181	358	0.895	0.509
39 (35)	172	163	656	172	163	310	1.000	0.473
40 (35)	200	178	605	197	181	242	0.985	0.400

Table 1 (continued)

Problem # (No. of vessels)	Method 1 (Solve-Primal-1)			Method 2 (Solve-Primal-2)			Method 2 / Method 1	
	UB*	LB*	CPU time (sec.)	UB	LB	CPU time	UB	CPU time
41 (40)	183	176	520	181	176	294	0.989	0.565
42 (40)	220	212	1186	219	216	463	0.995	0.390
43 (40)	313	242	1642	315	250	676	1.006	0.412
44 (40)	234	230	560	234	230	291	1.000	0.520
45 (40)	333	261	1514	339	265	473	1.018	0.312
46 (40)	269	240	939	278	250	546	1.033	0.581
47 (40)	271	239	1654	273	245	611	1.007	0.369
48 (40)	233	181	1510	215	195	464	0.923	0.307
49 (40)	250	228	705	255	236	307	1.020	0.435
50 (40)	359	255	2064	377	264	787	1.050	0.381
Average	**162**	**144**	**617**	**163**	**146**	**276**	**1.008**	**0.490**

*: UB, LB = the upper bound (the minimum feasible objective value) and the lower bound of (1), expressed in units of $1,000

5 Summary and conclusions

This study presents a method for scheduling the Berth and Quay cranes in port container terminals. A mathematical formulation is provided. A two-phase solution method is suggested for the formulation. In the first phase, the Berthing times and the positions of vessels are determined. In the second phase, the detailed operating schedule for individual cranes is constructed.

For the first phase, the subgradient optimization technique is applied. Two versions – a complete enumeration and a heuristic procedure – of the algorithms used for solving the decomposed subproblems were compared with each other by using randomly generated sets of data. The purpose of the first version was to optimize the decomposed primal problem by enumerating the solution space for the Berthing locations and times of each vessel (Solve-Primal-1). The purpose of the second version was to apply a heuristic to solve the decomposed primal problem (Solve-Primal-2). It was found that the final costs calculated by the method using Solve-Primal-2 were similar to those using Solve-Primal-1. However, the method using Solve-Primal-2 requires only one-half of the computational time required by the method using Solve-Primal-1 does. Thus, the method with Solve-Primal-2 can be recommended as the best solution method for the Berth-scheduling problem.

The dynamic programming technique is used for the crane-assignment problem. On average, the computation time of the dynamic programming algorithm for the second phase was 5 seconds, which is within the range for practical use. In the future, algorithms need to be developed to further reduce the computational time for the Berth-scheduling problem.

Appendix

An integer programming model for the Berth-scheduling problem

The following additional notations will be used for the detailed formulation of the Berth-scheduling problem:

Z_{ijk} : 1, if the reference point of vessel k is located at (i, j); 0, otherwise. The reference point of a vessel is the lower-left corner point of the rectangle corresponding to the vessel

v_{jk} : 1, if Y_{jk} is positive; 0, otherwise

M : A large positive real number

Based on the above notations, the Berth-scheduling problem can be formulated as follows:

$$(P)\ \ Min \sum_{k=1}^{l}\sum_{i=1}^{m}\sum_{j=1}^{n} Z_{ijk}\left\{c_{1k}\left|i-s_k\right| + c_{2k}(e_k - j)^+ \right. \tag{A.1}$$
$$\left. +c_{3k}(j-e_k)^+ + c_{4k}(C_k - d_k)^+\right\}$$

$$\text{subject to} \quad \sum_{k=1}^{l} X_{ijk} \leq 1 \text{ for } i = 1,\dots,m, j = 1,\dots,n \tag{A.2}$$

$$\sum_{k=1}^{l} Y_{jk} \leq c \text{ for } j = 1,\dots,n \tag{A.3}$$

$$\sum_{j=1}^{n} Y_{jk} \geq a_k \text{ for } k = 1,\dots,l \tag{A.4}$$

$$Y_{jk} \leq M\ v_{jk} \text{ for } j = 1,\dots,n, k = 1,\dots,l \tag{A.5}$$

$$v_{jk} \leq Y_{jk} \text{ for } j = 1,\dots,n, k = 1,\dots,l \tag{A.6}$$

$$(j+1)\cdot v_{jk} \leq C_k \text{ for } j = 1,\dots,n, k = 1,\dots,l \tag{A.7}$$

$$(j'' - j' + 1) \leq \sum_{j=j'}^{j''} v_{jk} + M(2 - v_{j'k} - v_{j''k}) \tag{A.8}$$
$$\text{for } j' = 1,\dots,n-1, j'' = 2,\dots,n, j' < j'', k = 1,\dots,l$$

$$v_{jk} \leq \sum_{i=1}^{m} \sum_{j'=1}^{j} Z_{ij'k} \text{ for } j = 1, \ldots, n, k = 1, \ldots, l \tag{A.9}$$

$$\sum_{i=1}^{m} \sum_{j=1}^{n} Z_{ijk} = 1 \text{ for } k = 1, \ldots, l \tag{A.10}$$

$$\sum_{i=1}^{i'-1} \sum_{j=1}^{n} X_{ijk} + \sum_{i=i'+b_k}^{m} \sum_{j=1}^{n} X_{ijk} \leq M \left(1 - \sum_{j=1}^{n} Z_{i'jk}\right)$$
$$\text{for } i' = 2, \ldots, m - b_{k,}, k = 1, \ldots, l \tag{A.11}$$

$$\sum_{i=1+b_k}^{m} \sum_{j=1}^{n} X_{ijk} \leq M \left(1 - \sum_{j=1}^{n} Z_{1jk}\right) \text{ for } k = 1, \ldots, l \tag{A.12}$$

$$\sum_{i=1}^{m-b_k} \sum_{j=1}^{n} X_{ijk} \leq M \left(1 - \sum_{j=1}^{n} Z_{(m-b_k+1)jk}\right) \text{ for } k = 1, \ldots, l \tag{A.13}$$

$$b_k - \sum_{i=1}^{m} X_{ijk} \leq M\left(1 - v_{jk}\right) \text{ for } j = 1, \ldots, n, k = 1, \ldots, l \tag{A.14}$$

$$X_{ijk}, v_{jk}, Z_{ijk} = 0 \text{ or } 1 \tag{A.15}$$

$$Y_{jk} = 0,\ l_k,\ \ldots, u_k \tag{A.16}$$

Constraints (A.5) and (A.6) define variable v_{jk} by relating it with Y_{jk}. Constraint (A.7) relates the departure time C_k to v_{jk}. Because v_{jk} is 1 if at least a Quay crane serves vessel k during time segment j that is between time j and $j+1$, the departure time of vessel k must be greater than equal to $j+1$. Constraints (A.8) and (A.9) eliminate invalid values of v_{jk}'s. Constraint (A.10) implies that only one grid can be chosen as the reference Berthing location and time for a vessel. Constraints (A.11) through (A.14) imply that only X_{ijk} within a rectangle can have the value of 1.

References

1. Brown GG, Lawphongpanich S, Thurman KP (1995) Optimizing ship Berthing. Naval Research Logistics 41: 1–15
2. Daganzo CF (1989) The crane scheduling problem. Transportation Research 23B (3): 159–175
3. Geoffrion M (1974) Lagrangean relaxation for integer programming. Mathematical Programming Study 2: 82–114
4. Held M, Wolfe P, Crowder HP (1974) Validation of sub-gradient optimization. Mathematical Programming 6: 62–88
5. Imai A, Nishimura E, Papadimitriou S (2001) The dynamic Berth allocation problem for a container port. Transportation Research (Part B) 35: 401–417
6. Lai KK, Shih K (1992) A study of container Berth allocation. Journal of Advanced Transportation 26 (1): 45–60
7. Li C-L, Cai X, Lee C-Y (1998) Scheduling with multiple-job-on-one-processor pattern. IIE Transactions 30: 433–445
8. Lim (1998) The Berth planning problem. Operation Research Letters 22: 105–110
9. Murty KG (1992) Network programming. Prentice Hall, Englewood Cliffs, NJ
10. Nishimura E, Imai A, Papadimitriou S (2001) Berth allocation planning in the public Berth system by genetic algorithms. European Journal of Operational Research 131: 282–292
11. Peterkofsky RI, Daganzo CF (1990) A branch and bound solution method for the crane scheduling problem. Transportation Research 24B (3): 159–172

A beam search algorithm for the load sequencing of outbound containers in port container terminals*

Kap Hwan Kim[1]**, Jin Soo Kang**[1]**, and Kwang Ryel Ryu**[2]

[1] Department of Industrial Engineering , Pusan National University, Changjeon-dong, Kumjeong-ku, Pusan 609-735, South Korea (e-mail: kapkim@pusan.ac.kr)
[2] Department of Computer Engineering, Pusan National University, Pusan, South Korea (e-mail: krryu@pusan.ac.kr)

Abstract. A beam search algorithm was applied to solve the load-sequencing problem in port container terminals. The algorithm was used to maximize the operational efficiency of transfer cranes and quay cranes (QCs) while satisfying various constraints on stacking containers onto vessels. The load-sequencing problem consisted of two decision-making subproblems. In the first subproblem, a pickup schedule was constructed in which the travel route of a transfer crane (TC) as well as the number of containers it must pick up at each yard-bay are determined. In the second subproblem, the load sequence for individual containers was determined. This study suggested a search scheme in which an algorithm to solve the second subproblem is imbedded into the algorithm for the first subproblem. Numerical experiments using practical data were performed to test the performance of the developed algorithm.

Keywords: Load sequencing – Container terminal – Beam search

1 Introduction

The container handling system considered in this study consists of QCs to load (unload) onto (from) containerships, TCs for transferring containers within a marshaling yard, and yard trucks (YTs) for delivering containers between the marshaling yard and QCs. A container terminal yard is divided into multiple blocks (see Fig. 1). A block consists of 20 to 30 yard-bays, each of which usually has four tiers and six stacks. To load a container in a yard onto a ship, a TC moves to a target

* This study was supported in part by the Korea Science and Engineering Foundation through the Center for Intelligent and Integrated Port Management Systems (CIIPMS) at Dong-A University. Authors also appreciate Director In-Tae Baek in Korea Logistics Network Co. for his help during the data collection.
Correspondence to: K. H. Kim

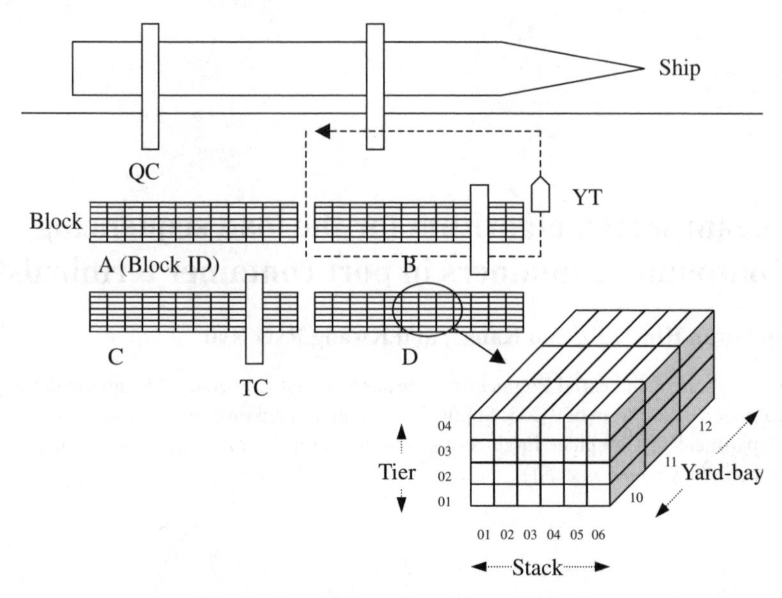

Fig. 1. An overview of container terminal

yard-bay, then its hoist picks up a selected container and takes it to one side of the block and loads it onto a waiting YT. A TC picks up containers in an order specified by a load planning process. Next, the YT transports the container to a QC. Finally, the QC picks up the container and loads it onto a ship. TCs usually move along the yard-blocks that are laid out parallel to the berth, while QCs can move on the rail that also runs parallel to the berth. A TC or a QC cannot pass another TC or QC.

Figure 2 shows a containership that has 28 ship-bays each of which consists of many stacks in hold and on deck. Hatch covers separates stacks on deck from those in hold. Thus, for each ship-bay, containers for slots in hold must be completely loaded before the loading operation into slots on deck can begin.

The ship operation consumes a large portion of the turnaround time of containerships in ports. The ship operation of a container ship consists of unloading inbound containers and loading outbound containers. Because inbound containers are usually unloaded onto a designated open space and there are fewer requirements to be satisfied in case of the unloading operation than in case of the loading operation, the sequencing of unloading operations is relatively easier than that of loading operations. In the loading operation, containers to be loaded into slots of a ship must satisfy various constraints on the slots pre-specified by a stowage planner. Also, locations of outbound containers may be scattered over a wide area in a marshaling yard. The time required for loading operations depends on the cycle time of QCs and TCs. Also, the cycle time of a QC depends on the loading sequence of slots, while the cycle time of a TC is affected by the loading sequence of containers in the yard. This study assumed that a transfer quay crane is exclusively assigned to each quay crane during the ship operation.

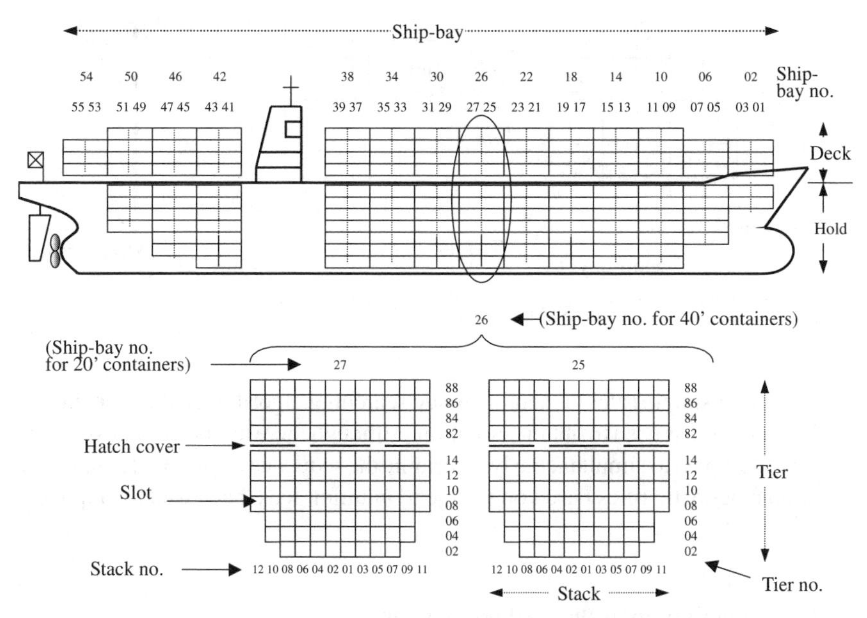

Fig. 2. Cross-sectional view of a containership

Research on load sequencing can be classified into three types according to its problem-solving approach: mathematical programming approaches (Cho [2], Kim [6]), heuristic algorithms (Beliech [1], Cojeen [3], Gifford [4]), and meta-heuristic approaches (Kim [5], Kozan [7], Ryu [10]). Research can also be classified by the scope of the problem. Some research has addressed the pickup scheduling problem in which the travel route of each yard crane and the number of containers to be picked up at each yard-bay on the route are determined during the loading process of a vessel (Kim [5], Kim [6], Narasimhan [7], Ryu [10]). Ryu et al. [10] suggested an algorithm based on "the ant system" for solving the pickup scheduling problem for TCs, which is a sub-problem of the load-sequencing problem. The performance of their algorithm will be compared with that of the algorithm in this study. Other research has attempted to determine the loading sequence of individual containers in the marshaling yard and slots in the vessel, a process that requires more detailed scheduling than does the pickup scheduling (Beliech [1], Cho [2], Cojeen [3], Gifford [4], Kozan [7]).

This study is different from previous studies in the following three aspects:

1. Many practical constraints and objective functions of the load-sequencing problem are considered in the algorithm. Examples are the travel distance of TCs, the handling convenience of TCs and QCs, the maximum height of a stack in hold, the maximum total weight of containers on a hatch cover, and the conformity of weights of loaded containers to the weight class specified in a stowage plan.
2. The loading sequence of slots in a vessel and containers in a marshaling yard are simultaneously determined.

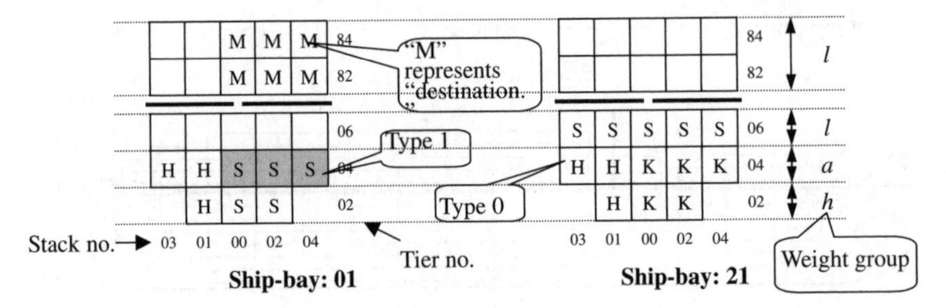

Fig. 3. An illustration of a stowage plan

3. Instead of simple rules, a meta-heuristic searching algorithm, called the filtered beam search, is used to obtain a solution. Although there has been some research that attempted to simultaneously determine the sequence of slots and containers, simple heuristic rules based on planners' intuition have been usually applied.

2 Problem definition of the load sequencing

By using stowage plans such as the one shown in Figure 3, shipping companies specify the port of destination (H, M, S, K), the size (20', 40', or 45'), the type (dry full container, refrigerated containers, empty containers, containers with dangerous cargo, etc.) and the weight group (light (l), medium (a), heavy (h)) of the container allowed to be loaded into each slot of a vessel. Containers of the same size and type and bound for the same destination is said to be in the same class. Because ship-bay numbers in Figure 3 are odd, all the outbound containers to be loaded into the ship-bays are 20' containers. Also, before the load sequencing for a vessel begins, planners usually construct a work schedule of QCs, which shows the sequence of ship-bays that each quay crane should perform discharging and loading operations, for the vessel, as shown in Figure 4. Then, load planners determine the loading sequence of slots in the vessel and containers in the yard. In the process, load planners use the yard map such as the one in Figure 5, which shows the distribution of containers in the yard. The yard map also shows the destination, the weight group, and the type of each container stacked in each position in the yard. An illustrative example of a load sequence list is provided in Table 1, which shows the load sequence of containers, the locations of the containers in the yard before loading, and their locations in the vessel after loading.

The following summarizes what load planners must consider during the load sequencing process. Some considerations are related to the operation of QCs, while others are related to the operation of TCs. Because many requirements must be satisfied, the load sequencing process is very time-consuming for planners and requires intensive computer support. In the load-sequencing algorithm in this study, some of the considerations are treated as constraints – which it was attempted to satisfy by imbedding them into the search procedure, while others were treated as factors in the objective function, as follows:

Work schedule for QC 1

Ship no: AA Voyage no.: 05 Date: June 25, 2002

Sequence	Ship-bay no.	Hold/deck	Unload/load	No. of containers		
				20ft	40ft	45ft
1	01	H	L	8		
2	01	D	L	6		
3	03	D	L	14		
4	05	H	L	27		
5	05	D	L	3		

Fig. 4. An illustration of a work schedule for QC 1

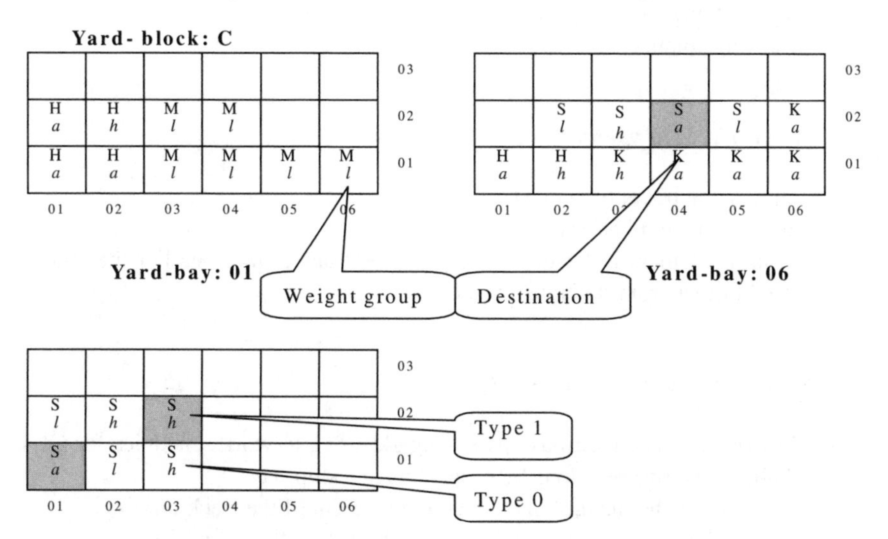

Fig. 5. An illustration of a yard map

Table 1. An example of a load sequence list

QC no.	Sequence	Container number	Location in yard	Location in vessel
1	1	MFU8408374	C-06-03-02*	01-00-02**
1	2	DMU2975379	C-06-02-02	01-02-02
1	3	DMU2979970	C-06-01-02	01-01-02
1	4	OLU0071308	C-01-02-02	01-01-04
1	5	MTU4015162	C-01-01-02	01-03-04
...	...	...	...	...

* Block no. – yard-bay no. – stack no. – tier no., ** Ship-bay no. – stack no. – tier no.

Objectives related to operation of QCs

1.1 First fill slots in the same stack in the hold. Filling slots in the same hold consecutively speeds to the loading operation of QC, because, in that case, automatic positioning function may used to move the spreader of QC.
1.2 First stack containers onto the same tier on deck. As objective 1.1, stacking containers onto the same tier on deck consecutively speeds up the lashing operation. Because preferences of QC operators are considered in objectives 1.1 and 1.2 and the preferences may be different from a terminal to another, the objectives may be modified if preferences of QC operators in a terminal are different from 1.1 and 1.2.
1.3 Stack containers of weights included in the same weight group as specified in the stowage plan.

Objectives related to operation of TCs

2.1 Minimize the travel time of TCs.
2.2 Minimize the number of rehandles.
2.3 Pick up containers in locations nearer to the transfer point earlier than those located farther from the transfer point.

Constraints related to operation of QCs

3.1 Follow precedence relationships among slots due to work schedules for QCs and due to relative positions between slots in a ship-bay.
3.2 Do not violate the maximum allowed total weight of the stack on deck.
3.3 Do not violate the maximum allowed height of the stack of a hold.
3.4 Load the same class, which is defined by the destination port, the size, and the type, of containers as specified in the stowage plan.

Constraints related to operation of TCs

4.1 Maintain the distance between adjacent TCs by at least 5 yard-bays.

For the problem formulation, the following notations are introduced:

Indices

i, j = Indices for containers in the yard.
s, t = Indices for slots in the vessel.
k = Index for load sequences for QCs.
c = Index for QCs.
u = Index for stacks in the vessel.

Problem data

m = The number of QCs.
n_c = The number of containers (slots) to be loaded (filled) by QC c.
n = The total number of containers to be loaded into the vessel. $n = \sum_{c=1}^{m} n_c$.
c_w = Penalty for the difference between the weight group of a container assigned to a slot and the weigh group planned for the slot in the stowage plan.
c_d = Penalty for the inconvenience of the loading operation for slots on deck by QCs. This is due to changing tiers on deck during the loading operation.
c_h = Penalty for the inconvenience of the loading operation for slots in hold by QCs. This due to changing stacks in hold during the loading operation.
a_t = Penalty for unit travel time by TCs.
a_r = Penalty of re-handling a container by TCs.
a_h = Penalty for the inconvenience of the transfer operation by TCs. This penalty is applied when a container is picked up before a container – which is located nearer to the transfer point than the container – is picked up.
g_s^o = Planned weight group – which is specified in the stowage plan – for slot s.
g_i = Weight group of container i.
w_i = Weight of container i.
h_i = Height of container i.
α_{st} = 1, if slot s and t are located in the same tier on deck; 0, otherwise. $a_{ss} = 0$.
β_{st} = 1, if slot s and t are located in the same stack in hold; 0, otherwise. $\beta_{ss} = 0$.
λ_{st} = 1, if slot s and t are located in the same stack on deck; 0, otherwise. $\lambda_{ss} = 0$.
t_{ij} = Travel time of TCs from the location of container i to the location of container j.
t_i = Transfer time of container i by a TC for picking up and putting down it on a YT.
γ_{ij} = 1, if container i and j are in the same stack of the yard and container i is located below container j; 0, otherwise.
δ_{ij} = 1, if container i is located farther from the transfer point than container j in a yard-bay; 0, otherwise.
θ_{is} = 1, if the class of container i is the same as the container class of slot s specified in the stowage plan; 0, otherwise.
w_u^m = The maximum allowed total weight of stack u on deck.
h_u^m = The maximum allowed height of stack u in hold.
M = A very large positive number.

Sets of indices

P	=	The set of pairs of slots with a precedence relationship between slots due to relative positions in a ship-bay or due to the work schedule specified for each QC. If $(s,t) \in P$, then slot s must be filled before slot t is filled with a container.
W_c	=	The set of slots assigned to QC c in the work schedule.
W_c^D	=	The set of slots on deck among slots in W_c.
W_c^H	=	The set of slots in hold among slots in W_c.
V_u	=	The set of slots in stack u.
T^D	=	The set of stacks on deck.
T^H	=	The set of stacks in hold.
U	=	The set of pairs of containers that cannot be transferred by TCs at the same time because of interferences between TCs.

Decision variables

X_{isk}^c	=	1, if container i is picked up in the kth order and stacked into slot s in the vessel by QC c; 0, otherwise.
S_i	=	The transfer starting time for container i by a TC.
T_i	=	The transfer completion time for container i by a TC.
Z_{ij}	=	1, if the transfer of container i by a TC is completed before starting the transfer of container j; 0, otherwise.

Then, the load-sequencing problem can be formulated as follows:

$$\begin{aligned} Min \quad & \Bigg(c_d \sum_{i=1}^{n}\sum_{j=1}^{n}\sum_{c=1}^{m}\sum_{s\in W_c^D}\sum_{t\in W_c^D}\sum_{k=1}^{n_c-1}(1-\alpha_{st})X_{isk}^c X_{jt(k+1)}^c \\ & + c_h \sum_{i=1}^{n}\sum_{j=1}^{n}\sum_{c=1}^{m}\sum_{s\in W_c^H}\sum_{t\in W_c^H}\sum_{k=1}^{n_c-1}(1-\beta_{st})X_{isk}^c X_{jt(k+1)}^c \\ & + c_w \sum_{i=1}^{n}\sum_{c=1}^{m}\sum_{s\in W_c}\sum_{k=1}^{n_c}|g_s^o - g_i| X_{isk}^c \\ & + a_t \sum_{i=1}^{n}\sum_{j=1}^{n}\sum_{c=1}^{m}\sum_{s\in W_c}\sum_{t\in W_c}\sum_{k=1}^{n_c-1} t_{ij} X_{isk}^c X_{jt(k+1)}^c \\ & + a_r \sum_{i=1}^{n}\sum_{j=1}^{n}\gamma_{ij}Z_{ij} + a_h \sum_{i=1}^{n}\sum_{j=1}^{n}\delta_{ij}Z_{ij} \Bigg) \end{aligned} \tag{1}$$

subject to

$$\sum_{i=1}^{n}\sum_{s\in W_c} X_{isk}^c = 1 \text{ for } c = 1,2,\ldots,m \text{ and } k = 1,2,\ldots,n_c, \tag{2}$$

$$\sum_{i=1}^{n}\sum_{k=1}^{n_c} X_{isk}^c = 1 \text{ for } c = 1, 2, \ldots, m \text{ and } s \in W_c, \tag{3}$$

$$\sum_{c=1}^{m}\sum_{s\in W_c}\sum_{k=1}^{n_c} X_{isk}^c = 1 \text{ for } i = 1, 2, \ldots, n, \tag{4}$$

$$\sum_{p=1}^{n}\sum_{q=1}^{n}\sum_{s\in W_c}\sum_{t\in W_c}\sum_{r=1}^{k-1}(t_p + t_{pq})X_{psr}^c X_{qt(r+1)}^c - S_i \le M(1 - X_{iuk}^c)$$
$$\text{for } i = 1, 2, \ldots, n, c = 1, 2, \ldots, m, u \in W_c, k = 1, 2, \ldots, n_c, \tag{5}$$

$$S_i + t_i = T_i \text{ for } i = 1, 2, \ldots, n, \tag{6}$$
$$S_j - T_i \le MZ_{ij} \text{ for } i, j = 1, 2, \ldots, n, \tag{7}$$

$$\sum_{i=1}^{n}\sum_{k=1}^{p} X_{isk}^c - \sum_{i=1}^{n}\sum_{k=1}^{p} X_{itk}^c \ge 0 \text{ for } c = 1, 2, \ldots, m, \text{ all}(s,t) \in P,$$
$$\text{and } p = 1, 2, \ldots, n_c, \tag{8}$$

$$\sum_{i=1}^{n}\sum_{c=1}^{m}\sum_{s\in V_p}\sum_{k=1}^{n_c} w_i X_{isk}^c \le w_p^m \text{ for all } p \in T^D \tag{9}$$

$$\sum_{i=1}^{n}\sum_{c=1}^{m}\sum_{s\in V_p}\sum_{k=1}^{n_c} h_i X_{isk}^c \le h_p^m \text{ for all } p \in T^H, \tag{10}$$

$$X_{isk}^c \le \theta_{is} \text{ for } i = 1, 2, \ldots, n, c = 1, 2, \ldots, m,$$
$$s \in W_c, k = 1, 2, \ldots, n_c, \tag{11}$$
$$Z_{ij} + Z_{ji} = 1 \text{ for all } (i,j) \in U \tag{12}$$
$$X_{isk}^c = 0 \text{ or } 1 \text{ for } i = 1, 2, \ldots, n, c =, 1, 2, \ldots, m,$$
$$s \in W_c, k = 1, 2, \ldots, n_c, \tag{13}$$
$$Z_{ij} = 0 \text{ or } 1 \text{ for } i, j = 1, 2, \ldots, n, \tag{14}$$
$$S_i, T_i > 0. \tag{15}$$

The terms of (1) correspond to objectives, 1.1, 1.2, 1.3, 2.1, 2.2, and 2.3, respectively. Each place in the loading sequence, each container, and each slot are assigned the value of one once and only once in the feasible solution by constraints (2), (3), and (4). Constraints (5), (6) and, (7) define variables S_i, T_i, and Z_{ij}. Constraints (8), (9), (10), (11), and (12) correspond to constraints 3.1, 3.2, 3.3, 3.4, and 4.1. Precisely defining, the value of t_i usually depends on the sequence of transfer, but, for the simplicity of the formulation, this study assumed that t_i is independent of the transfer sequence and has a constant value.

The objective function has quadratic terms as well as linear terms, and some decision variables are 0-1 binary variables. Considering loading containers numbers up to higher than 1000, developing a heuristic algorithm for near optimal solutions is a practical approach. Thus, a heuristic algorithm is proposed in the next section.

Table 2. Constructed yard-clusters

Cluster-ID	Port of destination	Size (ft)	Type	Stack location	Number of containers
H201*	H	20	0	C-01	4
M201	M	20	0	C-01	6
S204	S	20	0	C-04	4
S214	S	20	1	C-04	2
H206	H	20	0	C-06	2
S206	S	20	0	C-06	3
S216	S	20	1	C-06	1
K206	K	20	0	C-06	5

* H201 : H (port of destination), 2 (20ft), 0 (type), 1 (location).

3 A beam search algorithm for load-sequencing

This section introduces a beam search algorithm for the load-sequencing problem. The beam search method is similar to the branch and bound method in that both methods reject unpromising nodes in a large search tree, and thus save time and effort to search for the branches of the search tree growing from the rejected nodes. The filtered beam search does this by using a total cost evaluation function (a cost estimate projected from the current partial solution to a complete solution) and one-step priority evaluation function (a cost estimate only to the next step chosen by a simple priority rule). The calculation of a total cost evaluation function usually takes much longer than that of a one-step priority evaluation function. Thus, the filtered beam search first selects candidate solutions (called filtered nodes), whose number is the same as the filter-width, of the next stage by using a one-step priority evaluation function. And then, the filtered beam search procedure evaluates a total cost evaluation function of each filtered node and selects beam nodes, whose number is the same as the beam-width, among filtered nodes (Ow and Morton [8]).

To apply the beam search algorithm, first, a list of yard-clusters of containers is constructed. A yard-cluster is defined as a collection of containers of the same size and type (dry container, refrigerated container, empty container, container with dangerous cargo, etc.) that have the same destination port and which are stacked in the same yard-bay. Considering the example in Figure 6, the list of yard-clusters can be as shown in Table 2.

Two types of beam search are used to search for solutions. The load-sequence of yard-clusters is determined by the first search algorithm, which is called the filtered beam search. The load-sequence of individual containers is determined by the second beam search. The first beam search procedure starts from constructing initial beam nodes. For each initial beam node, nodes in the next stage are generated and filtered by using a total cost evaluation function to select a beam node. For the selected beam node, the sequence of individual containers is determined by the second beam search procedure. The second beam search procedure follows the normal beam search procedure in which, at each stage, beam nodes of the next stage are selected by one-step evaluation function.

Sequencing yard-clusters is equivalent to constructing the pickup schedule (Kim [5], Kim [6], Narasimhan [7], Ryu [10]) that specifies the visiting sequence of yard-bays and the number of containers to pick up at each visiting yard-bay. To determine the sequence of yard-clusters, the first search algorithm needs work schedules of QCs, a list of yard-clusters, and a stowage plan for vessels. The first search algorithm attempts to minimize the total travel time of TCs and to satisfy constraints related to the sequence of yard-clusters. That is, the first search algorithm solves the problem with objective 2.1 and constraints 3.1, 3.4, and 4.1 in the previous section.

The second beam search determines the sequence of individual containers to maximize the handling convenience of QCs and TCs and the degree of satisfaction of the weight requirement. The second beam search also attempts to find solutions to satisfy constraints on the maximum weight of a stack on deck and the maximum height of a stack in hold. Once again, the loading sequence of individual slots must obey the precedence relationships among slots (e.g.,the rule that slots in the bottom must be filled before slots on the top are filled with containers). Thus, the second search procedure solves the problem with objectives 1.1, 1.2, 1.3, 2.2, and 2.3 in the previous section and constraints 3.1, 3.2, and 3.3.

One approach to solving the load-sequencing problem is to sequentially solve two subproblems: sequencing yard-clusters and then sequencing individual containers for the resulting sequence of clusters. One of the difficulties of the sequential approach is that, because the constraints of the second subproblem are not considered when solving the first subproblem, the final solution of the first subproblem may result in an infeasible solution to the second subproblem. Therefore, the two subproblems must be solved simultaneously. Thus, in this study, the search procedure for the second subproblem (the second beam search) is imbedded within the search procedure for the first subproblem (the first beam search). During the second beam search for sequencing individual containers in a yard cluster which corresponds to a filtered node selected in the first beam search procedure, if no feasible sequence can be found, then the filtered node is removed from the set of filtered node of the first beam search tree. And then, the second beam search procedure starts again from the next-best filtered node in the first beam search tree.

For the description of the search algorithm, the following notations are introduced:

C	$=$	The set of yard-clusters.
N_e^o	$=$	The eth initial beam node in the first beam search procedure.
b	$=$	The number of initial beam nodes.
N_e	$=$	The current beam node connected to N_e^o.
F	$=$	The set of filtered beam nodes in the first beam search procedure.
f	$=$	The filtered beam width in the first beam search procedure.
$x(N_e)$	$=$	The partial solution – which can be represented by a sequence of yard-cluster for each QC – corresponding to the path from the root node to N_e in the first beam search tree.
$t(x(N_e))$	$=$	The total travel time of $x(N_e)$.

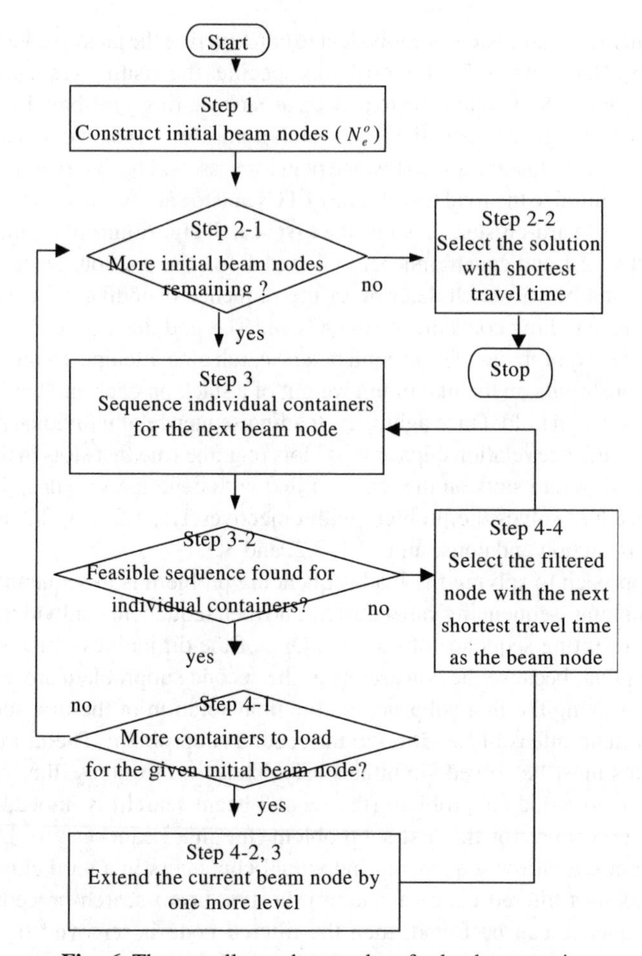

Fig. 6. The overall search procedure for load-sequencing

The overall search procedure is suggested in Figure 6, and a detailed explanation is provided in the following discussion. Note that the main flow of Figure 6 is related to the first subproblem. The beam search procedure for the second subproblem corresponds to Step 3 in Figure 6, which will be described in more detail in Figure 7.

Step 1. (Construct initial beam nodes)

In this step, initial beam nodes are constructed.

Step 1-1. (List candidate yard-clusters for each QC)

For each QC, list all the yard-clusters, from C, which have at least one container that satisfies constraints 3.1 and 3.4. That is, each listed yard-cluster must have

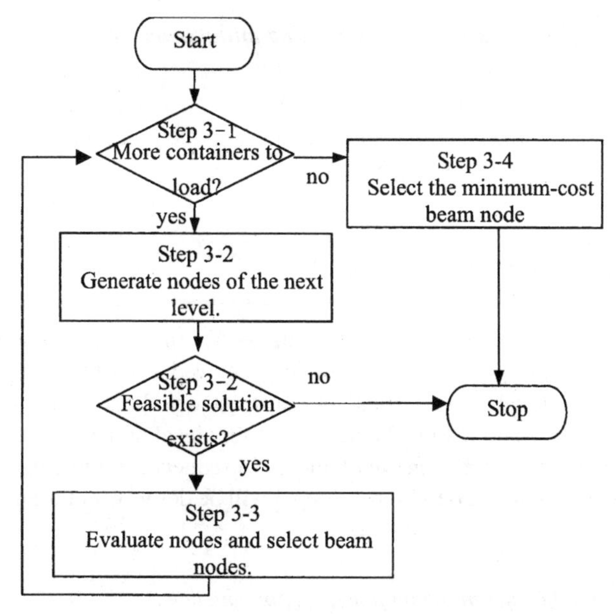

Fig. 7. The second beam search procedure for sequencing individual containers (Step 3)

at least one container that can be loaded into the lowest tier of stacks of the first ship-bay in the work schedule for the QC. Let C_c be the set of yard-clusters listed for QC c.

Step 1-2. (Globally evaluate candidate yard-clusters for each QC)

For each $v \in C_c$, by a neighborhood search, construct a complete pickup schedule which specifies the visiting sequence of yard-bays and the number of containers to pick up at each visiting yard-bay, for all the slots to be filled in the current hold or deck. The neighborhood search procedure constructs the complete pickup schedule by sequentially selecting the nearest yard-cluster, starting from yard-cluster v, and loading containers, satisfying constraints 3.1 and 3.4 in each selected yard-cluster, as many as possible. The complete pickup schedule for each $v \in C_c$ is evaluated by the travel time of the TC.

Step 1-3. (Construct and select the initial beam nodes)

List all the possible combinations of m elements – one element from each $C_c, c = 1, 2, \ldots, m$. Delete combinations that violate constraints 4.1. For each remaining combinations, sum all the travel times of m pickup schedules (of Step 1-2) corresponding to m elements (yard-clusters) in the combination. Among all the combinations, b combinations with the shortest total travel times are selected as initial beam nodes $(N_e^o, e = 1, 2, \ldots, b).e = 0$.

Step 2. (Check for the existence of remaining initial beam nodes)

$e = e + 1$. If $e > b$, then select $x(N_t), t = 1, 2, \ldots, b$, with the minimum $t(x(N_t))$ (the total travel time) as the final solution and stop the procedure (Step 2-2). Otherwise, $N_e = N_e^o$ and go to Step 3.

Step 3. (Sequence individual containers)

In this step, for yard clusters corresponding to N_e, individual containers are sequenced, and the sequenced containers are removed from the yard map. If no sequence of individual containers can be constructed without violating constraints 3.1, 3.2 and 3.3, then go to Step 2 when the current level of the first beam search is 1 or go to Step 4-5 when it is greater than 1. If a sequence of individual containers can be constructed, then go to Step 4. Step 3 will be described in more detail later.

Step 4. (Extend the current beam node, N_e, by one level)

Step 4-1. (Check for the existence of remaining containers)

For the current beam node, N_e, if more containers to be loaded exist, go to Step 4-2. Otherwise, go to Step 2.

Step 4-2. (List candidate yard-clusters)

Based on $x(N_e)$, select a QC, among the QCs that have remaining containers to load, which completed the previous work the earliest. Let the selected QC be QC c. Construct C_c by using candidate yard clusters for QC c as in Step 1-1.

Step 4-3. (Select filtered nodes)

To construct F, select f elements (yard-clusters) from C_c with the shortest travel time from the last location of the TC in $x(N_e)$ to the locations of the candidate yard-clusters. This evaluation process is called a "local evaluation." The selected f elements in C_c are called "filtered nodes."

Step 4-4. (Perform the global evaluation)

Perform a global evaluation for all the filtered nodes by the same procedure as the one in Step 1-2.

Step 4-5. (Select a beam node from filtered nodes)

Select the filtered node with the shortest travel time from F. Delete the selected node from F. The selected node becomes the new beam node, N_e. Go to Step 3.

The following describes how to sequence individual containers in Step 3. If Step 3 is performed during level 1 of the first beam search procedure, then the following procedure will be repeated as many times as the number of QCs, while, otherwise, the following procedure is performed once.

Before beginning Step 3, a yard-cluster has already been determined for sequencing individual containers for a QC. In the search tree for sequencing individual containers, one container for loading is determined at each level. Thus, the depth of the search tree is the same as the maximum number of containers that can be transferred from the current yard cluster.

For a more detailed explanation of Step 3, the following notations are used:

d = The beam width for the second beam search procedure.
r = Index representing the search level for the second beam search procedure.
B_r = The set of beam nodes at level r for the second beam procedure.
G_r = The set of all the generated nodes in level r.
M_r = A beam node in B_r.
$q(M_r)$ = The total penalty of objectives 1.1, 1.2, 1.3, 2.2, and 2.3 from the root node to M_r.
z = The number of containers to be sequenced.

Figure 7 shows the overall procedure of Step 3 that is also explained in the following:

$r = 0$.

Step 3-1. (Check for the existence of additional containers to load)

If $r = z$, then select, as the final solution, the beam node with the minimum $q(M_s)$, $s = 1, 2, \ldots, d$, and stop. Otherwise, $r = r + 1$ and go to Step 3-2.

Step 3-2. (Generate nodes for the next level)

For each M_{r-1}, generate all the feasible combinations, which satisfy constraints 3.1, 3.2, and 3.3, of the next candidate slots in the ship-bay and the next candidate containers in the yard-cluster. All the generated feasible combinations become elements of G_r. If no combination, which does not violate the constraints, can be found in the current yard-cluster, then the next filtered node with the next shortest travel time in Step 4-4 is selected as a beam node and repeat Step 3 again.

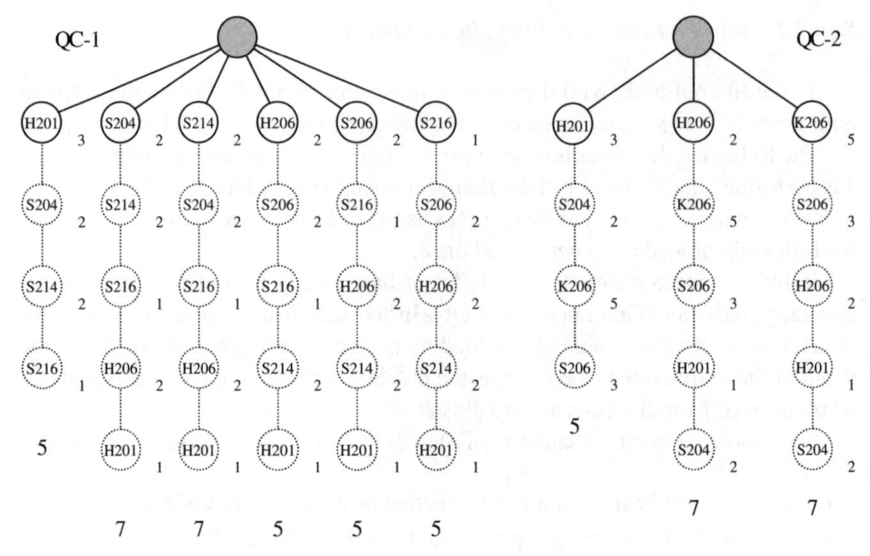

Fig. 8. The result of global evaluation for candidate yard-clusters

Step 3-3. (Evaluate nodes and select beam nodes)

Evaluate $q(y)$ for all the elements $y \in G_r$. Select d nodes with the minimum values of $q(y)$. The selected nodes are included in B_r. Go to Step 3-1.

A numerical example

By using the stowage plan, the yard map, and yard-clusters in Figures 3 and 5 and Table 2, the algorithm in this paper is illustrated in the following.

Step 1-1: Let QC-1 and QC-2 start the loading operation from ship-bay 01 and ship-bay 21 in Figure 3, respectively. $C_1 = \{$H201, S204, S214, H206, S206, S216$\}$ and $C_2 = \{$H201, H206, K206$\}$.

Step 1-2: As shown in Figure 8, to complete all the tasks in the current hold, three 20-foot containers bound for port "H" and five 20-foot containers bound for port "S" must be transferred for QC-1. The numerical value at each node represents the number of containers picked up from the corresponding yard-cluster. The numerical value at the end of each branch represents the travel time required for loading all the containers in hold. For example, let the first yard-cluster be H201. Then, three 20-foot containers can be picked up from H201. Next, the nearest yard-cluster that has 20-foot containers of type 0 bound for port "S" is S204. From yard-cluster S204, two 20-foot containers of type 0 (S204) and two 20-foot containers of type 1 bound for port "S" (S214) are picked up. Because one more 20-foot container of type 1 bound for port "S" is necessary, it is picked up from S216.

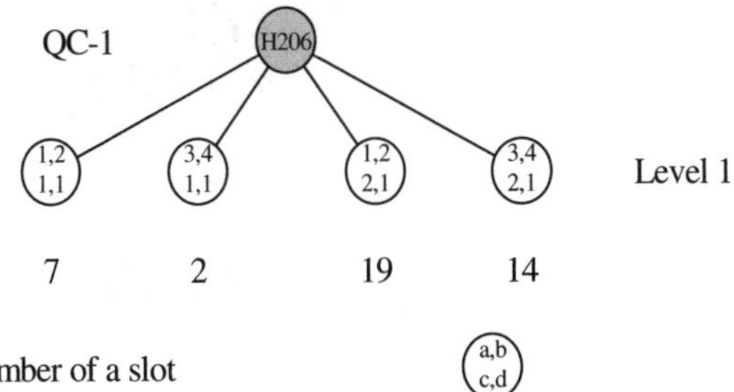

Fig. 9. Constructed nodes of level 1 in the second beam search tree

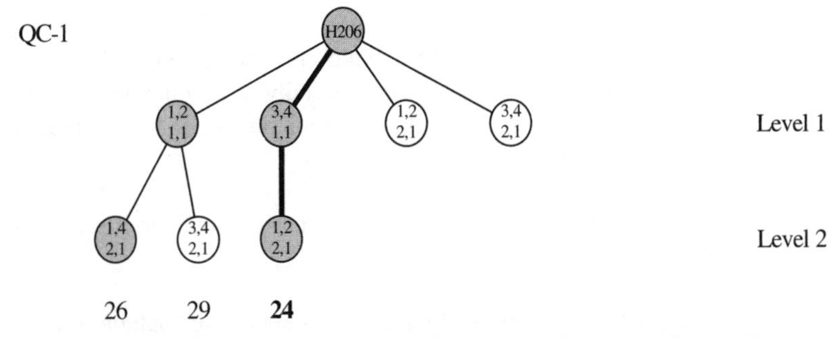

Fig. 10. Nodes constructed to level 2 in the second beam search procedure

Step 1-3: Let b be 5. Then, combinations of {<H206>, (H201)}, {<S206>, (H201)}, {<H201>, (H206)}, {<H201>, (K206)}, and {<S216>, (H201)} are selected as the initial beam nodes, where the bracket and the parentheses represent the first yard-cluster for QC-1 and QC-2, respectively.

Step 2: $N_1 = N_1^0$ {<H206>, (H201)}.

Step 3-1: Yes, there are containers to be loaded.

Step 3-2: Slots that can be selected as the first slot for QC-1 are (1, 2) and (3, 4), where slots are represented by (stack number, tier number). Containers that can be loaded first are (1, 1) and (2, 1), where containers are also denoted by (stack number, tier number). By combining all the candidate slots and containers, nodes in level 1 (G_1) are constructed as shown in Figure 9.

Step 3-3: Because $d = 2$ in this example, the first two nodes are selected as the beam nodes (elements in B_1). By applying Steps 3-1 through 3-3 once more, the tree as shown in Figure 10 is obtained.

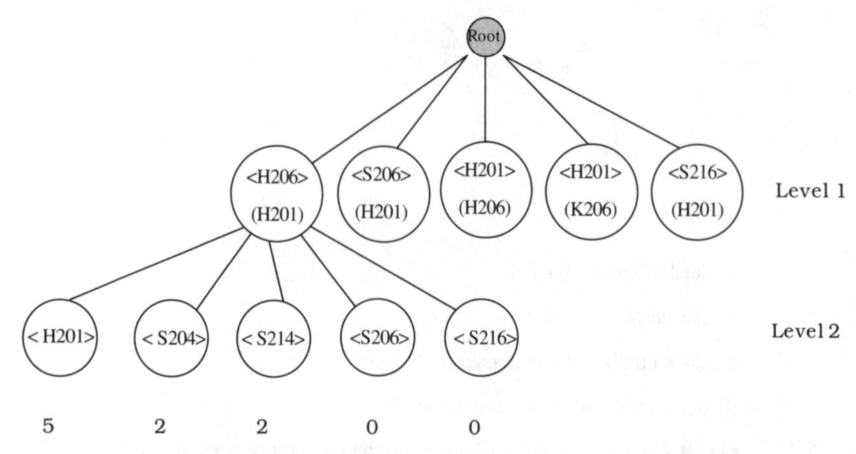

Fig. 11. Local evaluation for filtering nodes constructed at level 2 in the first beam search

Step 3-1: r equals to z. The best solution, which is represented by bold type in Figure 10, is selected as the loading sequence for individual containers for QC-1. The same procedure follows for QC-2 to obtain the loading sequence for individual containers of N_1, {<H206>, (H201)}.

Step 4-1. For N_1, {<H206>, (H201)}, because more containers remain to be loaded, go to Step 4-2.

Step 4-2. The cumulative travel time of the TC for transferring containers of QC-1 is shorter than that of QC-2. Thus, Figure 11 results. $C_1 =$ {<H201>, <S204>, <S214>, <S206>, <S216>}.

Step 4-3: f is set to be 2. Thus, $F =$ {<S206>, <S216>}.

Step 4-4: A global evaluation is performed for the two filtered nodes. It is found that the results are the same as those shown in Figure 12.

Step 4-5: Either of the two filtered nodes can be selected as the next beam node (N_1). This process is repeated until a feasible solution is obtained for the first initial beam node, {<H206>, (H201)}. Then, the algorithm moves to the second initial beam node, {<S206>, (H201)}.

4 Numerical experiments

Two numerical experiments were conducted to test the performance of the beam search algorithm suggested in this paper. The first experiment was conducted to test the sensitivity of algorithm's performance to changes in search parameters b, d, and f. The second experiment was for comparing the performance of the algorithm with two other approaches, the ant system approach (Ryu, 2001), and the neighborhood search algorithm.

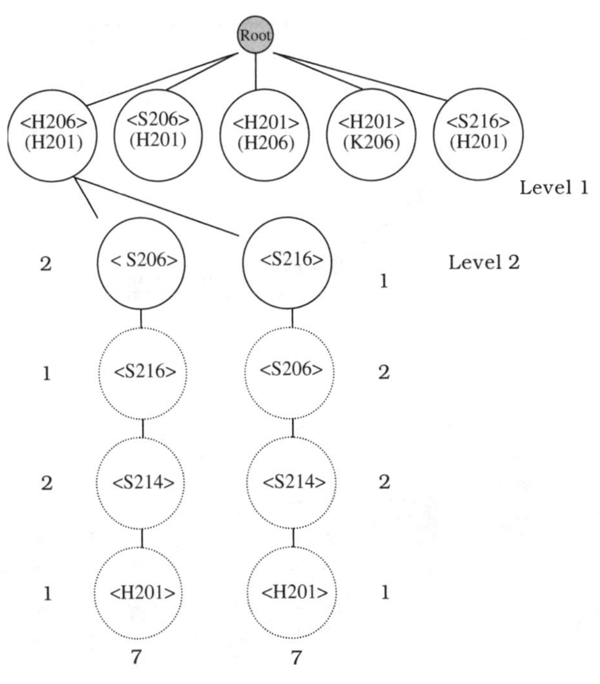

Fig. 12. Global evaluation for nodes filtered at level 2 in the first beam search procedure

The first experiment used a problem with 624 containers, 118 container groups, 32 ship bays, 79 yard-bays, and 4 QCs. Travel time of TCs between adjacent bays, travel time of TCs between different blocks, and travel time of TCs between blocks in different rows were set to be 1, 5, and 25 seconds, respectively. And, the values of the parameters were $c_w = 5$, $c_d = 3$, $c_h = 3$, $a_r = 10$, and $a_t = 1$.

The size of the search space for sequences of yard-clusters depends on the values of b and f, while that for sequences of individual containers is determined by the value of d. Figure 13 shows that the total travel time of TCs is affected significantly by the values of b, while the total travel time is insensitive to the values of f.

Figure 14 shows that both b and f contribute to the reduction of the total weighted penalty in which the penalty of the travel time was included. Figure 15 shows that a larger d results in a smaller total weighted penalty. However, the total travel time of TCs did not change for different values of d, which coincides with our intuition. The computational time was sensitive to the value of b (see Fig. 16) and d (see Fig. 17), while it was insensitive to the values of f.

Note that the algorithm enumerates whole solutions on the sub-tree below one initial beam node and then proceeds to the sub-tree below the next initial beam node. Thus, after the enumeration of the sub-tree below the first initial beam node is completed, the best so far known solution, which is feasible, is obtained. Thus, after then, at any time when the search process is terminated, one or more feasible solutions are available and the best so far feasible solution can be used as the final solution. This is a very useful property of the algorithm in this study. Figures 18 and

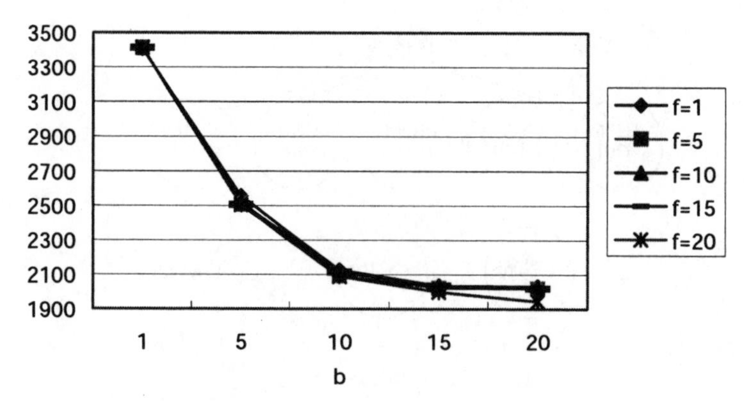

Fig. 13. The total travel time of TCs for different values of b and f ($d = 1$)

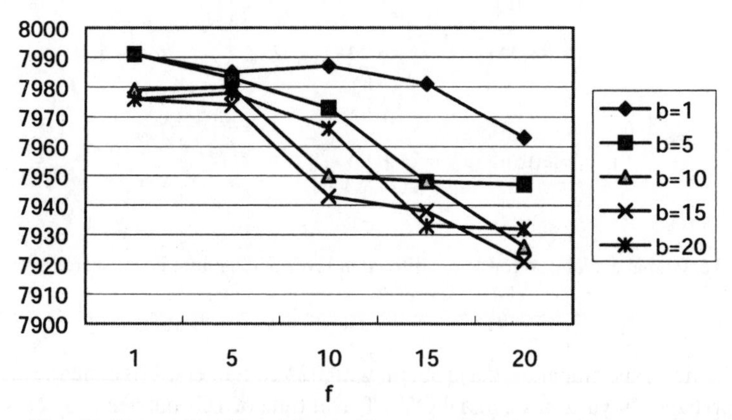

Fig. 14. The total weighted penalty for various values of b and f ($d = 1$)

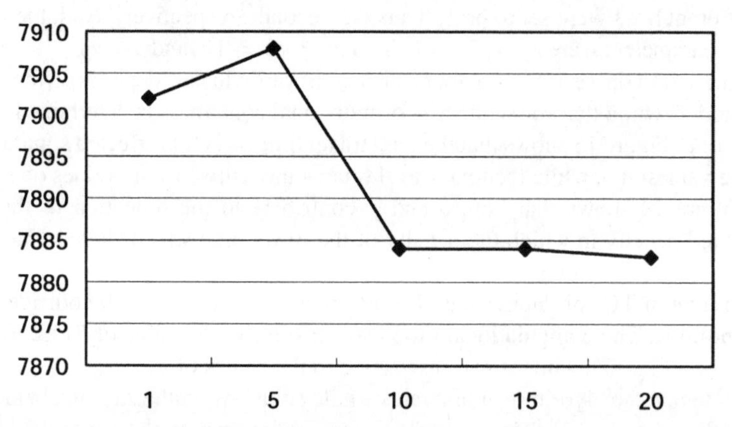

Fig. 15. The total weighted penalty for various values of d ($b = 10$, $f = 15$)

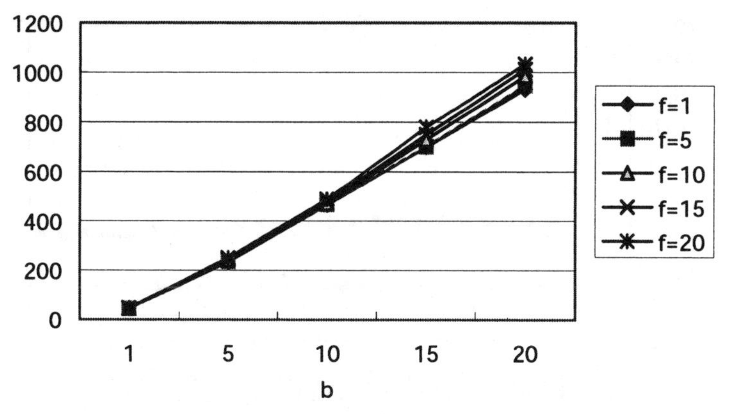

Fig. 16. Computational time (in seconds) for various values of b and f ($d = 1$)

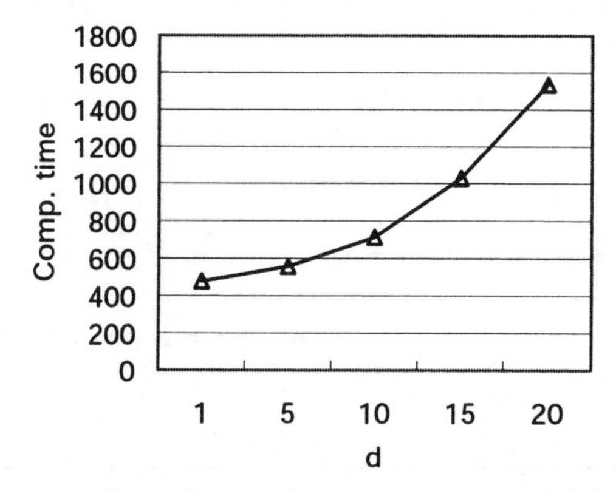

Fig. 17. Computational time (in seconds) for various values of d ($b = 10$, $f = 15$)

19 show how the minimum total travel time of TCs and the minimum total penalty change as the stopping time increases for a sample problem with the number of containers = 624, $b = 30$, $d = 10$, and $f = 15$. That is, these graphs show the trade-off between the quality of the final solution and the computational time.

The second numerical experiment was conducted with six sets of data collected from Pusan Eastern Container Terminal (PECT) in Korea. The size of the problems are listed in Table 3. Problems in Table 3 are representative of real problems in PECT.

Table 4 compares the performance of three solution algorithms: the neighborhood search, the algorithm in this study ($b = 10$, $d = 10$, $f = 15$), and an algorithm based on the ant system (ant algorithm) (Ryu [10]). Note that when values of parameters are set to $b = d = f = 1$, the algorithm in this study reduces to the neighborhood search. In the ant algorithm, the number of repetitions and the number of ants were set to 300 and 600, respectively. A personal computer with Pentium III-600 and 128 Mb-RAM was used for the numerical experiment. The algorithms in this study and in the neighborhood search were programmed by using

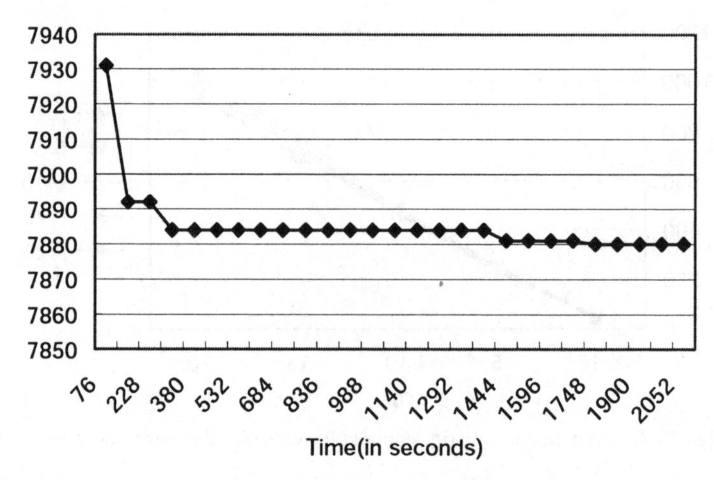

Fig. 18. The change in the minimum total travel time of TCs with respect to the stopping time

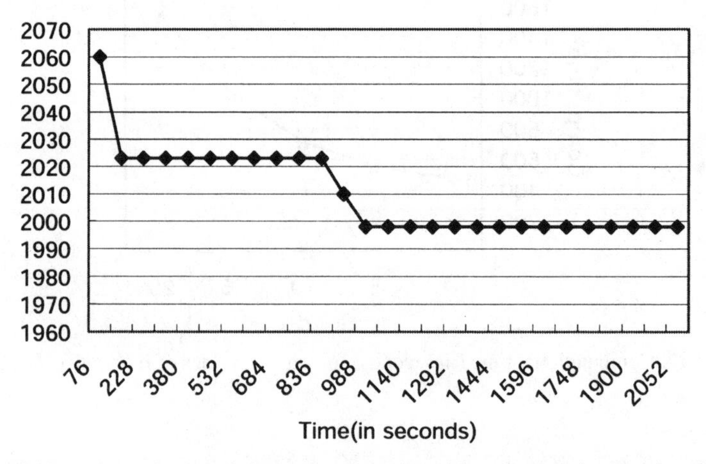

Fig. 19. The change in the minimum total penalty with respect to the stopping time

Table 3. Size of sample problems used in the second experiment

Problem number	Number of containers	Number of container groups	Number of ship bays	Number of yard bays	Number of QCs
1	313	36	18	21	2
2	624	118	32	79	4
3	653	117	21	59	3
4	1012	223	36	130	4
5	1304	242	23	126	3
6	1340	352	43	185	4

Table 4. Performance of three algorithms

	Performance	Neighborhood	Beam search	Ant algorithm
	Total travel time	170	168	326
Problem 1	Total penalty*	1,029	981	1,132
	Comp. time (sec)	13	173	1,410
	Total travel time	3,413	2,023	3,280
Problem 2	Total penalty	7,956	7,892	9,920
	Comp. time	45	712	4,482
	Total travel time	1,573	1,467	1,549
Problem 3	Total penalty	10,324	10,127	10,603
	Comp. time	43	689	4,917 tab4
	Total travel time	5,866	3,278	5,483
Problem 4	Total penalty	12,863	12,603	14,232
	Comp. time	74	1,289	10,552
	Total travel time	6,671	6,532	6,690
Problem 5	Total penalty	19,388	17,832	22,954
	Comp. time	115	2,513	12,554
	Total travel time	9,295	6,677	8,740
Problem 6	Total penalty	17,865	17,646	20,787
	Comp. time	111	2,498	11,234

* The penalty of the travel distance was excluded from the total penalty.

JAVA, while the ant algorithm was programmed by using C++. It is known that the processing speed of C++ is 4 to 5 times faster than that of JAVA.

The beam search algorithm obtained solutions higher in quality than those found by the neighborhood search, but at a cost of higher computational time. Note that the computational time can be adjusted by adjusting the values of b, d, and f, or by specifying the stopping time. The beam search algorithm in this study outperformed the ant algorithm in all three measures of performance. The difference between the two algorithms is due to the fact that, in the ant algorithm, the search procedure is hierarchically divided into two stages: sequencing yard-clusters, and sequencing individual containers in a yard and slots in a vessel. Note that, in this study, the two decisions are integrated and made simultaneously.

5 Conclusion

This paper discusses the load-sequencing problem for outbound containers in port container terminals in which TCs and YTs are used in the marshaling yard. Various constraints and objectives of the load-sequencing problem were introduced. A beam search algorithm was suggested to minimize the handling time of TCs and QCs, and to satisfy various constraints for loading containers. The algorithm in this paper has the following strength: various additional constraints and objectives can be considered without significantly modifying the algorithm, the pickup sequence by TCs in the yard and the loading sequence of slots in the vessel are determined simultaneously, the computational time can be adjusted by users, and the relative importance of elements in the objective function can be modified by users by adjusting the values of parameters of the objective elements.

A sensitivity analysis was performed to show how various performance measures are related to the values of parameters of the beam search algorithm. It was shown that the beam search in this paper outperforms the ant algorithm in the values of objective functions and the computational time.

References

1. Beliech Jr. D E (1974) A proposed method for efficient pre-load planning for containerized cargo ships. Master's Thesis, Naval Postgraduate School, Monterey, California
2. Cho D W (1982) Development of a methodology for containership load planning. Ph.D. Dissertation, Oregon State University
3. Cojeen H P, Dyke P V (1976) The automatic planning and sequencing of containers for containership loading and unloading. In: Pitkin, Roche, Williams (eds) Ship operation automation, pp 415–423. North Holland, Amsterdam
4. Gifford L (1981) A containership load planning heuristic for a transtainer-based container port. Master's Thesis, Oregon State University
5. Kim K H, Kim K Y (1999) Routing straddle carriers for the loading operation of containers using a beam search algorithm. Computers & Industrial Engineering 36: 109–136
6. Kim K H, Kim K Y (1999) An optimal routing algorithm for a transfer crane in port container terminals. Transportation Science 33 (1): 17–33
7. Kozan K, Preston P (1999) Genetic algorithms to schedule container transfers at multimodal terminals. International Transactions in Operational Research 6: 311–329
8. Ow P S, Morton T E (1988) Filtered beam search in scheduling. International Journal for Production Research 26 (1): 35–62.
9. Narasimhan A, Palekar U S (2002) Analysis and algorithms for the transfer routing problem in container port operations. Transportation Science 36 (1): 63–78
10. Ryu K R, Kim K H, Lee Y H, Park Y M (2001) Load sequencing algorithms for container ships by using metaheuristics. Proceedings of 16th International Conference on Production Research (CD-ROM), Prague, Czech Republic

A general framework for scheduling equipment and manpower at container terminals

Sönke Hartmann

OR Consulting, Bornstraße 6, 20146 Hamburg, Germany, and
Institut für Betriebswirtschaftslehre, Lehrstuhl für Produktion und Logistik,
Christian-Albrechts-Universität zu Kiel, 24098 Kiel, Germany
(e-mail: soenke.hartmann@epost.de)

Abstract. In this paper, we propose a general model for various scheduling problems that occur in container terminal logistics. The scheduling model consists of the assignment of jobs to resources and the temporal arrangement of the jobs subject to precedence constraints and sequence-dependent setup times. We demonstrate how the model can be applied to solve several different real-world problems from container terminals in the port of Hamburg (Germany). We consider scheduling problems for straddle carriers, automated guided vehicles (AGVs), stacking cranes, and workers who handle reefer containers. Subsequently, we discuss priority rule based heuristics as well as a genetic algorithm for the general model. Based on a tailored generator for experimental data, we examine the performance of the heuristics in a computational study. We obtain promising results that suggest that the genetic algorithm is well suited for application in practice.

Keywords: Container logistics – Container terminal – Optimization – Scheduling – Heuristics – Genetic algorithm

1 Introduction

In the 1960s, the container was introduced as a universal carrier for various goods. It soon became a standard in worldwide transportation. The success of the container is associated with the increasing containerization (which means that the number of goods transported in containers has steadily grown) and with increasing world trade. Container terminals are continuously facing the challenge of strong competition between ports and of turning around more and larger ships in shorter times. This leads to the necessity to use the highly expensive terminal resources such as quai cranes, straddle carriers, automated guided vehicles, and stacking cranes as efficiently as possible. A key factor of success is the optimization of the logistic processes. Therefore, many researchers and practitioners have developed optimization approaches for container terminal logistics.

Important optimization problems include the assignment of berths to arriving vessels (see Guan and Cheung [12], Imai et al. [18,19], Lim [29]) and the assignment of quai cranes to vessels or ship-bays (see Daganzo [4], Peterkofsky and Daganzo [34]). A berth assignment approach which simultaneously considers quai crane capacities has been developed by Park and Kim [33]. Scheduling the container transport on the terminal has been studied for two different types of equipment, namely straddle carriers (see Böse et al. [3], Kim and Kim [26], Steenken et al. [37]) and automated guided vehicles (AGVs, see Bae and Kim [1]). The case of multi-load AGVs (i.e., AGVs which can carry more than one container at a time) has been discussed by Grunow et al. [11]. Also the problem of allocating and scheduling stacking cranes has been considered (see Zhang et al. [44]). An integrated scheduling approach for automated stacking cranes and automated guided vehicles has been proposed by Meersmans and Wagelmans [31]. A method for sequencing the containers to be loaded onto a ship has been developed by Kim et al. [22]. Strategies for locating containers in the yard have been discussed for several problem settings (see de Castilho and Daganzo [5], Kim and Kim [23], Kim et al. [24], Taleb-Ibrahimi et al. [39], Zhang et al. [43]). In order to study the complex processes at container terminals with their dynamic nature, simulation models have been developed (see Gambardella et al. [8,9], Kim et al. [25], Legato and Mazza [28], Yun and Choi [42]). A simulation study to compare different automated vehicle types at a container terminal has been provided by Vis and Harika [41]. Finally, a method to generate data for experiments with optimization and simulation approaches has been suggested (see Hartmann [15]). Detailed literature surveys have been given by Meersmans and Dekker [30] as well as Vis and Koster [40].

In this paper, we introduce a general model for scheduling container terminal resources such as different types of equipment and manpower. This way, we take an approach that is different from the scheduling papers listed above because we provide a model that is not designed for a single application. Our model consists of a set of jobs (which could be transportation tasks or other activities) that must be scheduled and assigned to a resource. For the temporal arrangement of the jobs, sequence-dependent setup times have to be observed. They can be used to cover, e.g., the empty times of the equipment between two container transports. The objective is to minimize the average lateness per job as well as the average setup time. In order to demonstrate the generality of our model, we present four applications from the port of Hamburg (Germany), namely straddle carriers, automated guided vehicles (AGVs), stacking cranes, and workers. Subsequently, we propose priority rule heuristics and a genetic algorithm for the general model and analyze them in a computational study. The paper closes with a summary of the results and discusses opportunities for future research.

2 A general model for container terminal scheduling problems

In this section, we propose a general model for scheduling problems in container terminal logistics. After the definition of the model, we outline how several real-world scheduling problems can be captured by the general model.

2.1 Model formulation

We consider a set $J = \{1, \ldots, n\}$ of jobs that have to be carried out using a set $R = \{1, \ldots, m\}$ of identical resources. Each job $j \in J$ has to be executed on one of the resources. Since the resources are identical, each job can be performed on any of the resources. Each resource $r \in R$ can be occupied with only one job at a time.

Each job $j \in J$ has a processing time $p_j > 0$. Before job j can actually be processed, a setup time is required. Let $i_j \in J$ denote the job that is executed on resource $r \in R$ immediately before job j. A sequence-dependent setup time $s_{i_j,j} \geq 0$ has to be taken into account before processing job j. Once started, a job must not be interrupted, that is, both the setup and the processing phase of a job must be carried out as a non-preemptive whole. These requirements can be summarized as $f_{i_j} + s_{i_j,j} + p_j \leq f_j$ where f_j is the finish time of job $j \in J$. For the setup times, the following triangle inequality must hold: We assume $s_{ik} \leq s_{ij} + s_{jk}$ for any three jobs i, j, and k.

Some of the jobs may be related by precedence constraints. The predecessors of job j are given by the set P_j (analogously, the successors of job j are given by the set S_j). The precedence relation between a job j and a predecessor $i \in P_j$ is specified by a time lag $\tau_{ij} \geq 0$. The latter implies that job j must finish at least τ_{ij} units later than job i, that is, $f_i + \tau_{ij} \leq f_j$. $\tau_{ij} = 0$ represents a so-called end-to-end constraint because job j must not finish before job i finishes. $\tau_{ij} = p_j$ is a so-called end-to-start constraint since job j must not start before the finish time of job i. Note that also other types of precedence constraints can be expressed with the general time lags employed here (further details can be found in Bartusch et al. [2]).

Each job $j \in J$ is related to two specific time instants. First, time d_j is the due date of job j. That is, job j should be completed at or before time d_j. Completing job j later than time d_j is allowed, but it will lead to penalty costs in the objective function. Second, time e_j denotes the earliest time at which resource $r \in R$ that carries out job j is available again after the execution of job j. Hence, if job j is completed earlier than time e_j, then resource r becomes available at time e_j. Otherwise, resource r becomes available immediately after the completion of job j. Consider the finish time f_j of a job j and the earliest availability time e_{i_j} of the job i_j that is executed immediately before job j on the same resource. We can summarize the restriction imposed by the earliest availability time by $e_{i_j} + s_{i_j,j} + p_j \leq f_j$.

We assume that the model is applied repeatedly in a rolling planning horizon. Therefore, the scheduling problem must take the initial state of each resource into account, that is, the availability time and the setup state of that resource. The initial state of a resource is characterized by the last job carried out by that resource. For each resource $r \in R$, we denote the last job as j_r^+. This last job is a dummy job in the sense that it is assumed to be fixed, that is, it cannot be (re-)scheduled. The last jobs are comprised in the set $J^+ = \{j_r^+ | r \in R\}$, for which we assume $J \cap J^+ = \emptyset$. Each last job j_r^+ is associated with two types of information. First, it is related to a finish time $f_{j_r^+}$ that reflects the time at which resource r is available. Note that $f_{j_r^+} \geq t_{\text{now}}$ must hold, where t_{now} denotes the current time (without loss of generality, we assume $t_{\text{now}} = 0$). Second, job $j_r^+ \in J^+$ contains the initial setup state of resource $r \in R$ (hence the setup time s_{ij} must be given for all $i \in J \cup J^+$ and $j \in J$). Let us emphasize again that only the jobs in J (and not those in J^+) are considered for scheduling.

The problem now is to find a schedule for the jobs $j \in J$, that is, a resource r_j and a finish time f_j for each job $j \in J$ such that all constraints given above are observed. The objective is to minimize the weighted sum of average lateness per job and average setup time per job. The weight for the average lateness per job is denoted as α_L while α_S is the weight for the average setup time per job. Formally, the objective function can be given as

$$\text{minimize} \qquad \alpha_L \cdot \frac{1}{n} \cdot \sum_{\substack{j \in J \\ f_j > d_j}} (f_j - d_j) \quad + \quad \alpha_S \cdot \frac{1}{n} \cdot \sum_{j \in J} s_{i_j,j} \; .$$

Note that lateness and setup times should be measured on the same scale (e.g., seconds). Lateness minimization and hence observing the due dates is one of the most important goals in practice. In the applications, setup times correspond to empty times of resources which should be minimized especially if the due dates are not tight.

The model contains several features that are well known from other scheduling problems in the literature. The concept of jobs with precedence constraints and resource requests can also be found in the classical resource-constrained project scheduling problem (see Pritsker et al. [35]). Time lags and different types of precedence constraints have been studied in connection with project scheduling as well (see Bartusch et al. [2]). Moreover, the resource assignment part of the new model is a special case of the multi-mode extension of the resource-constrained project scheduling problem (see Elmaghraby [7], Talbot [38]). The objective of minimizing the lateness with respect to due dates has also been considered in the context of project scheduling (see, e.g., Kapuscinska et al. [21]). The concept of sequence-dependent setup times occurs in many problems in the field of lotsizing and batching (see Jordan [20]). We will make use of the similarity between the new model and resource-constrained project scheduling when designing heuristics for the new model.

Of course, the model outlined above is a rather abstract formulation. The generality of the approach enables us to apply it to different problem settings. The jobs reflect tasks that have to be carried out in container handling. The resources can reflect the technical equipment for container handling or manpower. The setup and processing times can correspond to moving a resource to some location, moving a container, or other tasks. Several practical applications will be given in the following subsections.

In practice, all applications are embedded in an overall terminal control system. Once a resource (e.g., straddle carrier, automated guided vehicle, stacking crane, reefer worker) has completed a job, the related scheduling procedure should be executed, and the resource should be assigned its next job according to the new schedule. This means that rescheduling should be frequently done such that all decisions are based on the current data. Despite rescheduling before each job assignment, it should be useful to compute schedules with a horizon of several successive jobs. This is because a longer horizon increases the degrees of freedom on when to execute a job the due date of which is not tight. While short horizons would lead to a first-come-first-serve-like approach, longer horizons allow to exploit the possibility of starting a job at a time which leads to a shorter setup time.

Finally, note that the model structure with only a single resource type implies that we have a separate problem for each resource type. This allows to define a general model that is independent of the terminal configuration (integrated models with more than one resource type are of course possible and might be promising, but they would be much more dependent on the actual terminal and thus less general).

2.2 Application to straddle carriers

Many container terminals employ straddle carriers (sometimes also called van carriers) for transportation of containers on the terminal. Straddle carriers are used for transportation between the stack on the one hand and other locations such as a quai crane, the area for external trucks, or the train area on the other hand. Since straddle carriers are able to unload themselves and high enough to stack containers on top of each other, they can also be employed to transport containers to their positions within a yard block. Thus, they can also carry out shuffle moves within a block, that is, they can move a container to another position in case it stands on top of another container that must be moved out. If stacking cranes serve the blocks, however, straddle carriers transport containers from and to hand-over positions at a block, and they do not carry out shuffle moves.

Of course, the set of the transportation tasks of the straddle carriers defines the set of jobs for scheduling, and the number of resources is given by the number of straddle carriers. The processing time p_j of a job j is the time that a straddle carrier needs for the transportation of a container between two locations on the terminal. It is defined as the time between picking up the container and putting it down. As the locations are fixed, the transportation times are assumed as fixed as well. The setup time s_{ij} between two jobs i and j is given by the time that a straddle carrier requires to get from the position where the container of job i was put down to the pick-up position of job j. Hence, the setup time models the empty travel time of a straddle carrier. For any two positions, an estimate of the processing and setup time is usually available for scheduling. Of course, scheduling decisions can affect only the empty (or setup) times but not the transportation (or processing) times.

Precedence relations exist between jobs that correspond to containers that stand on top of each other in the stack. The job i related to an upper container is a predecessor of the job j related to the lower one (if the upper container is not demanded by a ship, truck, or train, job i is a shuffle move). The time lag τ_{ij} is given by a buffer time which assures that the lower container can be picked up after the upper container has been moved away (although the latter job needs not necessarily be completed before the lower container is picked up).

The due date d_j of a job j is determined as follows. Let us first consider the case that a straddle carrier is supposed to bring a container from the stack to another location such as a quai crane or an external truck. In this case, the due date reflects the latest time at which the container should have arrived there. For a quai crane waiting for a container, the due date corresponds to the time at which the quai crane requires the container in order to keep its loading sequence on time. Considering an external truck, the due date reflects the acceptable waiting time of the truck driver. Moreover, keeping the waiting times for trucks short helps to avoid congestions in the truck area. In the second case, a straddle carrier is supposed to bring a container to the stack. Here, the due date is used to model the latest time at which a straddle carrier has to pick up a container at some location. For example, an arriving container must be picked up from a quai crane or an external truck at or before the due date. In this case, we have a due date for pick-up, say $\bar{d}_j$. The latter can be transformed into a due date related to the completion of the job by setting $d_j = \bar{d}_j + p_j$. This is possible because the time between pick-up and completion is a constant, namely the transportation or processing time p_j. Recall that the due date is always related to the finish time of a job (see the objective function).

The earliest availability time e_j of a resource after carrying out a job j is used to model the blocking of straddle carriers by other resources. Consider a job j that requires a straddle carrier to transport a container from a quai crane to a yard block. Let $\bar{e}_j$ denote the time at which the container is available for the straddle carrier (i.e., the time at which it has been put on the ground by the quai crane). The earliest release time of the straddle carrier is $e_j = \bar{e}_j + p_j$ which reflects the earliest possible time to complete this job. Now let us consider a job j to transport a container from the stack to a truck which is already waiting for the container. In this case, we have $e_j = 0$ because the straddle carrier is available immediately after completing the job. These two cases should be sufficient to illustrate the use of the earliest release times. Generally speaking, we have $e_j > 0$ if the straddle carrier will have to wait for another resource, whereas we have $e_j = 0$ if it is not blocked by another resource.

The use of the initial state and of the objective function is obvious. The lateness component of the objective function attempts to serve the quai cranes (and also trucks) on time such that they do not have to wait. The setup time component leads to short empty travel times. Summing up, we have employed all features of the general scheduling model of Subsection 2.1 to capture the straddle carrier scheduling problem.

2.3 Application to automated guided vehicles

On modern container terminals with a high degree of automation, often automated guided vehicles (AGVs) are employed to carry containers between the quai and the yard blocks. Unlike straddle carriers, AGVs are unable to unload themselves (automated vehicles with loading and unloading capability are often referred to as automated lifting vehicles, see Vis and Harika [41]—note that those vehicles can be modeled in the same way as straddle carriers). Therefore, stacking cranes are needed to serve the yard blocks. Hence, the jobs to schedule are the transportation of containers from a quai crane to a stacking crane and from a stacking crane to a quai crane. While the former case corresponds to the discharging of a vessel, the latter case occurs when loading a vessel.

The application of the general model to the AGV case is similar to the straddle carrier case with only a few differences. Again, the processing time p_j reflects the transportation of a container while the setup time s_{ij} models the empty time between two successive jobs. The due dates d_j reflect the latest hand-over times at the quai cranes. If an AGV exceeds its due date, this will lead to a waiting time of the quai crane. Thus, keeping the due dates is crucial for a high quai crane productivity and hence for a short time in port for the vessels. The earliest release times e_j consider the fact that AGVs can be blocked by the other terminal resources. Since AGVs cannot unload themselves, an AGV arriving before its due date will have to wait for the crane related to this job. Therefore, the earliest release time is equal to the due date, that is, $e_j = d_j$. Finally, precedence relations are needed to control the order of AGVs arriving at a quai crane with a container. This AGV order is important because the sequence of containers to be loaded onto a vessel is often fixed. The time lag τ_{ij} between two successive jobs i and j related to the same quai crane is given by a small buffer time that allows the first AGV to leave the quai crane position (typically, one will have $0 < \tau_{ij} \leq d_j - d_i$). Note that there are no precedence relations between jobs related to different quai cranes.

2.4 Application to stacking cranes

Usually, the stack is organized in several yard blocks. Many container terminals employ stacking cranes to serve these blocks. Normally, there is one crane for each block (occasionally, two cranes per block are used). So-called rail-mounted gantry cranes cannot be moved to another block. On the other hand, so-called rubber-tyred gantry cranes can be moved, but this takes a long time and is not done very often (decisions to transfer a crane are based on workload estimations of the blocks, see Zhang et al. [44]). Thus, for both crane types, crane assignments to blocks can be assumed to be fixed for the detailed scheduling problem considered here. Moreover, the assignment of containers to blocks is given (it is not a part of this scheduling problem). Therefore, we have a separate crane scheduling problem for each block. That is, we consider the set of jobs associated with a single block, and we have a single crane resource (or, in case of double cranes, two resources). The jobs include transportation moves between positions within the block on the one hand and AGVs, straddle carriers, or external trucks on the other hand. Furthermore, jobs representing shuffle moves have to be taken into account.

As for the straddle carrier case, we have precedence relations between jobs that correspond to containers that stand on top of each other. The processing time p_j of a job j is the transportation time that a crane needs between picking up the container and putting it down. Again, the setup time s_{ij} between two jobs i and j is given by the time that the crane requires to get from the position where the container of job i was put down to the pick-up position of job j. The due date d_j of a job j determines the latest acceptable completion time and hence also the waiting time of the other resource (AGV, straddle carrier) or the external truck. The earliest release times again model the blocking of the crane by other resources. For example, if a crane serves an AGV which arrives in a just-in-time fashion, the crane will not be available before the due date, that is, we have $e_j = d_j$. On the other hand, if a crane is to serve an external truck which has already arrived at the block, it will be available after completing the job, that is, we have $e_j = 0$.

2.5 Application to reefer workers

Reefer containers are used to carry goods that require a controlled temperature (such as refrigerated or frozen goods). They have a device for temperature regulation which requires electricity. Hence, they are stored in specific yard blocks or areas of blocks that provide electricity connections. Reefer containers require special handling by a manpower resource, the reefer workers.

The jobs carried out by the reefer workers are connecting arriving containers, disconnecting departing ones, doing small repair tasks, and controlling the temperature of the reefer containers in the stack. For each of these job types, a processing time p_j is given. Note that, unlike the previously described applications, here the processing time does not correspond to moving to another location. The use of the setup times, however, is similar to equipment scheduling. The setup time s_{ij} between a job i and a job j is the time that a reefer worker needs to move from the container associated with job i to the container related to job j. An estimate of this time can be assumed to be available. This estimate is based on the container positions in the stack.

For connection jobs, the due date d_j reflects the time that a reefer container is allowed to be without electricity. Of course, a container can keep its temperature in the allowed range only for a certain time if it is without electricity. For disconnection jobs, the due date reflects the latest time the container must be disconnected such that the stacking crane or straddle carrier can pick it up on time (the terminal control system might include a buffer time when computing the due date). For repair jobs and temperature control jobs, a due date is given as well. After a worker has completed a job, he can immediately move on to the next job. Therefore, we can set the earliest availability times to $e_j = 0$. A definition of precedence relations is not necessary.

In practice, the reefer workers are often equipped with portable radio data sets. When they have completed a job, they transmit this information to the terminal control system via the radio data set. In return, they get the next job.

3 Priority rule based heuristics

3.1 Single pass dispatching method

In this subsection, we present a simple dispatching heuristic for the container terminal scheduling problem. This method is a straightforward way to build job sequences for the resources. The following steps are repeated until all jobs have been scheduled:

- **Compute eligible jobs.** Compute the set E of eligible jobs as the set of the currently unscheduled jobs of which all predecessors have already been scheduled.
- **Select job.** Select the job j^* to be scheduled next as the eligible job with the smallest due date.
- **Select resource.** Select the resource r^* that leads to the smallest increase in the objective function, that is, r^* produces the smallest weighted sum of lateness and setup time for job j^* among all resources.
- **Update schedule.** Schedule job j^* at the end of the current job sequence of resource r^*.

The criteria for selecting a job and a resource can be seen as priority rules. According to the classification of Kolisch and Hartmann [27], this approach is a deterministic single-pass priority rule method. An alternative (and equally straightforward) rule for resource selection could be to select the resource with the earliest availability time in the current partial schedule. In Subsection 6.2, we carry out some computational experiments to compare both rules.

3.2 Multi pass sampling method

The single pass priority rule based method described above produces only one schedule. We now attempt to improve the results by allowing for multiple passes. In each pass, a schedule is constructed. In order to produce different schedules, we randomize the minimum due date priority rule in the job selection process. We obtain a multi-pass biased random sampling method (see Kolisch and Hartmann [27]). Hence, the sampling heuristic is essentially a repeated application of the single pass method in which the job selection mechnism is modified.

The job selection mechanism is adapted as follows. We define a parameter δ which determines how many of the jobs in the eligible set E are considered. Let E_δ denote the set of the δ eligible jobs with the smallest due dates in E. Now we select the eligible job $j^* \in E_\delta$ to be scheduled next on a biased random basis. The probability for eligible job $j \in E_\delta$ to be selected is given by

$$p(j) = \frac{d_{\max} - d_j + 1}{\sum_{i \in E_\delta}(d_{\max} - d_i + 1)},$$

where $d_{\max} = \max\{d_j \mid j \in E_\delta\}$ is the maximal due date. The resource selection mechanism is the same as in the single pass method. Once a schedule has been completed, the sampling heuristic proceeds to compute the next one. This is repeated until a time limit is reached, and the best schedule found is reported.

The priority rule based definition of the probabilities $p(j)$ implies that jobs with tighter due dates are more likely to be selected. This way, these jobs will be scheduled earlier, and the risk of exceeding their due dates is reduced. The parameter δ excludes jobs with non-tight due dates from consideration. Therefore, using small values for δ puts a focus on the jobs with the tightest due dates.

Two special cases of this sampling approach should be mentioned. $\delta \geq n$ induces conventional sampling in which the eligible set is not restricted. $\delta = 1$ implies that only the job with the smallest due date can be considered. Thus, in each pass, the same schedule is computed which is equal to the schedule found by the deterministic single pass approach.

4 Genetic algorithm

Genetic algorithms (see Goldberg [10], Holland [17]) adopt the principles of biological evolution to solve hard optimization problems. For our genetic algorithm (GA), we exploit the similarities of the general container terminal scheduling problem to the resource-constrained project scheduling problem. These similarities allow us to use the project scheduling GA of Hartmann [13] as a starting point. This GA will be adapted to solve the problem introduced in this paper. In particular, the representation, the decoding procedure, and the construction of the initial population require problem-specific knowledge.

4.1 Basic scheme

We apply the generational management framework of Eiben et al. [6]. The GA starts with the computation of an initial population, i.e., the first generation. The number of individuals in the population is referred to as POP. The GA then determines the fitness values of the individuals of the initial population. After that, we apply the crossover operator to produce new individuals ("children") from the existing ones ("parents"). Subsequently, we apply the mutation operator to the newly produced children. After computing the fitness of each child, we add the children to the current population. Then we apply the selection operator to reduce the population to its former size POP. Doing so, we obtain the next generation to which we again apply the crossover operator and so on. This process is repeated until a time limit is reached. Of course, other stopping criteria can be employed as well, e.g., a maximal number of generations or a number of generations without improvement of the best objective function value found so far. More formally, the GA scheme can be summarized as follows. Here, $\mathcal{P}$ denotes the current population (i.e., a set of individuals), and $\mathcal{C}$ is the set of children.

generate POP individuals for the initial population $\mathcal{P}$;
apply decoding procedure to compute fitness for individuals $I \in \mathcal{P}$;
while time limit is not reached **do**

begin
produce a set $\mathcal{C}$ of children from $\mathcal{P}$ by crossover;
apply mutation to children $I \in \mathcal{C}$;
apply decoding procedure to compute fitness for children $I \in \mathcal{C}$;
$\mathcal{P} := \mathcal{P} \cup \mathcal{C}$;
reduce population $\mathcal{P}$ to size POP by means of selection;
end.

4.2 Problem representation

Genetic algorithms for scheduling problems often do not operate directly on schedules but on representations of schedules. The latter are then transformed into schedules by means of a problem-specific decoding procedure. The advantage of such an indirect approach is that it allows to employ standard representations together with standard genetic operators (crossover, mutation).

In our genetic algorithm, a schedule is represented by a precedence feasible job list $(j_1, \ldots, j_n)$ and three additional genes β_L, β_S, and β_W, such that an individual I is given as

$$I = (\,(j_1, \ldots, j_n)\,, \beta_L, \beta_S, \beta_W\,).$$

In a job list, each job appears exactly once, that is, we have $J = \{j_1, \ldots, j_n\}$. A job list is precedence feasible if all predecessors of a job j appear in the list before job j, that is, $P_{j_i} \subseteq \{j_1, \ldots, j_{i-1}\}$ for $i = 1, \ldots, n$. A job list determines the order in which the jobs are scheduled by the decoding procedure. The job list representation is a generalization of the classical permutation based representation (see Reeves [36]). It has been shown to be superior to other representations for resource-constrained scheduling problems (see Hartmann [13, 14]).

While the representation controls the scheduling order, the task of assigning resources to jobs will be left to the decoding procedure. The remaining genes β_L, β_S, and β_W are parameters that are used by the decoding procedure when selecting a resource for a job (see the following subsection). Note that using algorithm parameters in the genetic representation implies that they are exposed to evolution and survival-of-the-fittest (see Goldberg [10]). This is often referred to as self-adaptation.

4.3 Decoding procedure and fitness

The decoding procedure employs problem-specific features to transform the representation into a solution for the container terminal scheduling problem. It scans the job list from left to right and successively schedules the jobs $j_1, \ldots, j_n$ using the following steps:

- **Select job.** Select the next job j_i from the job list.
- **Select resource.** Evaluate the partial schedules arising from assigning job j_i to resource $r \in R$ as last job by computing the resulting lateness $l_{j_i}(r)$, setup time $s_{j_i}(r)$, and waiting time $w_{j_i}(r)$ of job j_i. A waiting time occurs if a resource completes a job earlier than the earliest release time, i.e., the resource would have to wait until the earliest release time before the next job can be started. These three times are weighted with the related genes of the individual, leading to $y_{j_i}(r) = \beta_L \cdot l_{j_i}(r) + \beta_S \cdot s_{j_i}(r) + \beta_W \cdot w_{j_i}(r)$. Now select a resource r^* with the best evaluation $y_{j_i}(r^*) = \min\{y_{j_i}(r) \mid r \in R\}$.
- **Update schedule.** Schedule job j_i at the end of the current job sequence of resource r^*.

While the scheduling order of jobs is prescribed by the job list representation, the decoding procedure takes care of the resource assignment part of the problem. The usage of the weight genes β_L, β_S, and β_W for resource evaluation is based on two ideas. First, in the iterations of the decoding procedure, other settings than the overall objective function parameters might be useful. Second, they allow to penalize possible waiting times which might be worth avoiding during schedule computation because waiting times reduce the capacities that will be left for the jobs to be scheduled next. (Note, however, that considering waiting times in the overall objective function would not make much sense.) The fitness of an individual is defined as the objective function value of the schedule related to the individual.

4.4 Initial population

Let us now define how to determine the first generation containing POP individuals. For the construction of job lists, we employ the sampling strategy of Subsection 3.2. That is, in each step, we determine the restricted eligible set E_δ and draw a job $j \in E_\delta$ using the due date based probability $p(j)$. The selected job j is then added at the end of the job list. Next, we draw the genes β_L, β_S, and β_W using a parameter $\epsilon \geq 0$ which defines a range for the genes:

$$\beta_L \in [\max\{0, \alpha_L - \epsilon\}, \min\{\alpha_L + \epsilon, 1\}],$$

$$\beta_S \in [\max\{0, \alpha_S - \epsilon\}, \min\{\alpha_S + \epsilon, 1\}],$$

$$\beta_W \in [0, \epsilon].$$

Observe that the genes related to lateness and setup time are drawn from the ϵ-neighborhood of the respective objective function weights (for the waiting time, there is no related weight in the objective function). Having determined these three genes, their values are normalized such that we obtain $\beta_L + \beta_S + \beta_W = 1$, that is, we set $\beta_X := \frac{\beta_X}{\beta_L + \beta_S + \beta_W}$ for $X \in \{L, S, W\}$.

Considering the job list construction, one might as well think of the following straightforward approach: The latter selects jobs on a pure random basis without due date bias, that is, with equal probabilities $p(j) = \frac{1}{|E|}$, where E is the unrestricted eligible set. Using the sampling method instead should lead to more promising job lists while still maintaining reasonable genetic variety. The parameter ϵ controls the deviation of the weight genes from the objective function weights. Note that $\epsilon = 0$ reduces the resource evaluation to the respective objective function weights ($\beta_L = \alpha_L$ and $\beta_S = \alpha_S$) without considering waiting times ($\beta_W = 0$). The benefit of biased sampling and of the weight genes will be examined further in the computational tests of Subsection 6.3.

4.5 Crossover

For crossover, the current population is randomly partitioned into pairs of individuals. From each pair of individuals (parents), two new individuals (children) will be produced. Let us assume that two individuals of the current population have been selected for crossover. We have a mother individual $M = \left(\left(j_1^M, \ldots, j_n^M\right), \beta_L^M, \beta_S^M, \beta_W^M\right)$ and a father individual $F = \left(\left(j_1^F, \ldots, j_n^F\right), \beta_L^F, \beta_S^F, \beta_W^F\right)$. Now two child individuals have to be constructed, a daughter D and a son S.

Let us start with a definition of the daughter $D = \left(\left(j_1^D, \ldots, j_n^D\right), \beta_L^D, \beta_S^D, \beta_W^D\right)$. Combining the parent's job lists, we have to make sure that each job appears exactly once in the daughter's job list. We make use of a general crossover technique presented by Reeves [36] for permutation based genotypes. For the one-point crossover, we draw a random integer q with $1 \leq q \leq n$. The daughter's job list is determined by taking the job list of the positions $i = 1, \ldots, q$ from the mother, that is,

$$j_i^D := j_i^M .$$

The remaining positions $i = q+1, \ldots, n$ are derived from the father. However, the jobs that have already been selected must not be considered again. The remaining jobs are taken in their relative order in the father's job list, that is, we set for $i = q+1, \ldots, n$

$$j_i^D := j_k^F \quad \text{where} \quad k = \min\left\{1 \leq u \leq n \mid j_u^F \notin \left\{j_1^D, \ldots, j_{i-1}^D\right\}\right\}.$$

The three weight genes are taken from the mother, that is, we set

$$\beta_L^D := \beta_L^M, \quad \beta_S^D := \beta_S^M, \quad \beta_W^D := \beta_W^M.$$

The son individual is computed analogously. For the son's job list, the weight genes and the first part of the job list are taken from the father and the second part is taken from the mother.

In addition, we have tested a two-point crossover variant. Here, we draw two random integers q_1 and q_2 with $1 \leq q_1 < q_2 \leq n$. Analogously to the one-point variant, the weight genes and the first part of the job list until q_1 are taken from one parent. The second part of the job list between positions $q_1 + 1$ and q_2 is taken from the other parent, following the same logic as above. The third part of the list between $q_2 + 1$ and n is again taken from the first parent.

This crossover strategy creates job lists in which each job appears exactly once. Moreover, it has been proven by Hartmann [13] that the resulting job lists are precedence feasible, given that the parents' job lists were precedence feasible as well. Since this crossover operator produces feasible offspring, there is no need for a repair operator. This property leads to a good inheritance behavior of building blocks of solutions. Also note that taking the weight genes and the first part of the job list from the same parent implies that the first part of the related schedule is inherited. This means that good partial schedules can be passed on to the offspring without being destroyed by crossover.

4.6 Mutation

The mutation operator is applied to each newly produced child individual. The probability for each gene to be mutated is denoted as p_{mutation}.

In the first step, we consider the mutation of the job list for which two alternatives have been considered. The swap variant is defined as follows. We move through the job list from left to right. Consider a current position $i \in \{1, \ldots, n-1\}$ in the job list

$$(j_1, \ldots, j_i, j_{i+1}, \ldots, j_n).$$

If j_i is not a predecessor of j_{i+1}, we interchange these two successive jobs with probability p_{mutation}, which leads to a new job list

$$(j_1, \ldots, j_{i+1}, j_i, \ldots, j_n).$$

The shift variant can be described as follows. Again, we move through the job list from left to right. In this variant, however, we apply a right shift to each job with a probability of p_{mutation}. Consider a current position $i \in \{1, \ldots, n-1\}$ in the job list

$$(j_1, \ldots, j_i, \ldots, j_h, \ldots, j_z, \ldots, j_n).$$

Let z be the smallest index of the successors of job j_i, that is, $z = \min\{k \mid j_k \in S_{j_i}\}$. Now job j_i can be shifted behind some randomly drawn position $h \in \{i+1, \ldots, z-1\}$, which leads to job list

$$(j_1, \ldots, j_{i-1}, j_{i+1}, \ldots, j_h, j_i, j_{h+1}, \ldots, j_z, \ldots, j_n).$$

That is, job j_i is right shifted within the job list and inserted immediately after some job j_h. Clearly, the resulting job list is still precedence feasible.

The second step considers the three parameters for the decoding procedure. Using again parameter ϵ (see Subsect. 4.4), the operator randomly draws

$$\beta'_X \in [\max\{0, \beta_X - \epsilon\}, \min\{\beta_X + \epsilon, 1\}]$$

for $X \in \{L, S, W\}$. Again, each of these genes is mutated with probability p_{mutation}. Subsequently, the parameters are normalized as described in Subsection 4.4.

4.7 Selection

After the newly produced individuals have been added to the current population, the next step is to select the individuals that survive and make up the next generation. Following the study of Hartmann [13], we decided to employ the deterministic ranking method which follows a survival-of-the-fittest strategy (cf., e.g., Michalewicz [32]). This method sorts the individuals with respect to their fitness values and selects the POP best ones while the remaining ones are deleted from the population (ties are broken arbitrarily).

5 Generating experimental data

5.1 Generator

In order to demonstrate the applicability of the general scheduling model and the genetic algorithm, we carried out several computational experiments. These experiments required test instances as input data for the heuristic. In order to obtain a large number of test instances, we developed a data generator for our problem setting. It is controlled by parameters that allow to produce instance sets with specific characteristics. In particular, the parameters enabled us to produce quite realistic test sets for the container terminal applications mentioned in Section 2.

The generator works as follows. The number n of jobs and the number m of resources are specified by the user as parameters. In addition to the n regular jobs, the m dummy jobs that represent the initial setup states for the resources are generated. Each (non-dummy) job j is randomly assigned a processing time p_j from $\{p_{\min}, \ldots, p_{\max}\}$, where $p_{\min}$ and $p_{\max}$ are parameters that denote the minimal and maximal processing time, respectively. Next, we generate a setup time s_{ij} between two consecutive jobs i and j on the same resource. The following approach is designed to produce setup times which do not violate the triangle inequality. It employs two parameters S_1 and S_2, where S_1 can be interpreted as the minimal and $S_1 + S_2$ as the maximal setup time. First, each job j is randomly assigned a value $y_j \in \{0, \ldots, S_2\}$. Then the setup time between jobs i and j is defined as $s_{ij} = S_1 + |y_j - y_i|$. On the basis of a parameter T for the scheduling horizon, a due date d_j is drawn from $\{S_1 + p_j, \ldots, T\}$ for each job j. The earliest release times are determined using a parameter $\rho \in [0, 1]$. This parameter allows to control whether the earliest release time e_j of a job j is equal to its due date or zero (recall that these were the two typical cases in the applications of our model). With probability ρ, we set $e_j = 0$ and with probability $1 - \rho$, we set $e_j = d_j$.

5.2 Test sets

Using the generator described in the previous subsection, we generated a set of test instances for each of the four container terminal applications discussed in Section 2. That is, we have a straddle carrier set, an AGV set, a reefer worker set, and a stacking crane set. These four sets differ in the settings of the generator

Table 1. Characteristics of the four generated test sets

Test set	Number of jobs	Number of resources	Scheduling horizon
Straddle carrier	380	75	30 min
AGV	100	50	15 min
Reefer worker	120	5	60 min
Stacking crane	8	1	30 min

parameters and hence in their characteristics. We attempted to generate realistic instances which may similarly occur in practice at peak time on a medium-sized container terminal. The characteristics of the test sets can be summarized as follows (see also the main parameter settings displayed in Table 1):

- The straddle carrier set contains the largest instances. This is because we assume the straddle carriers to carry out transportation jobs both on the seaside and on the landside as well as shuffle moves. This set includes the largest number of resources and jobs and a medium scheduling horizon.
- The AGV case considers only transportation jobs between quai and stack. Therefore, we have selected smaller numbers of jobs and resources than for the straddle carrier set. Moreover, we have a shorter scheduling horizon because shuffle moves are not considered.
- In the reefer worker set, we have fewer resources than in the previous two cases but a longer scheduling horizon. The latter results from substantially longer time windows for reefer jobs.
- The stacking crane case is the smallest problem setting with only one resource (corresponding to one yard block served by a single crane). Due to shuffle moves and landside operations with less tight due dates, the horizon is the same as for the straddle carrier set.

For each of the test sets, the parameter settings were done in a way that leads to scarce resources. Scarce resources make it difficult to find schedules with small lateness—thus, we obtain challenging test problems. In fact, the parameters were adjusted such that we have an average lateness > 0 even for the best heuristic. This allows for a meaningful comparison of heuristics on the basis of the objective function values. Moreover, the case of scarce resources is particularly important in practice. In this case, minimization of lateness is crucial for successfully carrying out the logistic processes. For each of the four cases, 250 instances were generated. Hence, the test set consists of 1,000 instances altogether which should be a reasonable basis for a thorough computational analysis.

6 Computational results

6.1 Comparison of the heuristics

In order to evaluate the heuristics, they were coded in ANSI C and compiled with the lcc compiler. The experiments were carried out on a Pentium 4-m based computer with 1.6 GHz running under Windows XP. In order to consider the real-time

Table 2. Comparison of the heuristics (time limit: 1 s on Pentium 4 with 1.6 Ghz)

Problem	Heuristic	Objective	Lateness	Setup time	Late jobs
Straddle carrier	dispatching	31.0	27.3 s	63.9 s	27.5 %
	sampling	26.2	22.4 s	61.9 s	24.8 %
	GA	10.4	4.6 s	62.1 s	15.9 %
AGV	dispatching	22.8	17.6 s	69.2 s	20.5 %
	sampling	18.4	13.0 s	67.3 s	18.6 %
	GA	13.7	7.4 s	69.8 s	19.6 %
Reefer worker	dispatching	87.4	89.4 s	69.1 s	46.3 %
	sampling	13.1	9.5 s	45.5 s	10.3 %
	GA	8.8	5.8 s	34.9 s	7.4 %
Stacking crane	dispatching	30.1	28.1 s	47.6 s	30.1 %
	sampling	12.4	9.4 s	38.8 s	13.0 %
	GA	12.4	9.5 s	38.4 s	13.1 %

application of the scheduling model which requires very short computation times, a time limit of one second has been selected. Recall that rescheduling should be done after a resource has completed its last job and before it is assigned its next job, such that the computation time is waiting time for the resource.

The weights of the objective function have been set to $\alpha_L = 0.9$ and $\alpha_S = 0.1$ which should be a reasonable choice in practice. Meeting a due date is considered to be more important than setup time minimization (also, minimizing lateness will probably lead to small setup times). In case of small workload and thus plenty resource capacities, the lateness criterion becomes less critical and the setup time minimization plays a more important role.

Table 2 gives the results for the single pass dispatching method of Subsection 3.1, the sampling heuristic of Subsection 3.2, and the genetic algorithm of Section 4. The results are given separately for each of the four problem sets of Table 1. For each heuristic, we report the average objective function value together with the average lateness per job and the average setup time per job (both in seconds). The average percentage of late jobs is also displayed.

For the straddle carrier, AGV, and reefer worker cases, the sampling method performs better than the single pass approach while the GA outperforms both. In the stacking crane case, the sampling method and the GA produce similar results which is due to the small instance size. Also recall that we have an average lateness > 0 due to the design of the test instances. Considering the straddle carrier and the AGV cases, the GA reduces the lateness while all methods lead to similar setup times. In the reefer worker case, the GA reduces both the lateness and the setup times. Note that the long scheduling horizon in the reefer case leads to a greater optimization potential than in the other cases. The GA appears to be best suited for exploiting this optimization potential. Assuming that the test instances are quite realistic, we can state that these results suggest to apply the GA in practice.

Furthermore, recall that the GA uses priority rules to compute initial solutions and then proceeds with evolutionary inheritance. This leads to better results than

Table 3. Impact of priority rules for resource selection (deterministic single pass dispatching)

Problem	Priority rule	Objective	Lateness	Setup time	Late jobs
Straddle carrier	objective increase	31.0	27.3 s	63.9 s	27.5 %
	earliest available resource	360.5	375.4 s	226.4 s	86.3 %
AGV	objective increase	22.8	17.6 s	69.2 s	20.5 %
	earliest available resource	153.1	145.1 s	224.9 s	68.9 %
Reefer worker	objective increase	87.4	89.4 s	69.1 s	46.3 %
	earliest available resource	1367.5	1497.0 s	201.6 s	98.3 %

employing these priority rules over the entire computation time, even if the computation time is rather short. This observation is in line with studies on other scheduling problems (see [16]), but it should be noted that such a superior performance of the evolutionary inheritance mechanism is only possible with an appropriate genetic representation (see also [13]).

The following subsections deal with the configuration of the heuristics and with the effect of the algorithm parameters. The stacking crane test set will not be considered because the instances are too small for a meaningful analysis.

6.2 Configuration of the priority rule methods

We start the analysis with a comparison of the priority rules for resource selection in the deterministic single pass heuristic of Subsection 3.1. The results given in Table 3 show that the rule based on the increase in the objective function value leads to much smaller lateness and setup times than the alternative rule. This demonstrates that a seemingly reasonable rule as the earliest availability time criterion may in fact produce clearly inferior results. Due to this result, the objective increase rule is the standard rule in the following experiments with the sampling method (note that it has also been used to produce the results of Table 2).

Next, we examine the sampling approach in more detail. Again, a time limit of one second is used. We analyze the impact of parameter δ which determines the number of eligible jobs considered for sampling (cf. Subsect. 3.2). Table 4 displays the sampling results for different settings of δ. A value of $\delta = 10$ appears to be a good overall choice. Thus, it was selected as the standard setting which was used to produce the sampling results of Table 2. Also observe that this sampling approach is better than conventional sampling ($\delta = n$) and deterministic single pass scheduling ($\delta = 1$). Hence, the restriction of the eligible set for sampling is a good strategy for this problem setting.

6.3 Configuration of the genetic algorithm

In this subsection, we study the impact of the genetic algorithm parameters and provide the best configuration. Table 5 gives the computational results for various parameter settings. It is divided into three blocks corresponding to the three problem

Table 4. Configuration of the sampling heuristic (time limit: 1 s on Pentium 4 with 1.6 Ghz)

Problem	δ	Objective	Lateness	Setup time	Late jobs
Straddle carrier	n	133.7	141.5 s	63.6 s	43.5 %
	20	26.2	22.4 s	62.0 s	24.3 %
	10	26.2	22.4 s	61.9 s	24.8 %
	5	27.0	23.1 s	62.2 s	25.2 %
	1	31.0	27.3 s	63.9 s	27.5 %
AGV	n	28.3	23.9 s	68.1 s	23.4 %
	20	18.2	12.8 s	67.1 s	18.4 %
	10	18.4	13.0 s	67.3 s	18.6 %
	5	19.4	14.0 s	67.5 s	19.0 %
	1	22.8	17.6 s	69.2 s	20.5 %
Reefer worker	n	132.9	141.9 s	52.2 s	29.2 %
	20	21.0	18.1 s	47.4 s	12.9 %
	10	13.1	9.5 s	45.5 s	10.3 %
	5	13.3	9.7 s	45.4 s	11.7 %
	1	87.4	89.4 s	69.1 s	46.3 %

sets of straddle carriers, AGVs, and reefer workers. We examine the population size POP, the method to construct an initial population (simple random or priority based sampling), the number of crossover points C (1 or 2), the mutation rate p_{mut} (between 0.01 and 0.1), the mutation operator (swap or shift), and the range ϵ for the decoding procedure genes (between 0 and 1). The first row for each problem case provides the best parameter settings (this is the standard setting that was used to produce the of Table 2). The following rows of each block contain parameter variations which are underlined. Note that the values for the population size POP have been chosen with respect to the actual problem size (the more jobs and resources we have, the longer is the computation time for one schedule, which makes smaller populations within the same time limit reasonable). The standard settings for the other parameters are the same for all problem sets.

As shown in Table 5, some parameters have a significant impact on the GA results. First, the initial population should be produced by the sampling approach instead of a straightforward random assignment. The impact of sampling is stronger for longer scheduling horizons (straddle carrier and reefer worker cases). For long horizons, random job lists are likely to have jobs with early due dates in a position at the end of the list, which means that they are considerably late when they are scheduled. Second, the mutation operator should employ swaps rather than shifts. Shifting a job by many positions bears the risk that it will be scheduled much too late to meet its due date. Such a negative effect of the shift operator occurs particularly for the long horizon associated with the reefer worker case. Third, a range of $\epsilon > 0$ for the decoding procedure genes has a positive effect (recall that this means that resources are selected using adapted weights instead of the original weights of the objective function). This holds particularly for the straddle carrier and AGV cases where it pays to avoid waiting times which occur if a resource completes

Table 5. Configuration of the genetic algorithm (time limit: 1 s on Pentium 4 with 1.6 Ghz)

Problem	POP	Initial	C	p_{mut}	Mutation	ϵ	Objective	Lateness	Setup	Late jobs
Straddle	40	sampling	2	0.05	swap	0.5	10.4	4.6 s	62.1 s	15.9 %
carrier	20	sampling	2	0.05	swap	0.5	10.4	4.6 s	62.6 s	16.0 %
	60	sampling	2	0.05	swap	0.5	10.4	4.6 s	62.4 s	16.1 %
	40	random	2	0.05	swap	0.5	104.9	109.1 s	67.0 s	37.6 %
	40	sampling	1	0.05	swap	0.5	10.5	4.7 s	62.3 s	16.0 %
	40	sampling	2	0.01	swap	0.5	10.4	4.6 s	62.2 s	16.1 %
	40	sampling	2	0.10	swap	0.5	10.4	4.6 s	62.3 s	16.0 %
	40	sampling	2	0.05	shift	0.5	11.0	5.3 s	62.7 s	16.7 %
	40	sampling	2	0.05	swap	0.0	25.2	21.2 s	61.5 s	24.0 %
	40	sampling	2	0.05	swap	1.0	10.6	4.8 s	62.4 s	16.9 %
AGV	60	sampling	2	0.05	swap	0.5	13.7	7.4 s	69.8 s	19.6 %
	40	sampling	2	0.05	swap	0.5	13.7	7.5 s	69.9 s	19.7 %
	80	sampling	2	0.05	swap	0.5	13.7	7.6 s	69.7 s	19.7 %
	60	random	2	0.05	swap	0.5	19.3	13.1 s	75.5 s	26.3 %
	60	sampling	1	0.05	swap	0.5	13.7	7.4 s	69.8 s	19.7 %
	60	sampling	2	0.01	swap	0.5	13.8	7.5 s	70.0 s	19.7 %
	60	sampling	2	0.10	swap	0.5	13.7	7.4 s	70.0 s	19.7 %
	60	sampling	2	0.05	shift	0.5	13.7	7.4 s	69.9 s	19.8 %
	60	sampling	2	0.05	swap	0.0	16.8	11.3 s	66.1 s	19.6 %
	60	sampling	2	0.05	swap	1.0	14.1	7.8 s	69.8 s	19.7 %
Reefer	120	sampling	2	0.05	swap	0.5	8.8	5.8 s	34.9 s	7.4 %
worker	80	sampling	2	0.05	swap	0.5	8.8	6.0 s	34.6 s	7.6 %
	160	sampling	2	0.05	swap	0.5	8.8	5.9 s	35.4 s	7.6 %
	120	random	2	0.05	swap	0.5	137.5	148.5 s	38.4 s	28.6 %
	120	sampling	1	0.05	swap	0.5	9.2	6.3 s	35.6 s	7.7 %
	120	sampling	2	0.01	swap	0.5	9.1	6.2 s	34.9 s	7.6 %
	120	sampling	2	0.10	swap	0.5	8.8	5.8 s	35.2 s	7.4 %
	120	sampling	2	0.05	shift	0.5	11.4	8.0 s	42.1 s	9.1 %
	120	sampling	2	0.05	swap	0.0	8.9	6.1 s	34.6 s	7.4 %
	120	sampling	2	0.05	swap	1.0	8.9	6.0 s	35.2 s	7.5 %

a job before its earliest release time. Note that in the reefer worker case waiting times cannot occur because there are no earliest release times to be considered. The remaining parameters have only a very small impact. This means that the GA is robust in the sense that its performance does not deteriorate if those parameters are chosen according to a rule of thumb.

6.4 Further results

For the sampling method and the GA, Table 6 reports the average number of schedules that is computed within the time limit of one second (on a Pentium 4 with

Table 6. Average number of schedules computed within the time limit (1 s)

	Sampling	GA		
Problem	#Schedules	#Schedules	#Generations	Population size
Straddle carrier	321	320	8	40
AGV	1903	2047	35	60
Reefer worker	5172	7524	63	120

1.6 Ghz). For the GA, also the number of generations and the population size are given. Generally speaking, the GA appears to produce a sufficient number of schedules within the short time limit to allow for a successful evolution. This holds in particular since GAs with good initial solutions require less iterations to produce near-optimal solutions than GAs which start from random solutions (recall that our GA employs the priority rule based sampling method to compute the initial generation).

Furthermore, in both heuristics, the computation time for one schedule increases with the number of jobs and the number of resources. Hence, the number of schedules computed within the time limit depends of the problem set. It is also interesting to note that the number of computed schedules is different for the two approaches. The GA includes additional effort for crossover and selection. On the other hand, the job selection is faster in the GA where the next job is simply picked from the chromosome. In the sampling method, the eligible jobs have to computed, and a randomized selection mechanism is applied. Therefore, if the number of resources is small (as in the reefer worker case), the job selection makes up for a large part of the computational effort in the sampling heuristic. Thus, the GA produces more schedules within the same time limit.

7 Conclusions and research perspectives

In this paper, we proposed a general optimization model for scheduling jobs at container terminals. We showed that our model is applicable to straddle carriers, automated guided vehicles, stacking cranes, and reefer workers. The generality of our model is advantageous in practice because it allows to use the same model and optimization algorithms for several different scheduling problems. Furthermore, we developed priority rule heuristics and a genetic algorithm to solve the proposed problem. With a tailored instance generator, we generated several large sets of test instances for a computational analysis. The experiments showed that the genetic algorithm leads to better results than the priority rule methods. It appears to be well suited to solve test instances of realistic size within very short computation times. This makes it applicable to online scheduling within a terminal control system.

An important topic for future research is the integration of the optimization approach proposed in this paper into a simulation model. Simulation models cover the behavior of the equipment types as well as the control and optimization strategies. Providing a dynamic environment, they allow to capture the logistic processes in a realistic way. Simulation models have various applications ranging from the assessment of equipment capacities and layout alternatives to tests of optimization components. In particular, they provide a more realistic test bed for scheduling approaches than an offline study like the one that was carried out in this paper. On container terminals, the typical overall objectives are to maximize the productivity (i.e., the number of containers handled per hour) and to minimize the ships' times in port. Such objectives are not applicable to equipment scheduling. Therefore, an objective like the minimization of lateness and setup times is often chosen for scheduling. In order to analyze whether the scheduling objective really leads to good overall productivity and short times in port, a simulation model is needed. In addition, simulation models allow to examine various other points such as the impact of inaccurate estimates of processing times or delays (and resulting updates of due dates or availability times) as well as the coordination of different resource types. Thus, the scheduling approach proposed in this paper can only be fully evaluated by integrating it into a simulation model. While the experiments presented here showed that the GA produces good results in terms of the scheduling objective, the next step would be to examine whether the scheduling objective leads to a good terminal productivity. Note, however, that a simulation study always depends on the container terminal under consideration since it captures the actual equipment and the design of the terminal control system. Therefore, a simulation model cannot be used to evaluate a scheduling approach in a way that leads to general results (i.e., results which hold for any terminal configuration). The scheduling model suggested in this paper has been successfully tested in a simulation study for a real-world container terminal (for reasons of confidence, however, details on that study cannot be given).

References

1. Bae J W, Kim K H (2000) A pooled dispatching strategy for automated guided vehicles in port container terminals. International Journal of Management Science 6: 47–70
2. Bartusch M, Möhring R H, Radermacher F J (1988) Scheduling project networks with resource constraints and time windows. Annals of Operations Research 16: 201–240
3. Böse J, Reiners T, Steenken D, Voß S (2000) Vehicle dispatching at seaport container terminals using evolutionary algorithms. In: Sprague R H (ed) Proceedings of the 33rd Annual Hawaii International Conference on System Sciences, pp 377–388. IEEE, Piscataway
4. Daganzo C F (1989) The crane scheduling problem. Transportation Research B 23: 159–175
5. de Castilho B, Daganzo C F (1993) Handling strategies for import containers at marine terminals. Transportation Research B 27: 151–166
6. Eiben A E, Aarts E H L, van Hee K M (1990) Global convergence of genetic algorithms: a markov chain analysis. Lecture Notes in Computer Science 496: 4–12
7. Elmaghraby S E (1977) Activity networks: project planning and control by network models. Wiley, New York
8. Gambardella J M, Bontempi G, Taillard E, Romanengo D, Raso G, Piermari P (1996) Simulation and forecasting of an intermodal container terminal. In: Bruzzone A G, Kerckhoffs E J H (eds) Simulation in industry – Proceedings of the 8th European Simulation Symposium, pp 626–630. SCS, Ghent, Belgium

9. Gambardella L M, Rizzoli A E, Zaffalon M (1998) Simulation and planning of an intermodal container terminal. Simulation 21: 107–116
10. Goldberg D E (1989) Genetic algorithms in search, optimization, and machine learning. Addison-Wesley, Reading, MA
11. Grunow M, Günther H O, Lehmann M (2004) Dispatching multi-load AGVs in highly automated seaport container terminals. OR Spectrum 26: to appear
12. Guan Y, Cheung R K (2004) The berth allocation problem: models and solution methods. OR Spectrum 26: 75–92
13. Hartmann S (1998) A competitive genetic algorithm for resource-constrained project scheduling. Naval Research Logistics 45: 733–750
14. Hartmann S (2001) Project scheduling with multiple modes: a genetic algorithm. Annals of Operations Research 102: 111–135
15. Hartmann S (2004) Generating scenarios for simulation and optimization of container terminal logistics. OR Spectrum 26: to appear
16. Hartmann S, Kolisch R (2000) Experimental evaluation of state-of-the-art heuristics for the resource-constrained project scheduling problem. European Journal of Operational Research 127: 394–407
17. Holland H J (1975) Adaptation in natural and artificial systems. University of Michigan Press, Ann Arbor
18. Imai A, Nagaiwa K, Tat C W (1997) Efficient planning of berth allocation for container terminals in Asia. Journal of Advanced Transportation 31: 75–94
19. Imai A, Nishimura E, Papadimitriou S (2001) The dynamic berth allocation problem for a container port. Transportation Research B 35: 401–417
20. Jordan C (1996) Batching and scheduling – models and methods for several problem classes. Lecture Notes in Economics and Mathematical Systems, No. 437. Springer, Berlin Heidelberg New York
Grun
21. Kapuscinska T T, Morton T E, Ramnath P (1998) High-intensity heuristics to minimize weighted tardiness in resource-constrained multiple dependent project scheduling. Technical report, Carnegie Mellon University, Pittsburgh, PA
22. Kim K H, Kang J S, Ryu K R (2004) A beam search algorithm for the load sequencing of outbound containers in port container terminals. OR Spectrum 26: 93–116
23. Kim K H, Kim H B (1999) Segregating space allocation models for container inventories in port container terminals. International Journal of Production Economics 59: 415–423
24. Kim K H, Park Y M, Ryu K R (2000) Deriving decision rules to locate export containers in container yards. European Journal of Operational Research 124: 89–101
25. Kim K H, Won S H, Lim J K, Takahashi T (2001) A simulation-based test-bed for a control software in automated container terminals. In: Proceedings of the International Conference on Computers and Industrial Engineering, pp 239–243. Montreal, Canada
26. Kim K Y, Kim K H (1999) A routing algorithm for a single straddle carrier to load export containers onto a container ship. International Journal of Production Economics 59: 425–433
27. Kolisch R, Hartmann S (1999) Heuristic algorithms for solving the resource-constrained project scheduling problem: classification and computational analysis. In: Weglarz J (ed) Project scheduling: recent models, algorithms and applications, pp 147–178. Kluwer, Amsterdam
28. Legato P, Mazza R M (2001) Berth planning and resources optimisation at a container terminal via discrete event simulation. European Journal of Operational Research 133: 537–547
29. Lim A (1998) The berth planning problem. Operations Research Letters 22: 105–110

30. Meersmans P J M, Dekker R (2001). Operations research supports container handling. Technical Report EI 2001-22, Econometric Institute, Erasmus University Rotterdam
31. Meersmans P J M, Wagelmans A P M (2001) Effective algorithms for integrated scheduling of handling equipment at automated container terminals. Technical Report EI 2001-19, Econometric Institute, Erasmus University Rotterdam
32. Michalewicz Z (1995) Heuristic methods for evolutionary computation techniques. Journal of Heuristics 1: 177–206
33. Park Y M, Kim K H (2003) A scheduling method for berth and quai cranes. OR Spectrum 25: 1–23
34. Peterkofsky R I, Daganzo C F (1990) A branch and bound solution method for the crane scheduling problem. Transportation Research B 24: 159–172
35. Pritsker A A B, Watters L J, Wolfe P M (1969) Multiproject scheduling with limited resources: a zero-one programming approach. Management Science 16: 93–107
36. Reeves C R (1995) Genetic algorithms and combinatorial optimization. In: Rayward-Smith V J (ed) Applications of modern heuristic methods, pp 111–125. Alfred Waller, Henley-on-Thames
37. Steenken D, Henning A, Freigang S, Voß S (1993) Routing of straddle carriers at a container terminal with the special aspect of internal moves. OR Spectrum 15: 167–172
38. Talbot F B (1982) Resource-constrained project scheduling with time-resource tradeoffs: The nonpreemptive case. Management Science 28: 1197–1210
39. Taleb-Ibrahimi M, de Castilho B, Daganzo C F (1993) Storage space vs. handling work in container terminals. Transportation Research B 27: 13–32
40. Vis I F A, de Koster R (2003) Transshipment of containers at a container terminal: an overview. European Journal of Operational Research 147: 1–16
41. Vis I F A, Harika I (2004) A comparison of vehicle types for automated container terminals. OR Spectrum 26: 117–143
42. Yun W Y, Choi Y S (1999) A simulation model for container-terminal operation analysis using an object-oriented approach. International Journal of Production Economics 59: 221–230
43. Zhang C, Liu J, Wan Y W, Murty K G, Linn R J (2001) Storage space allocation in container terminals. Technical report, Hong Kong University of Science and Technology
44. Zhang C, Wan Y W, Liu J, Linn R J (2002) Dynamic crane deployment in container storage yards. Transportation Research B 36: 537–555

Dispatching multi-load AGVs in highly automated seaport container terminals

Martin Grunow, Hans-Otto Günther, and Matthias Lehmann

Department of Production Management, Technical University Berlin,
Wilmersdorfer Straße 148, 10585 Berlin, Germany
(e-mail: {m.grunow;ho.guenther;m.lehmann}@pm-berlin.net)

Abstract. This paper is concerned with AGV dispatching in seaport container terminals. Special attention is given to multi-load vehicles which can carry more than one container at a time. The characteristics of this complex application environment and the impact on the AGV dispatching problem are analyzed and various solution techniques considered. For practical application within an online logistics control system, a flexible priority rule based approach is developed, making use of an extended concept of the availability of vehicles. For evaluation reasons, this approach is complemented by an alternative MILP formulation. Finally, the performance of the priority rule based approach and the MILP model are analysed for different scenarios with respect to total lateness of the AGVs. The main focus of the numerical investigation is on evaluating the priority rule based approach for single and dual-load vehicles as well as comparing its performance against the MILP modelling approach.

Keywords: Multi-load AGVs – Dispatching – Container terminals – Online control

1 Introduction

Since the 70s automated guided vehicle (AGV) systems have been employed in many industrial environments. The use of AGVs in container ports, however, is a rather recent development and therefore far less investigated. In this paper, we consider a highly automated seaport container terminal, which consists of a berthing, an AGV, and a storage area. Figure 1 illustrates the layout of one of the latest seaport container terminals. The berthing area is equipped with quay cranes for the loading and unloading of vessels. When a vessel arrives at the port, it has already been determined at which position the vessel is berthed and which quay cranes will be working on the vessel. Equally, the unloading sequence of the containers is known

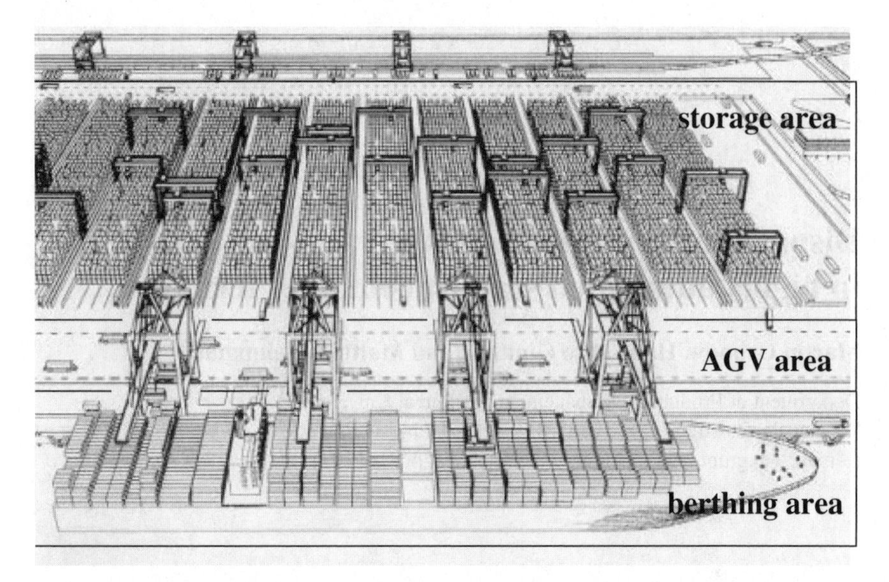

Fig. 1. Layout of a seaport container terminal[1]

in advance for each vessel. Thus, detailed schedules for the quay cranes can be derived from the given unloading sequence. At the same time, the final destination in the storage area is determined for each container. The storage area is divided into blocks each of which is serviced by one or more stacking cranes. After unloading a container, the stacking cranes at the affected block are scheduled to meet the estimated arrival time of the container. The transport of the containers from the berthing area to the storage yard is realized by AGVs, which can carry up to two 20ft containers or alternatively one 40 or 45ft container at a time. In the container terminal considered, AGVs are operated in single-carrier mode, but shall be used as multi-load carriers in the future. The particular difficulty of AGV dispatching in the container terminal environment considered is that AGV pick-up and delivery times for each container have to coincide with the schedules of the quay and stacking cranes to avoid idle times of this equipment and to guarantee short berthing of the vessels. The operations necessary to load a vessel are similar.

AGV dispatching usually consists of three sub-problems, namely assigning AGVs to transportation orders, routing the AGVs, and traffic control. Generally, algorithms for routing and traffic control are already included in the control software provided by the AGV manufacturer. Thus, only the assignment problem is investigated in this paper.

There are several characteristics that distinguish this problem from the assignment problem commonly found in manufacturing systems. First of all, operational conditions typically found in container ports involve numerous strongly stochastic elements, e.g., weather conditions, demolition status of the containers, and manual operations which influence the processing times. The planning horizon for the as-

[1] Source: http://www.hhla.de/C/cont.htm (05.03.2003).

signment problem therefore only covers the immediately upcoming events which may be forecasted with sufficient precision. The lack of buffers between different types of equipment units combined with the necessity of short berthing times lead to rigid constraints with respect to the pick-up and drop-off times for each transportation order which significantly increase the problem complexity. Finally, the enormous spatial dimensions of container ports with hundreds of equipments units are unique characteristics. Despite the complexity of the problem and the size of real-life problem instances, the logistics control software must allow for real-time reactions to the frequent changes in the container port environment.

In the case of single-load carriers, AGV dispatching can be reduced to the classical $m : n$ assignment problem with the objective of minimizing the costs associated with not meeting the target pick-up and delivery times. The corresponding linear optimization model can be solved very efficiently due to its pure binary nature. However, in the case of multi-load carriers the assignment problem is significantly more complex. In addition to the basic order-vehicle-assignment, the various pick-up and delivery operations have to be sequenced for each one AGV.

The remainder of this article is organized as follows. In the next section we give an overview of the relevant literature. In Section 3, a detailed analysis of the dispatching problem under consideration is given. A priority rule based algorithm for the dispatching of multi-load vehicles is presented in Section 4. This approach is complemented by an alternative MILP model formulation in Section 5. Finally, the performance of both approaches is compared in Section 6.

2 Relevant literature

Vehicle routing and scheduling problems have been studied extensively during the past two decades. Laporte and Osman (1995) provide a comprehensive literature overview of routing problems in general. The routing of vehicles is reviewed in Savelsbergh and Sol (1995). In their review, the authors discuss the general pick-up and delivery problem (GPDP) as well as its most important modifications, the vehicle routing problem (VRP), the pick-up and delivery problem (PDP) and the dial-a-ride problem (DARP). A recent review of real-time vehicle routing can be found in Ghiani et al. (2003).

Among the three special cases of the GPDP, the VRP has probably received the greatest attention due to its broad applicability. Laporte (1992) and Gambardella (2000) review different heuristic and exact solution techniques for this problem. A recent comparison of 10 typical heuristics for the VRP with and without side-constraints is given in Van Breedam (2002). The author provides a classification of solution approaches to the VRP, distinguishing between route-building methods, two-phase solution techniques, incomplete optimization algorithms, and improvement methods. Whereas route-building methods and two-phase solution techniques have received less attention, a variety of approaches has been proposed for the two latter types.

As a universal modelling tool, mixed-integer linear programming (MILP) has always been an obvious choice to model vehicle routing problems. MILP models have been solved using branch-and-cut or branch-and-bound procedures, as in Laporte et al. (1986), Achuthan et al. (1996) and Mingozzi et al. (1999) for the VRP or in Kohl et al. (1999) and Bard et al. (2002) for the VRP with time windows (VRPTW). The predominant numerical search method applied is tabu search, as in Golden et al. (1997) and Angelelli and Speranza (2002) for the VRP or in Gendreau et al. (1994) for the VRPTW. Other authors have proposed simulated annealing (e.g. Van Breedam, 1995), genetic algorithms (e.g. Hwang, 2002), or an extension of the well-known savings procedure (e.g. Fleischmann, 1998).

In contrast to the VRP, the PDP explicitly considers pick-up and delivery operations. Solution approaches for the dynamic and static PDP with several side-constraints are reviewed in Savelsbergh and Sol (1995). Many investigations on the PDP in a job shop environment have been performed. In this context, PDP is typically characterized as the vehicle dispatching problem. To solve the problem of AGV dispatching for single-load vehicles a variety of priority rule based approaches has been developed (see, e.g., Egbelu and Tanchoco, 1984; Lau and Liang, 2001; Nayyar and Khator, 1993; Schrecker, 2000). Moreover, extensions have been proposed for multi-load carriers (see, e.g., Bilge and Tanchoco, 1997; Lee et al., 1996; Sinriech and Kotlarski; 2002). Wallace (2001) developed a multi-agent system to control AGVs. Recently, the tabu search approach of Carlton and Barnes (1996) for the TSP with time windows has received much attention. Nanry and Barnes (2000) extend this approach to solve the PDP with time windows (PDPTW). Lau and Liang (2001) propose a further extension for the same problem. Another promising research direction in AGV dispatching is the application of bidding concepts (e.g., Hwang and Kim, 1998; Lim et al., 2003).

The DARP is a special case of the PDP, considering only single-unit orders. This additional feature permits the application of efficient dynamic programming approaches, as proposed in Desrosiers et al. (1989) and Psaraftis (1980). In Psaraftis (1986), the grouping-clustering-routing (GCR) method and the advanced dial-a-ride algorithm with time windows (ADARTW) introduced by Jaw et al. (1986) are compared for the DARP with time windows. Moreover, local search has been used by Healy and Moll (1995) to solve the DARP.

The main issue of vehicle scheduling problems (VSP) is to determine the sequence in which given trips are performed by a vehicle. In contrast to vehicle routing, different transportation orders are not allowed to be processed concurrently. For solution approaches to the VSP, see, e.g., Fischetti (2001).

As opposed to the huge interest in vehicle dispatching in a manufacturing environment (e.g., job shops and flexible manufacturing systems), vehicle dispatching in sea port container terminals has received less attention. So far AGVs are only used in a few highly automated terminals. In practice there is a growing interest in replacing manually driven carts by automatic unmanned carriers, since this approach offers the benefit of increased availability and lower operating costs. In the academic literature, however, very few papers have been published dealing with AGV systems in a container port environment (e.g., Evers and Koppers, 1996; Bae and Kim, 2000). Another recent application area is the use of AGVs in underground transportation systems (e.g., van der Heijden et al., 2002).

In this paper, AGV dispatching for multi-load carriers in seaport container terminals is considered. This problem is similar to the PDPTW and the DARPTW, but does not require disjoint vehicle routes. Furthermore, AGVs may differ with respect to their initial load and their location. These additional issues are of considerable importance and, therefore, must be explicitly taken into account by the dispatching algorithm.

3 Problem statement

3.1 Logistics control at a highly automated container terminal

In order to achieve a short berthing of the vessels, the complete handling chain consisting of quay cranes, stacking cranes and AGVs has to be optimized. The determination of an exact global optimum solution to this extremely complex logistics optimization problem is impractical even under the assumption of a deterministic application environment. Hence, a decentralized control scheme is required. Major constituents of an adequate logistics control architecture are modules which optimize the operations of the different types of handling equipment, i.e., quay cranes, stacking cranes and AGVs. While the scheduling of quay and stacking cranes has been the subject of numerous investigations (e.g., Kim and Kim, 1999; Chung et al., 2002; Park and Kim, 2003; Kim and Park, 2003), efficient approaches for dispatching multi-load AGVs are lacking. At the core of each logistics control system, a module for coordinating the different equipment units within the handling chain is needed.

In principle, there are two possible modes of coordination, which are determined by the organizational layout of the container terminal and the temporal sequence of the handling operations (see Fig. 2 a, b). In the sequential mode depicted in Figure 2 (a) the target times of each equipment unit in the handling chain are determined on the basis of the target times of its predecessor. Since the overall objective usually is to minimize the berthing time of a vessel, the quay cranes serve as a trigger in the entire handling chain. This is true not only for unloading but also for loading a vessel. In the latter case the respective target times of the AGV and the stacking crane are derived from those of the quay crane in a backward manner. These coordination principles lead to very small idle times of the quay crane, but some idle time may occur at the stacking crane, being the last link of the chain.

However, an efficient usage of the stacking cranes is vital for the overall productivity of a container port. The coordination is therefore organized in a triangular mode as depicted in Figure 2 (b) for both the loading and unloading of a vessel. When a container is loaded onto a ship, an AGV carrying the required container has to be positioned underneath the quay crane. The corresponding point in time, which has been derived from the loading sequence of the vessel, is propagated to the software module coordinating the AGV operations (drop-off time of the AGV's transportation order). Also, the container has to be supplied by one of the stacking cranes serving the block where it is stored. Based on the estimated transportation

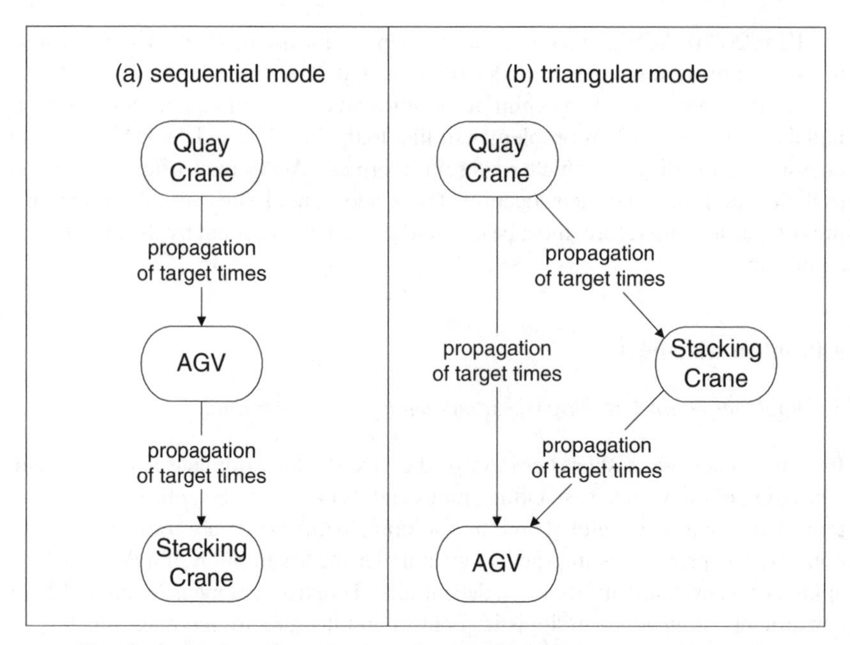

Fig. 2 a,b. Modes of coordination of equipment units in the handling chain

time, the approximate point in time at which this operation should be carried out is passed on to the controller of the stacking cranes. This controller determines the schedule of the stacking cranes and transmits the precise pick-up time to the AGV controller which based on both, pick-up and drop-off time, assigns the transportation order to an AGV. The unloading of containers is coordinated in a similar fashion.

Compared to the sequential mode (Fig. 2 a) in which the target times are propagated along the handling chain, the triangular mode (Fig. 2 a) significantly reduces the scheduling flexibility of the AGV controller. In this mode both, pick-up and delivery times, are predetermined input data which serve as constraints in the assignment of transportation orders to AGVs.

It should be noted that for the calculation of the estimated arrival time at the storage yard no individual AGV is considered. Instead, the distance between the corresponding locations of the quay crane and the storage block as well as average traffic conditions are considered to predict the real travel time. As it is impossible to exactly meet the target times of both, quay and stacking cranes, for the great number of transportation orders, it seems reasonable to soften the hard arrival time constraints for the AGVs and to introduce penalty costs for late arrivals. In the sequel, we focus on determining an efficient solution to the AGV dispatching problem for multi-load carriers, given the AGV target times induced by the schedules of quay and stacking cranes.

3.2 Event driven dispatching

Due to the rapidly changing and non-deterministic environment, re-scheduling has to be performed in real time. Moreover, the scheduling system must provide sufficient flexibility to react on frequent changes arising during the terminal operation. In order to reduce system nervousness and scheduling complexity, the dynamic scheduling problem is usually solved by repeatedly updating the assignment between transportation orders and vehicles. For this approach, it is important to define on which condition a new instance of the static scheduling problem has to be solved. Such a condition could be the expiry of a given time interval or the occurrence of a triggering event. The second approach requires more differentiation and is therefore more complex. Nonetheless, in the frequently changing container port environment it seems to be more adequate because of its higher flexibility.

The concept of event driven dispatching of AGVs has been introduced by Egbelu and Tanchoco (1984). A new dispatching request can be invoked, if a vehicle completes its current mission (vehicle-initiated dispatching) or a new transportation order emerges (order-initiated dispatching). In the former case, the vehicle, which is set idle, has to be assigned to a new order from the pool of waiting transportation orders or to be parked, if the pool is empty. On the other hand, if a new order emerges, an available vehicle has to be assigned to that order or the transportation order has to be added to the pool of waiting orders, if there is no vehicle available at that moment. Thus, depending on the triggering event, either a 1:m or an m:1 assignment between transportation orders and vehicles has to be determined.

In the case of a container terminal, due to the interaction of the material handling equipment, a higher degree of coordination is required. Given pick-up and delivery times for all loads, all vehicles becoming available within a given time window and all unassigned orders should be considered, independently of which event triggers a new dispatching request. The resulting problem is a many-to-many assignment, which requires more powerful solution techniques. Moreover, the dispatching strategies need to be modified so as to make use of the additional information provided. Surplus transportation orders are not assigned to a global pool of waiting orders any more, but to a list of pre-assigned orders for each vehicle. As a result, a new dispatching request does not need to be invoked if a vehicle completes a transportation order on time, as long as there are other orders previously assigned to that vehicle. Only if a vehicle does not complete an order on time, the assignment has to be updated by invoking a new dispatching request. Summarizing these considerations, in the container port environment five triggering events can be identified:

1) a new transportation order emerges within the given time window,
2) vehicle target times change due to a change in either the quay or the stacking cranes schedule,
3) a vehicle finishes its last transportation order,
4) a vehicle does not complete an operation, either pick-up or drop-off, on time,
5) none of the former four events occurs during a predefined time slice.

The last condition ensures that system information is updated after a reasonable time, thus allowing the control system to react properly on unexpected changes in the environment.

3.3 Total and partial availability of vehicles

Each time a dispatching request is initiated, a new assignment between waiting transportation orders and available AGVs has to be determined. In this context, a key issue is the definition of availability of the multi-load vehicles.

It should be noted that, at the moment a new dispatching request is invoked, the AGVs will usually not reside at a station or a park position, but will be on their way to a pick-up or a drop-off station. It is therefore especially important to consider the actual mission of each vehicle. To this end it will be assumed that an AGV will not be redirected from its current route to a different stop. The main reasons for this assumption are the reduction of system nervousness and scheduling complexity. In addition, due to a sparse grid of sensors in the AGV guide path system, it may not be possible to precisely locate a vehicle at every arbitrary moment.

As shown in Schrecker (2000), there are two views of the availability of a vehicle. It could be considered available if no container is physically loaded onto the vehicle or if not even an order is assigned to it. In this paper, we adopt the second view. In the case of multi-load AGVs, a vehicle is considered *not available*, if its capacity is fully occupied by the orders assigned to it. If on the other hand the vehicle provides unused capacity, it is said to be *available*. In this case, the vehicle may be either *totally available*, if no order at all is assigned to it, or *partially available*, otherwise. For instance, in the context of dual-load carriers, a vehicle is partially available at a pick-up station (quay crane or stacking crane) just after having loaded the first 20ft container. On the other hand it is totally available after starting its trip for delivery of the second 20ft container or a 40ft/45ft container. Figure 3 illustrates the concept of availability used in our investigation.

Assuming the actual mission of each AGV to be fixed, the status of the vehicle during a trip can be identified with that at the next stop. For instance, if a vehicle is not available at all at the next stop, it is also considered not available during the trip to this stop. This would be the case for the trip to the second pick-up of a 20ft container or the pick-up of a 40ft/45ft container, as indicated by the solid lines in Figure 3. On the other hand, during the trip to the second delivery of a 20ft container or the delivery of a 40ft/45ft container (dotted line in Fig. 3) the vehicle is already considered totally available, because the location and time it becomes available are already known. Similarly, during the trip to the first pick-up or drop-off of a 20ft container the vehicle is partially available (dashed line in Fig. 3).

A special feature of modern quay and stacking cranes is the so-called twin lift capability. Two coupled 20ft containers are simultaneously placed on and lifted from an AGV. The twin lift capability is always used for both, the pick-up and the drop-off operation of a transportation order. Hence, transportation orders involving twin lifts have the same effect on AGV availability as transportation orders for 40/45ft containers and are thus dealt with in an identical fashion.

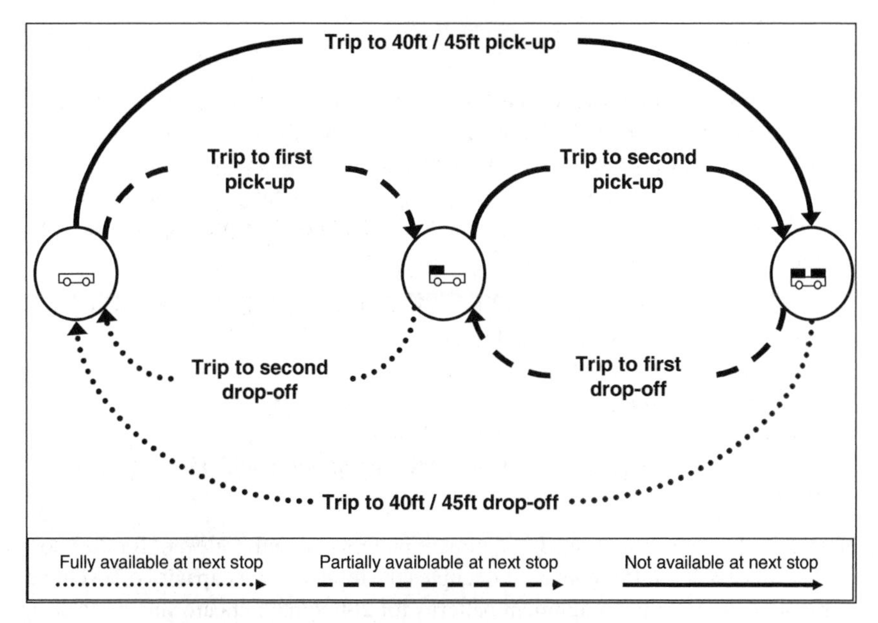

Fig. 3. Availability status of vehicles

From the dispatching point of view, a totally available multi-load carrier can be treated just like a single-load carrier. For obvious reasons, a vehicle being unavailable at all does not need to be considered. In contrast, if a vehicle is partially available, it has to be decided how to insert the pick-up and drop-off operations of a new transportation order into the already scheduled tour. To determine combined tours for the case of dual-load vehicles a heuristic dispatching algorithm and an MILP model based approach are presented in the subsequent sections.

4 Priority rule based dispatching algorithm

For implementation within an online logistics control system in a container terminal environment, a dispatching methodology is required which guarantees response times of merely a few seconds of CPU time (ideally, even less than a second). Thus, for real-time dispatching, priority rule based approaches seem appealing due to their flexibility and their low computational effort. However, simple dispatching rules, which have been developed for the application within manufacturing environments, are not directly applicable here. Specifically, the entire sequence of pick-up and delivery operations of multi-load vehicles must be considered. This is supported by the introduction of assignment patterns which express the feasible options of assigning a transportation order to an available vehicle.

4.1 Assignment patterns

In the following, we consider dual-load carriers as they are employed at some of the latest highly automated seaport container terminals. If a vehicle is partially available, i.e., only a single order for a 20ft container has been previously assigned to it, there are various options of assigning an additional (20ft container) order and integrating this order into the vehicle's schedule. The pick-up and delivery operations for the new order could be scheduled:

- prior to the already assigned one (assignment pattern "*nnaa*", read "new (pick-up) – new (delivery) – assigned (pick-up) – assigned (delivery)"),
- subsequent to the already assigned one ("*aann*"),
- in between ("*anna*"),
- enclosing the already assigned order ("*naan*"),
- or the respective pick-up and delivery operations could alternate ("*anan*", "*nana*").

Note, that the above mentioned assignment patterns are only relevant for 20ft containers, since a 40 or 45ft container occupies the total loading capacity of a vehicle. In Figure 4 all possible assignment patterns for 20ft containers are shown. Pick-up and delivery operations are indicated by an arrow pointing upwards or downwards, respectively. From the six theoretically possible assignment patterns only three are considered in our priority rule based algorithm. At the time the dispatching request is initiated, the vehicles might already be on their way to the service point of the next operation. Thus, in order to avoid re-routing of a vehicle's mission and to prevent that an already assigned transportation order is infinitely delayed, assignment patterns "*nnaa*", "*nana*", and "*naan*" are not considered here.

4.2 Heuristic algorithm

To deal with the complexity of the dispatching problem at hand, the original $m : n$ order-vehicle-assignment problem is reduced to repeatedly determining $1 : n$ assignments, i.e. assigning a single order to one of the available vehicles. Moreover, a look-ahead time window is defined. Only orders initiated within this time window are considered for dispatching. It is assumed that their service times are known with certainty. Actually, orders beyond the time window are known in advance. Their timing, however, is due to frequent changes of the individual terminal operations and exact due-dates are hardly predictable. Hence, the length of the look-ahead time window should be reasonably short and comprise only a limited number of orders with reliable service times.

At every dispatching request, a priority rule based procedure is called up, consisting of the following steps:

1) From all unassigned transportation orders within the given time window, determine the one with the earliest service time.
2) For the selected transportation order, consider all feasible assignments to available AGVs. Evaluate the assignments according to a given cost function, e.g.,

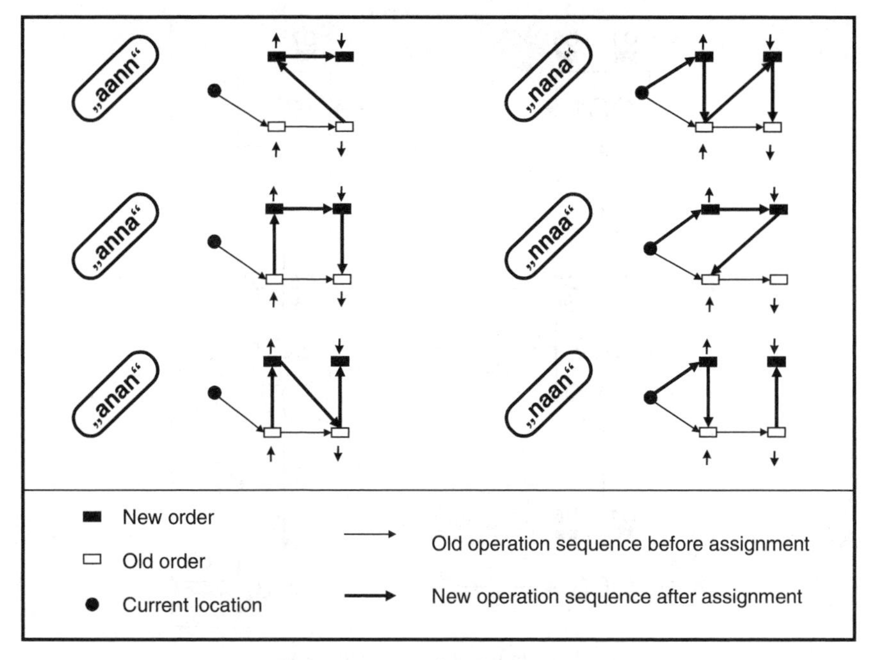

Fig. 4. Possible assignment patterns

penalizing late deliveries. In this case, penalties also apply for delays of transportation orders, which have already been assigned to the vehicle. Such additional delays may be caused by inserting new pick-up and delivery operations into the schedule, cf. using assignment patterns "*anan*" and "*anna*".

3) Select the order-vehicle-assignment showing the lowest assignment costs, remove the order from the list of unassigned orders, update ready times and prospective locations of the AGV, and go to 1).

Note that the algorithm is designed to solve the general case of assigning 20ft containers to multi-load carriers. However, for the assignment of 40 or 45ft containers to dual-load carriers or dealing with single-load carriers, the complexity of step 2) is significantly reduced. In this case, only totally available vehicles are considered in step 2) and assignment patterns become irrelevant. In the remainder of this paper, the original version of the heuristic algorithm is referred to as priority rule for multi-load carriers, while the simplified version is called priority rule for single-load carriers.

4.3 Example

The priority rule based dispatching algorithm is explained using an elementary example. In Figure 5 an example of a bi-directional guide path network is shown. At the given point of time, three empty AGVs (V1 to V3) are positioned at stacking cranes SC2 and SC3 and quay crane QC3. The distances (measured in travel time)

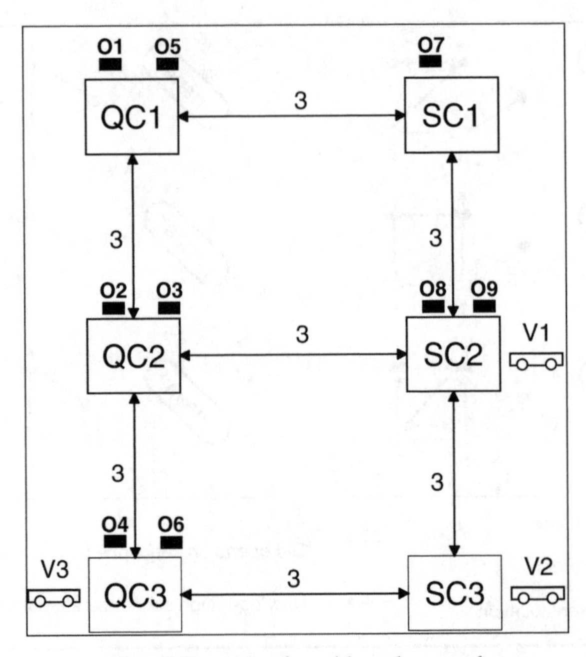

Fig. 5. Example of a guide path network

between the locations are given as well as nine orders (O1 to O9) waiting for pick-up. In this example we assume that all transportation orders comprise 20ft containers.

Not all of the transportation orders shown in Figure 5 are known to the AGV controller at the time of the dispatching request. This issue is illustrated in Figure 6, which indicates the schedule after the first three order-vehicle-assignments. In the figure, each location is represented along with a vertical time bar. A pentagon next to the time bar indicates the pick-up and delivery times and locations of the orders, either pointing upwards (pick-up) or downwards (drop-off). The routes of the AGVs are trajectories in the graph, indicating the change of space and time. The state of availability of each AGV is shown at the bottom of the figure, indicating an assigned order by parentheses, which are omitted as soon as the container is physically loaded.

At the beginning, only orders 2, 3, 4, and 6 are to be picked up within the given look-ahead time window of six periods (marked in white in the figure). After three iterations of the priority rule based algorithm, the most urgent orders O4, O6 and O2 are assigned to vehicles V3, V2 and V1, respectively. The vehicles V1 and V3 arrive early at the pick-up locations of O2 and O4 and hence have to wait until the respective quay cranes are ready for transfer. Still, none of the vehicles is fully loaded. Vehicle V2, e.g., is partially available after picking up O6 at QC3 in period 3 and totally available after delivering this order at SC2 in period 9. Thus, in the next step of the algorithm, order O3 can be assigned to each vehicle according to the assignment patterns "*aann*", "*anan*", "*anna*" introduced in Section 4.1. From

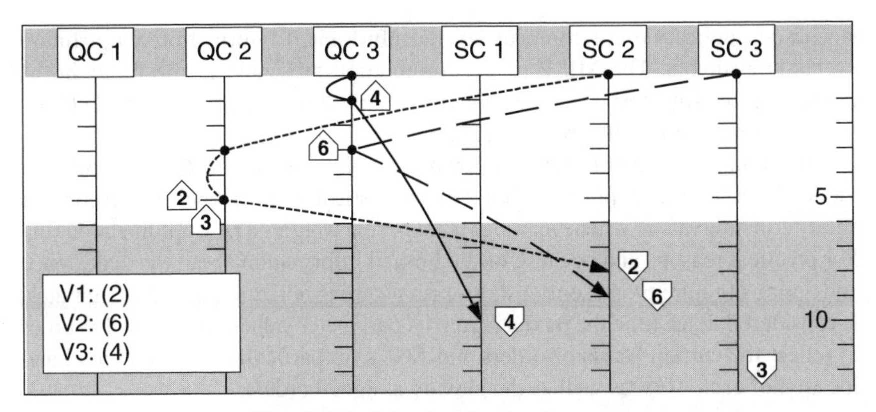

Fig. 6. Schedule after three assignments

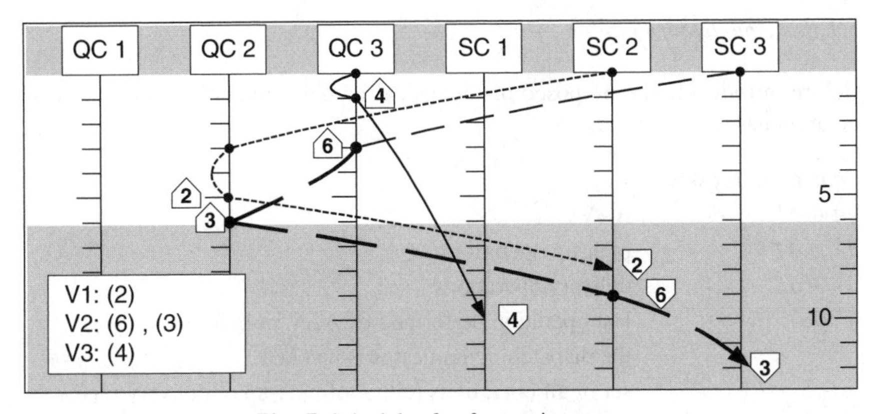

Fig. 7. Schedule after four assignments

the nine feasible options, the assignment to V2 according to pattern "*anan*" appears best (see Fig. 7). The former itinerary of V2 has to be modified in order to execute O3, too. After this step, no unassigned order is left within the look-ahead time window. The current dispatching request is completed and a new request is only invoked if a new order is generated or a vehicle completes a delivery operation.

5 MILP model formulation

The AGV dispatching problem for multi-load carriers can be formulated as a mixed-integer linear programming (MILP) model. However, instead of assigning complete orders to AGVs, multi-load AGVs require the assignment of entire pickup and delivery sequences. Hence, additional constraints have to be introduced in order to ensure the consistency of these sequences. The major drawback of the MILP modelling approach is that, for realistic scenarios, computational times easily exceed the allowable response time of an online logistics control system in a container terminal environment. Moreover, MILP models are based on a finite scheduling horizon and deterministic data. Thus they do not provide optimal solutions in a stochastic application environment where only limited information about future events is available. The MILP model formulation presented in the following is therefore primarily motivated as a benchmark for evaluating the priority rule based approach presented in the previous section.

In a container terminal application, it is practically impossible to solve an MILP model for a long-term planning horizon. Realistically, the model is solved after short-term intervals or whenever a significant event triggers a new optimization run. For practical reasons and because of the limited information about the occurrence and timing of future events, only a short-term planning horizon, e.g., 10–20 minutes, is considered. Each time the model is run, its parameter values are updated in order to reflect the current status of orders and AGVs, in particular the actual load and location of each AGV as well as the already assigned orders.

5.1 Basic model formulation

Before introducing the proposed mathematical model formulation, the following notation has to be defined.

Indices, index sets

$m \in M$	AGVs
$i^+ \in I^+$	pick-up operations
$i^- \in I^-$	delivery operations
$i^L(m)$	last operation performed by AGV m before the dispatching request was invoked
$i \in I = I^+ \cup I^-$	set of all operations (either of pick-up or delivery type, with each tuple (i^+, i^-) representing a transportation order)
$i \in I^L$	last operations of all AGVs ($I^L = \bigcup_{m \in M} I^L(m)$)
$i^N(m)$	next operation of AGV m as scheduled in the last dispatching request
$i \in I^N$	next operations of all AGVs ($I^N = \bigcup_{m \in M} I^N(m)$)
$i \in I^{loaded}(m)$	pick-up operations already performed for containers physically loaded onto AGV m (i.e., the corresponding drop-off operations are to be scheduled)
$i \in I^{loaded}$	pick-up operations already performed for containers physically loaded ($I^{loaded} = \bigcup_{m \in M} I^{loaded}(m)$)
$s \in S$	set of service locations (i.e. all quay and stacking cranes) in the guide path network

Parameters

t_i	target time for operation i
t_i^h	hand-over time for operation i

s_i	service location (pick-up or drop-off) for operation i
q_i	AGV load capacity requirement of operation i ($q_i \in \{1,2\}$ for the pick-up of 20 or 40ft/45ft containers, respectively, and $q_i \in \{-1,-2\}$ for the corresponding drop-off)
Q_m	load capacity of AGV m (=2 for dual-load carriers)
$d\left(s_1, s_2\right)$	(temporal) distance (in time units) between locations s_1 and s_2
B_1	sufficiently large number ($B_1 >$ *maxDepTime* + *maxDistance*, where *maxDepTime* is the maximal possible departure time of an order and *maxDistance* the largest distance in the guide path network)
c_l	costs per unit of waiting time of the stacking or quay crane caused by late arrival of the AGV (identical for both equipment types)
Decision variables	
$x_{m,i1,i2}$	$= 1$, if operation $i1$ precedes operation $i2$ by AGV m (0, otherwise)
t_i^{arr}	arrival time at s_i before performing operation i ($i \notin I^L$)
t_i^{dep}	departure time from s_i after performing operation i, ($i \notin I^L$, with $t_0^{dep}(m)$ being the departure time at last location)
l_i	lateness in performing operation i
y_{mi}	used capacity of AGV m after performing operation i, ($i \notin I^L$, with initial load $y_0(m)$)

Objective function. Total lateness costs are minimized.

$$\min \sum_{i \in I} c_l \cdot l_i \tag{1}$$

subject to the following constraints.

Initial values. The schedule of each AGV starts with the last operation performed before the dispatching request has been invoked (2). This operation is unique for each AGV and may not be assigned to other AGVs (3). The next operation of AGV m as scheduled in the last dispatching request directly follows this operation (4, 5). Finally, the initial load of each AGV (6) and the departure time at the last service location (7) is set according to the actual values.

$$x_{m,i1,i^L(m)} = 0 \qquad \forall m \in M, \forall i1 \in I \tag{2}$$

$$x_{m,i^L(m),i1} = 0 \qquad \forall m \in M, \forall i1 \in I^L - \left\{i^L(m)\right\} \tag{3}$$

$$x_{m,i^L(m),i^N(m)} = 1 \qquad \forall m \in M \tag{4}$$

$$x_{m,i1,i^N(m)} = 0 \qquad \forall m \in M, \forall i1 \in I - I^L \tag{5}$$

$$y_{m,i^L(m)} = y_0(m) \qquad \forall m \in M \tag{6}$$

$$t_{i^L(m)}^{dep} = t_0^{dep}(m) \qquad \forall m \in M \tag{7}$$

Assignment of operations to AGVs. No operation is allowed to be its own predecessor (8). Additionally, each operation not yet performed must be assigned to exactly one AGV, i.e., it must succeed operation $i^L(m)$ (9). Pick-up operations for containers already loaded onto an AGV have been performed in the past and, hence, need not to be assigned to an AGV anymore (10). The sequence of two corresponding operations must be clearly defined if and only if they are assigned to the same AGV (11, 13). Considering two arbitrary operations, at most one of them is allowed to precede the other (12). If two operations are assigned to different AGVs, no precedence relation exists between them at all. Finally, the pick-up and delivery operations of an order must be assigned to the same AGV (15, 16).

$$x_{m,i,i} = 0 \qquad \forall m \in M, \forall i \in I \tag{8}$$

$$\sum_{m \in M} x_{m,i^L(m),i1} = 1 \qquad \forall i1 \in I - \left(I^L \cup I^{loaded}\right) \tag{9}$$

$$\sum_{m \in M} x_{m,i^L(m),i1} = 0 \qquad \forall i1 \in I^{loaded} \tag{10}$$

$$x_{m,i^L(m),i1} + x_{m,i^L(m),i2} - 1 \leq x_{m,i1,i2} + x_{m,i2,i1} \qquad \forall m \in M,\ \forall i1,i2 \in I - I^L : i1 \neq i2 \tag{11}$$

$$\sum_{m \in M} \left(x_{m,i1,i2} + x_{m,i2,i1}\right) \leq 1 \qquad \forall i1, i2 \in I \tag{12}$$

$$2 \cdot x_{m,i1,i2} \leq x_{m,i^L(m),i1} + x_{m,i^L(m),i2} \qquad \forall m \in M,\ \forall i1,i2 \in I - I^L : i1 \neq i2 \tag{13}$$

$$x_{m,i^L(m),i^+} = x_{m,i^L(m),i^-} \qquad \forall m \in M,\ \forall \left(i^+, i^-\right) : i^+ \notin I^{loaded} \tag{14}$$

$$x_{m,i^L(m),i^-} = 1 \qquad \forall m \in M, \forall \left(i^+, i^-\right) : i^+ \in I^{loaded}(m) \tag{15}$$

Load balances. If operation $i2$ is assigned to AGV m, the load of m after performing $i2$ results from the initial load adjusted by the changes resulting from performing all operations prior to $i2$ including $i2$ (16). Moreover, the load of each AGV may not exceed its capacity (17).

$$y_{m,i2} = y_{m,i^L(m)} + \sum_{\substack{i1 \in I \\ i1 \neq i^L(m)}} \left(q_{i1} \cdot x_{m,i1,i2}\right) + q_{i2} \cdot x_{m,i^L(m),i2} \qquad \forall m \in M, \forall i2 \in I \tag{16}$$

$$y_{m,i1} \leq Q_m \qquad \forall m \in M, \forall i1 \in I \tag{17}$$

Sequence of pick-up and delivery operations. If pick-up and drop-off operations of an order are assigned to the same AGV m (which is ensured by (14)), the pick-up must precede the corresponding drop-off operation (18).

$$x_{m,i^L(m),i^+} + x_{m,i^L(m),i^-} - 1 \leq x_{m,i^+,i^-} \qquad \forall m \in M,\ \forall (i^+, i^-) : i^+ \notin I^{loaded} \tag{18}$$

Travel time balances. If operation $i1$ precedes $i2$ on AGV m, the arrival time of $i2$ must be greater or equal to the departure time of $i1$ plus travel time from the service location of $i1$ to that of $i2$ (19). Note that one important precondition for the correctness of this constraint is that the triangle inequality be valid, i.e. for $d(s_1, s_3) > d(s_1, s_2) + d(s_2, s_3)$ (19) does not hold. Also the departure time of an operation is not less than its arrival time plus hand-over time (20) and not less than its target time plus hand-over time (21).

$$t_{i1}^{dep} + d(s_{i1}, s_{i2}) \leq t_{i2}^{arr} + B_1 \cdot (1 - x_{m,i1,i2}) \qquad \forall m \in M,\ \forall i1, i2 \in I \quad (19)$$

$$t_i^{dep} \geq t_i^{arr} + t_i^h \qquad \forall i \in I \quad (20)$$

$$t_i^{dep} \geq t_i + t_i^h \qquad \forall i \in I \quad (21)$$

Lateness. The lateness in performing an operation is calculated as the difference between arrival time and target time (22).

$$l_i \geq t_i^{arr} - t_i \qquad \forall i \in I \quad (22)$$

Variable domains: Only the assignment variables are restricted to binary values (23). All other variables are defined as non-negative (24, 25). It should be noted that variables y_{mi} indicating the load of an AGV automatically take integer values in the optimal solution according to Equation (16).

$$x_{m,i1,i2} \in \{0, 1\} \qquad \forall m \in M, \forall i1, i2 \in I \quad (23)$$

$$y_{mi} \geq 0 \qquad \forall m \in M, \forall i \in I \quad (24)$$

$$l_i,\ t_i^{dep}, t_i^{arr} \geq 0 \qquad \forall i \in I \quad (25)$$

5.2 Extensions

In the practical application considered, minimizing the lateness of transportation orders is regarded at the predominant objective at the operational planning level. Issues of secondary importance are to minimize the earliness of transportation orders and empty travel times of AGVs.

If an AGV arrives too early at some destination, it has to wait for the quay or stacking crane to be served. Depending on the specific layout of the container terminal, buffer space for waiting AGVs may be limited. Therefore, minimization of AGV earliness can be of practical interest. To include this aspect in the basic model formulation, additional decision variables e_i and cost parameters $c_e$are introduced.

- e_i earliness in performing operation i
- c_e costs per unit of waiting time of AGV (early arrival of AGV)

Moreover, the following constraint is needed.

Earliness of AGV. The earliness in performing an operation is calculated as the difference between target time and arrival time (26).

$$e_i \geq t_i - t_i^{arr} \qquad \forall i \in I \quad (26)$$

Minimizing empty travel times tends to improve the overall availability of AGVs. This issue has some practical relevance in a highly stochastic terminal environment, where unforeseen dispatching requests frequently occur. To model empty travel times of AGVs, penalty costs are charged for each time unit an AGV travels unloaded or partially loaded. For dual-load carriers a distinction between empty and half-loaded travel of an AGV has to be made. Thus, corresponding weighting factors of 1.0 and 0.5, respectively, are used to measure empty travel costs. The consideration of empty travel times requires the corresponding decision variable λ_{mi} and additional parameters B_2, B_3 and c_{empty} to be included in the basic model formulation.

λ_{mi} weighted empty travel time of AGV m for travelling to the service location of operation i (distance between last location and service location of operation i weighted by the unused load capacity of AGV m)

B_2 sufficiently large number ($B_2 > Q_m \cdot$ *maxDistance*, where *maxDistance* indicates the largest distance in the guide path network)

B_3 sufficiently large number ($B_3 > B_2 \cdot$ *maxLengthOrderList*, where *maxLengthOrderList* is the maximum length of the AGV order list)

c_{empty} costs per unit of empty travel time of AGV

The following additional constraint has to be included in the basic model formulation in order to model empty travel times of AGVs.

Weighted empty travel times. If operation $i1$ *directly* precedes $i2$ on AGV m, the weighted empty travel time of m is calculated by multiplying the distance from the last location visited to the service location of $i2$ by the number of unused units of the load capacity of m during this mission (27). Note that the second and third terms on the right hand side of (27) equal zero, if $i1$ directly precedes $i2$ on AGV m (i.e. $i1$ precedes $i2$ on m and has one predecessor less than $i2$). It is important to set $B_3 > B_2 \cdot$ *maxLengthOrderList*, since otherwise the inequality would be binding, which is not desired if $i1$, but not $i2$, is assigned to AGV m.

$$\lambda_{m,i2} \geq (2 - y_{m,i1}) \cdot d(s_{i1}, s_{i2}) - B_3 \cdot (1 - x_{m,i1,i2}) + B_2 \cdot \left(1 - \sum_{i3 \in I} x_{m,i3,i2} - x_{m,i3,i1}\right) \quad \forall m \in M, \forall i1, i2 \in I \tag{27}$$

To consider issues of AGV utilization, additional terms for earliness and empty travel times of AGVs have to be included in the objective function. The extended objective function is given by:

$$\min \sum_{i \in I} c_l \cdot l_i + \sum_{i \in I} c_e \cdot e_i + \sum_{m \in M} \sum_{i \in I} c_{empty} \cdot \lambda_{mi} \tag{28}$$

In a real application, the objective function has to be adjusted according to the requirements of the specific application environment. In particular, appropriate values of cost parameters have to be determined. From a theoretical point of view, it is almost impossible to measure the costs of earliness, lateness, and empty travel times of AGVs. Therefore, penalties have to be used instead. As mentioned before, minimizing the lateness of AGVs is seen as the primary objective in order to prevent idle time of critical equipment units, e.g. quay and stacking cranes. Thus, earliness of AGVs and empty travel times should receive smaller weights or may be disregarded at all.

The final extension is aimed to reduce the computational effort required to solve the proposed MILP model. Since the AGV dispatching model is intended for real-time application, short computational times are of great importance. In order to reduce CPU times, redundant constraints may be added to the basic model formulation. Such constraints prevent the solver from exploring unnecessary branches in the branch-and-bound search tree. When dealing with precedence orders, transitivity constraints are an obvious choice.

$$x_{m,i1,i2} + x_{m,i2,i3} - 1 \leq x_{m,i1,i3} \qquad \forall m \in M, \forall i1, i2, i3 \in I \qquad (29)$$

6 Numerical investigation

6.1 Sample problems

Two main issues are addressed through our numerical investigation. First, we analyse the relative performance of dual-load AGVs against single-load carriers. This is achieved by comparing the results obtained by the priority rule based dispatching algorithm outlined in Section 4 for both types of AGVs. Second, we examine the performance of the priority rule based dispatching algorithm by comparing its results to those obtained from the MILP model presented in the previous section.

As a test bed, two simplified layouts are defined which still capture the main characteristics of real sea port container terminals. In the smaller layout, three quay cranes and three storage blocks (or stacking cranes, respectively) are served by a fleet of three AGVs (see Fig. 8 a). The quay cranes are lined up in a row along the berth, while all storage blocks are located in a row at the opposite side, a typical design in highly automated container ports. All distances (measured in travel time required by the AGVs) are normalized at 1 time unit for a trip between two adjacent handling units. Since modern AGV control systems allow flexible routing of the AGVs, it seems reasonable to assume symmetrical distances. In the larger layout the numbers of AGVs, quay and stacking cranes are doubled (see Fig. 8 b).

For both layouts we investigate two different workloads. Table 1 shows the number of transportation orders that have to be processed in each of the four resulting scenarios. The starting time of the transportation orders is drawn from the uniform distribution $U[1, 20]$. Equally, the source and the sink are randomly generated with equal likelihood given to all possible connections between quay and stacking cranes. As outlined in Section 4.3, not all jobs are known at the time the dispatching is invoked. It is assumed that only transportation orders with a pick-up operation within a look-ahead time window of six periods are known for a specific dispatching request. In accordance with many port settings, the share of 40ft/45ft containers is set to 30%. For each of the scenarios 10 problem instances were randomly generated.

6.2 Implementation

The scenarios were modelled with the aid of the simulation package eM-Plant 4.6.12 and the priority rule was implemented in eM-Plant's programming language

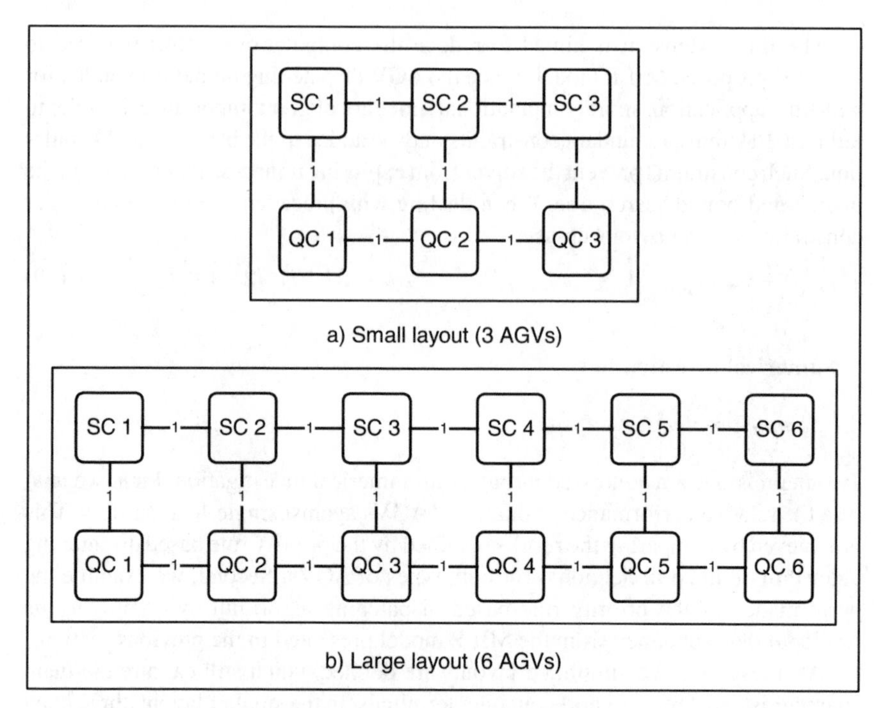

a) Small layout (3 AGVs)

b) Large layout (6 AGVs)

Fig. 8 a,b. Layouts of test scenarios

Table 1. Scenarios

Scenario	Layout	Workload
1	Small	Low (8 TOs)
2	Small	High (12 TOs)
3	Large	Low (14 TOs)
4	Large	High (18 TOs)

SIMTALK. The MILP was algebraically represented by use of ILOG's OPL Studio 3.1 and solved using CPLEX 7.0. All calculations were carried out on a PC with an AMD 1 GHz Athlon processor under Windows 2000.

The computational times per dispatching request are summarized in Table 2. Apparently, the priority rule based algorithm is well suited for a real-time application environment. For both, the single-load and the multi-load case, the computational times were less than a second. However, for the MILP solution this is only true for Scenario 1, which is characterized by the small layout and the low workload. On average, the CPU time for solving the MILP model is noticeably low. But in a few cases, the time required to prove the optimality of the solution obtained was not acceptable for the planning environment under consideration. Therefore, the maximum CPU time allowed was set to three minutes. This time limit was reached for one dispatching request in one instance in Scenario 2 and three dispatching requests in two instances in scenario 4. (Since CPLEX does not continuously check

Table 2. Computational times per dispatching request in seconds

		Scenario 1	Scenario 2	Scenario 3	Scenario 4
Priority rule					
	average	< 1	< 1	< 1	< 1
	maximum	< 1	< 1	< 1	< 1
MILP					
	average	< 1	3.42	3.18	10.40
	maximum	< 1	182.36	94.25	193.99

Table 3. Average lateness per handling operation in number of time units

	Priority rule SLC	Priority rule MLC	MILP MLC
Scenario 1	0.04	0.04	0.03
Scenario 2	0.35	0.30	0.14
Scenario 3	0.11	0.06	0.04
Scenario 4	0.35	0.19	0.14

the CPU time limit, computational times of slightly greater than 180 seconds are exhibited in Table 2.) For both layouts shown in Figure 8 average CPU times are considerably higher for the high-workload scenarios 2 and 4 compared to their low-workload counterparts 1 and 3.

6.3 Numerical results

In our numerical investigation, the average lateness per handling operation is considered as the main performance criterion for evaluating the different dispatching approaches. As pointed out in Section 3, idleness of quay and stacking cranes should be avoided due to the capital investment required for these types of equipment and the need to minimize the berthing time of vessels. Therefore, under a rational logistics control system, the lateness of the AGVs is generally small but does increase with a higher workload. A summary of the numerical results obtained is given in Table 3. Entries indicate the ratio of the total lateness observed in the simulation runs divided by the number of pick-up and a drop-off operations resulting form the generated workload.

The second and the third column of Table 3 show the average lateness observed when the priority rule based dispatching algorithm outlined in Section 4 is employed. One set of experiments was performed assuming that all AGVs operate in single-load carrier mode (second column). Similar experiments were carried out with all AGVs operated as dual-load carriers (third column). Apart from Scenario 1, the average lateness is smaller for the multi-load carrier mode. This can be attributed to the increased capacity of the vehicles which may carry up to two 20ft containers and to the enhanced flexibility of the multi-load carrier mode. Significant improvements can be observed particularly for the large layout (Scenarios 3 and 4). In these cases the average lateness could almost be halved. Obviously, the greater number of transportation orders generated in these scenarios and the increased fleet size of AGVs provide more flexibility for combining different transportation orders within a tour.

The fourth column of Table 3 gives the results of the MILP model application. For all scenarios, this dispatching approach achieves the smallest average lateness. However, the comparison is only fair between the MILP model and the multi-load version of the priority rule. As the MILP model represents the more elaborative solution approach, its superiority is not surprising. In particular, in the high-workload scenarios 2 and 4, the MILP model clearly outperforms the priority rule based dispatching algorithm. This superiority, however, is achieved at the expense of excessive CPU times, which are prohibitive in a large-sized real container terminal configuration, especially if dispatching has to be accomplished in real-time. Therefore, the MILP model approach merely serves as a benchmark for evaluating fast heuristic dispatching algorithms. Finally, it should be noted that, due to the limited look-ahead time window and the highly stochastic application environment, the MILP model by no means provides an exact optimal solution to the complex real-world scheduling problem.

7 Summary and conclusions

Due to the complexity of automated seaport container terminals, the dispatching methods utilized in the control software of the AGV subsystem must be capable to react rapidly to changes in the highly dynamic planning environment. In our investigation, the focus is on the development of fast dispatching methods suitable for real-time application. In particular, we examine multi-load AGVs which can carry either two 20ft containers or one 40ft/45ft container. A major concept upon which our analysis is based is the distinction of different degrees of AGV availability and the definition of assignment patterns which allow alternative AGV tours to be generated with very limited computational effort. This concept led to the development of a novel heuristic dispatching algorithm. Since this approach is based on a myopic priority rule and laid out as a single-pass heuristic, it can be efficiently applied to even large problem instances occurring in real-life container port settings. This is verified in our numerical tests, where the solution time of the heuristic was less than one second for all scenarios, even if a high workload on the AGV system was imposed. The heuristic algorithm developed thus seems to be suitable for integration into an AGV online control system.

Our numerical results also reveal that considerable reductions in AGV lateness may be obtained if the AGVs are operated in a multi-load instead of a single-load carrier mode. The performance was particularly superior for the large layout considered in our test scenarios. As configurations of real-world port container terminals are even larger in size, this result indicates that the use of multi-load AGVs will improve the overall performance of real world container terminals.

To evaluate heuristic approaches, we also developed an MILP model formulation of the multi-load AGV dispatching problem. The model application is based on a limited look-ahead time window and the MILP model is solved upon every single dispatching request. The model can easily be solved using readily available standard optimization software. Our analysis also shows that the MILP model application clearly outperforms the priority rule based dispatching algorithm in terms of AGV lateness. However, the superior solution quality comes at the cost of significantly increased computational effort. Thus a CPU time limit of three minutes was imposed to account for real-time requirements. However, this may still be too large for real container port applications. Since we observed excessive solution times only for a few of the dispatching requests, a hybrid solution scheme which replaces the MILP model solution by a solution obtained from a fast heuristic dispatching algorithm on just these occasions seems promising. In addition, more advanced heuristics, e.g. based on neighbourhood search, will be developed in order to fill the gap between the MILP model and the priority rule.

References

Achuthan N R, Caccetta L, Hill S P (1996) A new subtour elimination constraint for the vehicle routing problem. European Journal of Operational Research 91: 573–586

Angelelli E, Speranza M G (2002) The periodic vehicle routing problem with intermediate facilities. European Journal of Operational Research 137: 233–247

Bae J W, Kim K H (2000) A pooled dispatching strategy for automated guided vehicles in port container terminals. International Journal of Management Science 6: 47–67

Bard J F, Kontoravdis G, Yu G (2002) A branch-and-cut procedure for the vehicle routing problem with time windows. Transportation Science 36: 250–259

Bilge Ü, Tanchoco J M A (1997) AGV Systems with multi-load carriers: basic issues and potential benefits. Journal of Manufacturing Systems 16: 159–174

Carlton W B, Barnes J W (1996) Solving the traveling-salesman problem with time windows using tabu search. IIE Transactions 28: 617–629

Chung R K, Li C L, Lin W (2002) Interblock crane deployment in container terminals. Transportation Science 36: 79–93

Desrosiers J, Dumas Y, Soumis F (1989) The multiple vehicle DIAL-A-RIDE problem. Ecole de Hautes Études Commerciales, Montreal

Egbelu P J, Tanchoco J M A (1984) Characterization of automatic guided vehicle dispatching rules. International Journal of Production Research 22: 359–374

Evers J J M, Koppers S A J (1996) Automated guided vehicle control at a container terminal. Transportation Research A 30: 21–34

Fischetti M, Lodi A, Martello S, Toth P (2001) A polyhedral approach to simplified crew scheduling and vehicle scheduling problems. Management Science 47: 833–850

Fleischmann B (1998), Tourenplanung. In: Isermann H (ed) Logistik. Verlag Moderne Industrie, Landsberg/Lech, pp 287-301 (in German)

Gambardella L M (2000) Vehicle routing problems. Technische Universiteit Eindhoven Eindhoven

Gendreau M, Hertz A, Laporte G (1994) A tabu search heuristic for the vehicle routing problem. Management Science 40: 1276–1290

Ghiani G, Guerriero F, Laporte G, Musmanno R (2003) Real-time vehicle routing: Solution concepts, algorithms and parallel computing strategies. European Journal of Operational Research 151: 1–11

Golden B L, Laporte G, Taillard E D (1997) An adaptive memory heuristic for a class of vehicle routing problems with minimax objective. Computers Operations Research 24: 445–452

Healy P, Moll R (1995) A new extension of local search applied to the Dial-A-Ride problem. European Journal of Operational Research 83: 83–104

Hwang H (2002) An improved model for vehicle routing problem with time constraint based on genetic algorithm. Computers and Industrial Engineering 42: 361–369

Hwang H, Kim S H (1998) Development of dispatching rules for automated guided vehicle systems. Journal of Manufacturing Systems 17: 137–143

Jaw J J, Odoni A R, Psaraftis H N, Wilson N H M (1986) A heuristic algorithm for the multi-vehicle advance request dial-a-ride problem with time windows. Transportation Research B 20: 243–257

Kim K H, Kim K Y (1999) An optimal routing algorithm for a transfer crane in port container terminals. Transportation Science 33: 17–33

Kim K H, Park Y-M (2003) A crane scheduling method for port container terminals. European Journal of Operational Research (to appear)

Kohl N, Desrosiers J, Madsen O G B, Solomon M M, Soumis F (1999) 2-path cuts for the vehicle routing problem with time windows. Transportation Science 33: 101–116

Laporte G, Mercure H, Norbert Y (1986) An exact algorithm for the asymmetrical capacitated vehicle routing problem. Networks 16: 33–46

Laporte G (1992) The vehicle routing problem: an overview of exact and approximate algorithms. European Journal of Operational Research 59: 345–358

Laporte G, Osman I H (1995) Routing problems: a bibliography. Annals of Operations Research 227–262

Lau H C, Liang Z (2001) Pickup and delivery with time windows: algorithms and test case generation. Proceedings of the 13th IEEE International Conference on Tools with Artificial Intelligence (ICTAI'01), pp 333–340

Lee J, Tangjarukij M, Zhu Z (1996) Load selection of automated guided vehicles in flexible manufacturing systems. International Journal of Production Research 34: 3383–3400

Lim J K, Kim K H, Yoshimoto K, Lee, J H, Takahashi T (2003) A dispatching method for automated guided vehicles by using a bidding concept. OR Spectrum 25: 25–44

Mingozzi A, Giorgi S, Baldacci R (1999) An exact method for the vehicle routing problem with backhauls. Transportation Science 33: 315–329

Nanry W P, Barnes J W (2000) Solving the pickup and delivery problem with time windows using reactive tabu search. Transportation Research B 34: 107–121

Nayyar P, Khator S K (1993) Operational control of multi-load vehicles in an automated guided vehicle system. Computers and Industrial Engineering 24: 503–506

Park Y-M, Kim, K H (2003) A scheduling method for berth and quay cranes. OR Spectrum 25: 1–23

Psaraftis H N (1980) A dynamic programming solution to the single vehicle many-to-many immediate request dial-a-ride problem. Transportation Science 14: 130–154

Psaraftis H N (1986) Scheduling large scale advanced request dial-a-ride systems. American Journal of Mathematical and Management Sciences 6: 327–367

Savelsbergh M, Sol M (1995) The general pickup and delivery problem. Transportation Science 29: 17–29

Schrecker A (2000) Planung und Steuerung fahrerloser Transportsysteme (Planning and control of automated guided vehicle systems). Gabler Verlag & Deutscher Universitäts-

Verlag, Wiesbaden (in German)
Sinriech D, Kotlarski J (2002) A dynamic scheduling algorithm for a multiple load multiple carrier system. International Journal of Production Research 40: 1065–1080
Van Breedam A (1995) Improvement heuristics for the vehicle routing problem based on simulated annealing. European Journal of Operational Research 86: 480–490
Van Breedam A (2002) A parametric analysis of heuristics for the vehicle routing problem with side-constraints. European Journal of Operational Research 137: 348–370
Van der Heijden M, Ebben M, Gademan N, van Harten A (2002) Scheduling vehicles in automated transportation systems – algorithms and case study. OR Spectrum 24: 31–58
Wallace A (2001) Application of AI to AGV control - agent control of AGVs. International Journal of Production Research 39: 709–726

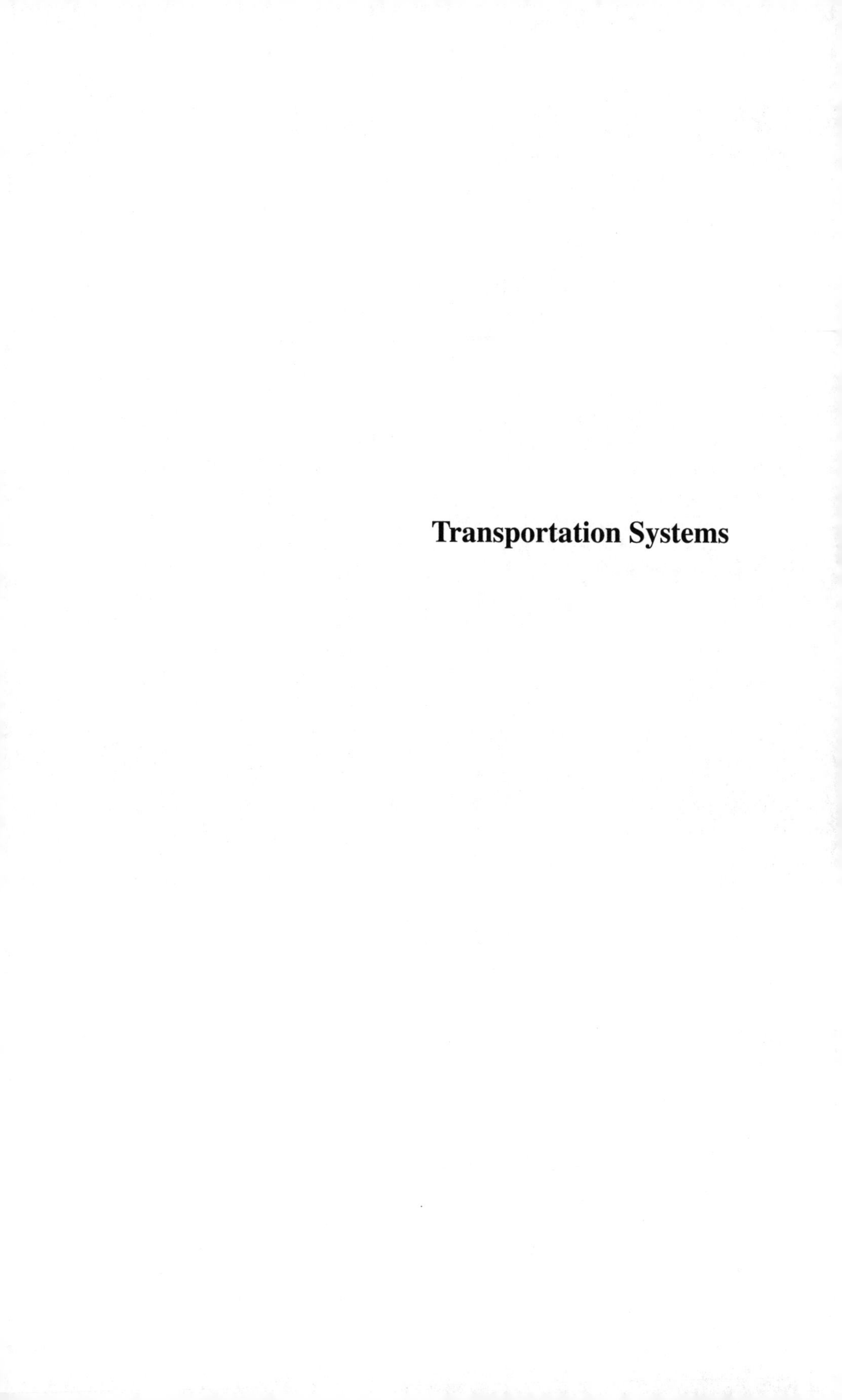

Transportation Systems

Scheduling vehicles in automated transportation systems

Algorithms and case study*

Matthieu van der Heijden, Mark Ebben, Noud Gademann, and Aart van Harten

University of Twente, Centre for Production, Logistics and Operations Management, Faculty of Technology and Management, P.O. Box 217, 7500 AE Enschede, The Netherlands (e-mail: m.c.vanderheijden@sms.utwente.nl)

Received: June 21, 2000 / Accepted: January 22, 2001

Abstract. One of the major planning issues in large scale automated transportation systems is so-called *empty vehicle management,* the timely supply of vehicles to terminals in order to reduce cargo waiting times. Motivated by a Dutch pilot project on an underground cargo transportation system using Automated Guided Vehicles (AGVs), we developed several rules and algorithms for empty vehicle management, varying from trivial First-Come, First-Served (FCFS) via look-ahead rules to integral planning. For our application, we focus on attaining customer service levels in the presence of varying order priorities, taking into account resource capacities and the relation to other planning decisions, such as terminal management. We show how the various rules are embedded in a framework for logistics control of automated transportation networks. Using simulation, the planning options are evaluated on their performance in terms of customer service levels, AGV requirements and empty travel distances. Based on our experiments, we conclude that look-ahead rules have significant advantages above FCFS. A more advanced so-called *serial scheduling method* outperforms the look-ahead rules if the peak demand quickly moves amongst routes in the system.

Key words: Freight transportation – Vehicle scheduling – Simulation

---

* We thank the Dutch Centre for Transportation Technology (CTT) for their funding of the simulation study that has been the basis of our research results. CTT is initiator and coordinator of the project to design and develop the underground logistics system around Amsterdam Airport Schiphol that has been used as a case study in this paper.

Correspondence to: M. van der Heijden

1 Introduction

As in many countries, traffic congestion is becoming a severe problem in the Netherlands, causing delay to both private and commercial transportation. Aiming to be a main hub for Western Europe, this forces the Netherlands to search for solutions in order to guarantee rapid processing of transportation orders within and through the country, particularly when the freight supply keeps growing as has occurred in the past decade.

One of the options currently considered is to move part of the freight underground via a fully automated transportation system. In several governmental studies, the technical and economical feasibility of such a system has been established. As a next step, a pilot system is currently developed, focussing on the processing of time-critical products between Amsterdam Airport Schiphol, the flower auction in Aalsmeer and a future rail terminal in Hoofddorp (cf. van der Heijden et al., 2000). All three locations are situated in the western part of the Netherlands. An example of time-critical transportation is the supply of flowers to the auction in Aalsmeer, which have to arrive in time in order to be processed the same day. Also, the transportation of export flowers to Amsterdam Airport Schiphol is critical, as delay may cause that the cargo arrives when the plane has already left. Because the automated transportation system should be rapid and especially reliable, a competitive advantage on other transportation modes, such as traditional door-to-door road transportation, can be obtained.

Such an automated transportation system consists of a number of terminals, connected by an underground tube system (see Fig. 1). The situation in Figure 1 is the primary layout option at the time of our research for the automated transportation network around Amsterdam Airport Schiphol. Automatic Guided Vehicles (AGVs) carry cargo between terminals in standardised load units (air pallets, flower mid boxes). Order patterns are usually time dependent and may vary over days in the week and over hours on a day. Each terminal consists of a number of docks where vehicles can be loaded or unloaded. Transportation is constrained by arrival times and due times. These due times can be met by a combination of sufficient resources (vehicles, terminals, docks) and a set of logistics planning and control rules. Of course, both issues are interrelated, because efficiency gained by clever planning and control rules leads to reduced capacity requirements. On the other hand, simple myopic or look-ahead rules are usually easier to implement, require less information exchange and are more robust to disturbances.

The network as shown in Figure 1 consists of three main locations, namely Amsterdam Airport Schiphol (AAS) in the north consisting of five terminals, Aalsmeer flower auction (VBA = "Verenigde Bloemenveiling Aalsmeer") in the east consisting of two terminals and Rail Terminal Hoofddorp (RTH) in the southwest consisting of one terminal. AGVs that are not needed for a while or for which there is no room in the local parking are dispatched to a central parking that is located just south of Schiphol Airport, close to the intersection. All terminals and the central parking will be located at the surface, while just the tube system is underground (about 15 meters below the surface). Therefore slopes are planned between each terminal entrance / exit and the tube system. AGVs drive slower on these slopes

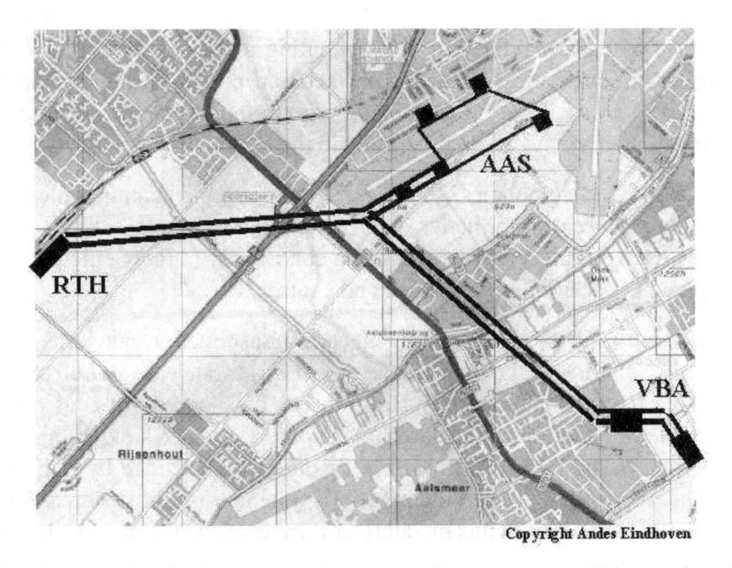

Fig. 1. A layout option for the automated transportation system around Amsterdam Airport Schiphol

than in the tube, which is included in the simulation model. The slopes have a length of approximately 150 meters each.

A large part of the system has two-way traffic, only the five terminals at Schiphol Airport are connected by a loop where only one-way traffic (counter clockwise) is possible. To give an indication for the scale, the distance between RTH and VBA is approximately 10 km. Because of legal speed limitations on the terminals, a small terminal size is more efficient regarding throughput times. Furthermore, the terminal size has significant impact on construction costs, in particular in the case of subterranean construction. Therefore, parking capacity in a terminal is small, both in local parkings and at docks.

In this paper we shall discuss various options for the logistics planning and control of an automated transportation network as described above. Key planning decisions include:

1) empty vehicle management, i.e. the pre-positioning of empty vehicles in the network to anticipate known and predicted demand;
2) order release, i.e. the assignment of transportation orders to load docks in the terminal of origin;
3) order scheduling, i.e. the assignment of empty vehicles to loads at docks when starting a transportation job and the assignment of loaded vehicles to docks after arrival at the terminal of destination;
4) task allocation to docks, i.e. the assignment of load and/or unload activities. Depending on the logistics control concept chosen, these decisions can be taken separately or (partly) simultaneously.

Although we shall discuss a decision structure containing all these questions, this paper focuses on the issue of empty vehicle management. The scheduling of empty AGVs is closely related to the scheduling of transportation jobs. Transportation jobs require empty AGVs and finished transportation jobs result in empty AGVs. Therefore we will study the planning of both empty vehicle travel and loaded vehicle travel. Especially, we are interested in the following subjects:

- How do look-ahead policies compare to simple myopic planning rules?
- What is the value of information about future transportation orders?
- To which extent can central coordination contribute to a more efficient system, compared to decentralised planning and control?

These issues will be discussed for general transportation networks. The pilot project as discussed above serves as a test case. For ease of reference, we will describe the methodology in relation to our case study. We constructed an object-oriented simulation model based on the object-oriented simulation software eM-Plant and on the Logistics Modeling Framework as defined by Van der Zee (1997). This framework gives guidelines for a systematic construction of logistics simulation models. We used this model to test and evaluate the various options for empty vehicle management.

The remaining part of this paper is organised as follows. In the next section, we present the model for the transportation network, discuss the assumptions and describe the case study to which we applied our decision structure. An overview of related literature is given in Section 3. Section 4 deals with a decision structure of the logistics planning and control. Various options for empty vehicle management, the focus of our paper, are presented in Section 5. The simulation study with numerical results, based on the case study, is the subject of Section 6. Finally, we summarise our main conclusions and give some directions for further research (Sect. 7).

2 Model and assumptions

We consider a closed transportation network consisting of a fixed number of origin / destination nodes (terminals), connected by an asymmetric road network, see Figure 1 for an example. Each terminal consists of a fixed number of docks, where vehicles can load or unload (one vehicle at a time), (see Fig. 2 for an example). The vehicles transport loads from the terminal of origin to the terminal of destination. Each vehicle can handle only one load at a time. A load may result from the consolidation of several small volume orders with the same origin and destination and similar due times. However, this is a separate decision that is not discussed in this paper. The route between a given origin-destination pair is fixed and determined in advance using a common shortest path algorithm. In the presence of uni-directional flow paths (as is the case in the pilot project), the distance matrix is asymmetric. When a vehicle is not needed for a while, it may be parked in the terminal (local parking) or on a separate parking lot (a central empty vehicle buffer) that can be considered as a network node.

Transportation jobs arrive in the system (e.g. as a result of train arrivals carrying multiple loads to be transported). The orders may be (partly) known some time before the actual load arrival in the system. Each load has a specific *due time* before

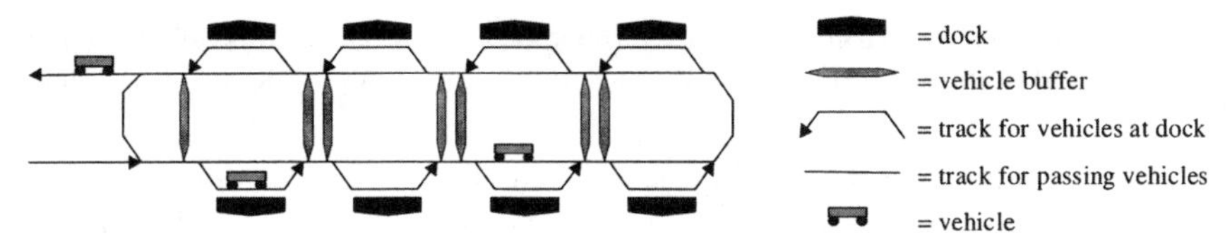

Fig. 2. Example of a terminal

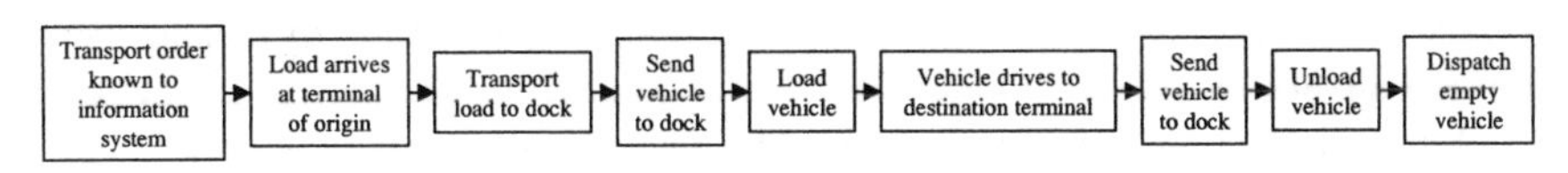

Fig. 3. Order processing

which it should be delivered at the destination terminal, defined as the time at which the unload operation has to be finished. Key performance measures considered are:

- the order throughput time, i.e. the time from order arrival at the origin until the job is unloaded at the destination;
- the fill rate, i.e. the fraction of orders that is delivered before the due time;
- the lateness, i.e. the difference between actual job completion time and due time
- average empty vehicle kilometres per day, being an indicator for energy waste and loss of vehicle capacity.

The processing of a transportation order is depicted in Figure 3. First an order is issued, which may occur some time before the actual arrival of the load. Next the load is moved to a dock (after the *order release* decision), where it waits for a vehicle to be assigned to (*order scheduling*). After arrival at the dock, the empty vehicle picks up the load and drives to the terminal of destination. There it is assigned to an unload dock (possibly after some delay in the case of high utilisation), based on a second order scheduling decision. Then the vehicle drives to the dock and unloads as soon as the dock becomes free. After unloading, the transportation job is considered as completed, so further order handling is not considered. Then the vehicle is available to transport another order. Note that the order release and both order scheduling decisions may be taken simultaneously or sequentially, depending on the logistics control concept chosen (cf. Sect. 4). With respect to the second option, it is relevant to note that there may be a significant time lag between the three actions corresponding to the three decisions (say 5-30 minutes in the Schiphol case). This may be an argument to postpone the second and/or third decision.
As mentioned before, an empty vehicle that cannot be used immediately can be moved to a parking to wait for orders, either locally (possibly short waiting time, limited space) or centrally (possibly longer waiting times, high capacity). So an empty vehicle that has just unloaded may receive one of the following directions:

1) Move to another dock at the same terminal to pick up a load.
2) Move to a local parking to wait until a load order is assigned.
3) Move to another terminal, because it is expected or known that a (high priority) transportation order is available over there.

4) Move to a central parking, because a transportation order for this vehicle is not expected in the short run and/or because there is no room in the local parking.

In our definition, empty vehicle management includes the decisions regarding these four actions. Note that additional actions may be defined, such as vehicle repair after failure, vehicle maintenance and battery changing or recharching. Although we skip these issues for convenience, we remark that the planning framework as discussed in this paper can be extended to include both failure and energy management (cf. Ebben, 2001).

Empty vehicle management is needed because transportation flows will generally be unbalanced over time, i.e. the number of vehicles arriving in a time interval at a terminal differs from the number of vehicles to be loaded and dispatched in the same interval and from the same terminal. It is especially critical in situations where due times are tight and resource utilisation (vehicles, docks) is high. Therefore, we shall pay special attention to empty vehicle management.

Another part of the logistics control structure is the assignment of tasks to docks (load or unload). Obviously, it is most convenient if a dock can load and unload in random order without capacity loss. In some cases however, separation of incoming and outgoing flows may be preferred for organisational and technical reasons. If the ratio between incoming and outgoing transportation flows is stable in time, docks can be dedicated to either load or unload activities. Otherwise, we may consider using dynamically dedicated docks whose function can periodically be switched (e.g. each hour). Then, the periodic decision how many docks should load and how many should unload is referred to as task allocation to docks.

For the model as described above, we further make the following technical assumptions:

1) All vehicles are identical, so we have a homogeneous fleet.
2) The number of vehicles is fixed.
3) All orders have to be executed, even when they are late.
4) Loading and unloading times at docks vary over jobs; therefore these times are represented by independent and identically distributed random variables.
5) Travel times between nodes are deterministic and known in advance.
6) At the terminal entrance there are two infinite capacity FIFO vehicle buffers, one for loaded and one for empty vehicles.

Regarding the last assumption, we note that queuing in front of the terminal entrance may occur if an arriving vehicle cannot be assigned to a free destination in the terminal (dock, parking) or if the destination to which it has already been assigned is occupied. In practice, these queues operate according to the First-In, First Out (FIFO) principle. Separate queues for loaded and empty vehicles offer flexibility to set priorities for either job types. Although infinite queuing capacity implies that terminal parking capacity may be omitted from vehicle planning, queuing may have impact on the logistics performance. The reason is that loaded vehicles cannot overtake in the queue whereas this may be desirable because of different job priorities.

In contrast to travel times, the *order handling times* may fluctuate considerably between orders having the same origin and destination. These handling times consist of

- variations in loading and unloading time,
- waiting times at the destination terminal (which depend on the time-dependent work load)
- dock-dependent distances between dock and terminal entrance/exit.

For a specific job, the handling time can only be observed *after* job completion, whereas this value is needed for planning *in advance*. As the actual handling time of a specific job is unknown at the time of planning, we treat these as random variables for which the mean and standard deviation can be estimated (cf. Sect. 4 for more details).

3 Literature

Related literature includes AGV scheduling in internal transportation systems and vehicle scheduling in other external transportation systems (rail, road) and container scheduling.

Vehicle selection and dispatching rules are important for a good performance in AGV systems. Vehicle dispatching involves a decision rule for selecting a vehicle or dock for a specific transportation order. Egbelu and Tanchoco (1984) classify the dispatching rules into two categories, vehicle and work center initiated dispatching, and evaluate some heuristic rules. Klein and Kim (1996) propose various multi-criteria dispatching rules and show better performance than single-criteria rules. These rules do not take into account the precise timing of operations. Because timing is important in the case of limited dock capacities, a scheduling approach can be used. Akturk and Yilmaz (1996) propose an analytical model to incorporate the AGV system into the overall decision-making hierarchy. Ulusoy and Bilge (1993) make the scheduling of AGVs an integral part of the overall scheduling activity; machines and AGVs are scheduled simultaneously.

Directly related to the vehicle dispatching problem is the problem of positioning idle vehicles. The most common objective is to minimize the maximum or mean response time. Kim and Kim (1997) propose a procedure to determine the home location of idle vehicles in a loop layout in order to minimize the mean response time, while Hu and Egbelu (2000) present a framework to determine the optimal home locations in a unidirectional AGV system. They evaluate an exact solution approach and a heuristic algorithm. Gademann and Van de Velde (2000) show that the problem of determining the home location of idle vehicles in a loop layout in order to minimize maximum response time is solvable in polynomial time for any number of AGVs. The same is true when the criterion minimum *average* response time is used. These papers deal with so-called dwell-point strategies, i.e. the positioning of vehicles that are currently not needed for transportation. The focus is on AGV systems for internal transport in a warehouse or production facility. Usually such networks are relatively small. On the other hand, we consider *external* transport between various facilities, in which case the networks are much larger. As a consequence, the response time for an empty vehicle to arrive at its destination is substantial, and the system status may change significantly in this period (high priority order arrivals, equipment failure or recovery, etc.). Hence the empty vehicle management in this paper differs from the usual idle vehicle positioning in AGV networks.

One of the first papers in the other related area, empty vehicle management, is by White and Bomberault (1969), who model the allocation of empty freight cars in a railroad system with deterministic demand as a transhipment problem. Later on, Jordan and Turnquist (1983) formulated a dynamic model for empty vehicle allocation with stochastic demand and stochastic travel times. A recent contribution is by Holmberg et al. (1998), who develop an optimisation model for a situation with deterministic supply and demand, pre-specified train schedules, heterogeneous fleet and limited train capacity. In the field of empty truck distribution, especially Powell has done a lot of work. Powell et al. (1988) describe the basic problem setting, where a practical solution is given for empty truck repositioning for a full truckload carrier in the United States under uncertain demand. Theoretical improvements and model extensions are presented amongst others in Cheung and Powell (1994), Powell (1996) and Powell and Carvalho (1998). A recent example of empty container allocation is by Cheung and Chen (1998), who construct a single-commodity stochastic network flow model for the distribution of empty containers under random demand and transportation capacities.

For a literature review of earlier work in the area of empty vehicle management, we refer to Dejax and Crainic (1987). They present an overview and classification of empty vehicle allocation models. We can relate our model to the classification scheme of Dejax and Crainic (1987) as follows: We have an operational model for empty vehicle dispatching with single transportation mode and homogeneous fleet in a dynamic setting with stochastic demand.

If we compare our model to available literature, we see that the model as discussed by Powell (1996) shows the most similarities. However, there are some important differences:

1) We face capacity restrictions at each network node, arising from a limited number of load docks and unload docks.
2) The time horizon is very short compared to applications in railcar distribution, truck distribution or container distribution. While in existing applications the planning horizon covers at least a couple of days, we have to plan for at most a few hours. Orders are known only a short period in advance and should be processed very rapidly. This implies that a relatively high planning frequency and hence fast planning methods are required. From a computational point of view, a heuristic approach seems to be more suitable in this situation than time-consuming optimisation models.
3) We focus on attaining customer service levels, measured by fill rates, instead of cost minimisation.
4) We allow an asymmetric road network, which is quite uncommon in other applications.
5) The presence of terminals requiring operational control rules for order release and scheduling asks for a proper planning decomposition. A logical choice

is to use a hierarchical empty vehicle planning procedure, namely a *global* empty vehicle manager to distribute vehicles between network nodes (terminals, central parkings) and a *local* empty vehicle manager covering the assignment of load orders or empty vehicle dispatch orders to the vehicles available within a node. For local empty vehicle management, we can use principles from AGV dispatching and positioning of idle vehicles.

To our knowledge, empty vehicle management in such a setting has not been discussed before in the literature.

4 Decision structure and options for planning and control

As stated in the introduction, the amount of information used and the level of planning coordination can affect the logistics performance significantly. It is likely that the best system performance can be attained if all major decisions (order release, order scheduling, task assignment to docks and empty vehicle management) are taken at a central level using all system information available. This implies that one central organisational unit should be responsible for integral system planning and that all relevant information should be available at this central level. The relevant information includes

1) order information (for known and forecasted orders):
   a. the origin and the destination
   b. the arrival time and the due time
   c. the current status (load in process, waiting to be processed or still to arrive)
2) vehicle information:
   a. the current status (driving loaded, driving empty and assigned to an order, empty and unassigned)
   b. the current (approximate) location and destination
   c. the expected finishing time of the current activity
3) dock information
   a. the current status (assigned loading or unloading task; free or occupied)
   b. the expected finishing time of the current activity

Order information and dock information are usually readily available at the local terminal level, but it takes more effort to collect and maintain this information centrally. As a consequence of frequent data alteration, extensive and reliable data exchange is essential, leading to an extensive (and possibly vulnerable) information and communication system. A compromise may be to exchange aggregate information only, such as the number of load jobs for the next period at each terminal (without knowledge of destination or due times). Furthermore, central authority causes additional communication with local authorities such as terminal managers. Also, it may be more difficult for a central authority to react quickly on unexpected events such as equipment failure and the arrival of rush orders. Local responsibility and authority can be more flexible in this respect. A prerequisite is that the two hierarchical layers should communicate in a simple, yet efficient way. A local control approach will therefore be the basis of our control framework. Still, we shall analyse the effect of local versus central decision making on the logistics performance.

When decisions are taken at a central level, it is possible to take the order release and order scheduling decisions at once. That is, the central authority may decide simultaneously at which dock an order should be loaded by which AGV and at which dock the AGV will unload at the terminal of destination. Because there can be a significant time lag between the load and unload activity (up to half an hour in our case study), it is not guaranteed that the second order scheduling decision is still optimal when the AGV arrives at the terminal of destination. If congestion and/or AGV failures cause (unpredictable) fluctuations in the travel times, the actual AGV queue length for each dock at the destination terminal can be different from the expectation at planning time. In such a situation, it can be preferable to postpone the second order scheduling decision. The latter will happen automatically in a local control concept.

As total system behaviour is mainly affected by empty vehicle management, we focus on this decision area. We choose some plausible decision rules for the other four decision areas. Further analysis of the impact of local decision rules is a separate research topic. In this section, we describe the rules for order release, order scheduling and task assignment. Further we give the general outline of empty vehicle management. Variants for the latter decision area are discussed in Section 5.

To define the decision rules in this section and the next section, we use the following notation:

i, j = node index (terminal, central parking)
n = order index
k = vehicle index
t_0 = current clock time
tK_n = time at which order n is known in the system
tA_n = arrival time of order n (load arrives at terminal of origin)
tD_n = due time of order n, i.e. the latest time at which order n should be unloaded at the terminal of destination
tL_n = latest departure time of order n (see below for the calculation)
sL_n = the latest time that an empty vehicle has to be dispatched in order to pick up load n before the latest departure time tL_n
τ_{ij} = the travel time of a vehicle from the exit of node i to the entrance of node j
H_{ij} = handling time for an order to be transported from node i to node j, consisting of loading time, travel time, waiting time and unloading time; for planning purposes, H_{ij} is modelled as a random variable with mean $E[H_{ij}]$ and standard deviation $\sigma[H_{ij}]$

Note that the order handling time is *not* equal to the throughput time, as the latter also includes the time that the order has to wait for an empty AGV.

4.1 Order release

By order release we mean the assignment of transportation orders to load docks in the terminal of origin. The procedure is as follows. At each dock, one or more loads can be placed in a small dock buffer, waiting to be loaded. When a load arrives at the terminal of origin, it is assigned to the load dock having the shortest queue of unprocessed orders. If all load dock buffers are occupied (which will generally be true in peak periods), the load waits at the terminal until a buffer position becomes free. At that time, the order with highest priority is released to the dock where a vacancy arises. This implies that the general load buffer at the terminal should have random access (orders can be released in any arbitrary order, independent of the arrival order). The load buffers at the *docks* can both be random access buffers or channel-like devices. In the latter case, orders cannot overtake. We use the latter assumption in our case study, but modification to random access buffers is straightforward within our logistics control framework.

Priority is assigned to the orders according to the minimum latest departure time $\mathrm{tL_n}$. For each order, this time $\mathrm{tL_n}$ is set such that the due time for order n is met with high probability:

$$tL_{\mathrm{n}} = tD_{\mathrm{n}} - E[H_{\mathrm{ij}}] - k\sigma[H_{\mathrm{ij}}] \tag{1}$$

where k is a safety factor. The term $\mathrm{k}\sigma[\mathrm{H_{ij}}]$ allows for (time-dependent) fluctuations in the handling time, caused by

- variation in load and unload times;
- possible variation in transportation times (congestion effects, variation in route length depending on the dock and/or local parking choice)
- (unpredictable) waiting time at the terminal of destination depending on the work load.

Note that the mean and standard deviation of the handling time can be estimated using a standard exponential smoothing procedure (see e.g. Silver et al., 1998).

Order release according to the priority rule based on minimum latest departure time is applicable to most look-ahead rules for empty vehicle management (Sects. 5.1 and 5.2). However, integral planning of empty vehicles may conflict with this priority rule and therefore the order release heuristic may be overruled (see Sect. 5.3 for more details).

4.2 Order scheduling

Order scheduling defines the start times for load orders and unload orders at a dock. That is, an empty vehicle is assigned to a specific order and dock to pick up the load at the origin terminal and a loaded vehicle is assigned to a dock at the destination terminal. Here we have to take into account that a limited number of positions may be available to park vehicles at each dock, e.g. one or two. If all positions are occupied, a vehicle cannot be sent to a dock, because it could block other traffic in the terminal. Then a vehicle can either wait in front of the terminal entrance or drive to the local parking if there is space left. For both parts of order scheduling (load orders and unload orders), we shall use simple decision rules:

- A vehicle that becomes available for loading at a terminal is assigned to the load dock with the highest priority order that is released but not yet scheduled (i.e.

no vehicle is assigned); it is dispatched to that dock if a free AGV position is available, otherwise it is sent to a local parking to wait until it can be dispatched to the dock.

- A vehicle arriving at the terminal of destination is assigned to the unload dock with the shortest queue if a free AGV position is available, otherwise it is sent to a local parking to wait.

4.3 Task allocation to docks

Assigning load and/or unload activities to docks (time-dependent docks dedication) should obviously be based on the expected number of load orders and unload orders in the system. Tasks are changed periodically only, so that (i) incoming and outgoing product flows can be clearly separated and (ii) the allocation of dock capacity can be fitted to the expected work load. The period during which the task allocation at a terminal is fixed should preferably be not too short and tuned on the frequency in which the balance between load and unload orders changes. For this decision, simple rules of thumb can be used. For example, choose the number of load docks and unload docks such, that the utilisation is balanced for orders that have to be handled in the next period (with latest departure time as selection criterion for load jobs that have to be handled). We refer to Ebben (2001) for more details.

4.4 Empty vehicle management

Empty vehicle management can be considered at two hierarchical levels, namely at network level (how to allocate empty vehicles amongst terminals and central parkings) and at terminal level (how to handle an empty vehicle in a terminal). Although it is possible to consider these decisions simultaneously in a central decision setting, a hierarchical decision structure seems to be more convenient. At a local level, it is easier to react on unexpected events, such as the arrival of rush orders and the failure of docks or vehicles. At a network level, we can focus on balancing vehicle flows without taking into account all detailed events at terminals. If local problems are solved locally and network problems centrally, we can construct a relatively simple and robust control structure without excessive information traffic. To this end, we distinguish between *global empty vehicle management* that covers the network level and *local empty vehicle management* that covers the terminal level. Such an approach is to be preferred from a practical point of view, but requires coordination between the two hierarchical levels. Variants for both levels are discussed in the next section.

5 Options for empty vehicle management

In this section, we shall discuss four options for empty vehicle management (EVM). We shall focus on global empty vehicle management and explain the connection to the local empty vehicle manager per option. The variants differ with respect

Table 1. Variants for empty vehicle management (EVM)

Variant for EVM	Central information	Coordination
EVM1: FCFS, myopic	1. release time and route for • orders present 2. vehicle status, location and time ready	local
EVM2: FCFS, look-ahead	1. release time and route for • orders present • future orders known 2. vehicle status, location and time ready	local
EVM3: Hierarchical coordination	1. release time, latest departure time and route for • orders present • future orders known 2. vehicle status, location and time ready	hierarchical
EVM4: Integral planning	1. release time, latest departure time and route for • orders present • future orders known 2. vehicle status, location and time ready	central

to the amount of information used and the level of coordination (see Table 1 for an overview). The names mentioned in the table primarily refer to the operation of the global empty vehicle manager. The first two variants are merely on-line dispatching rules that react on each order arrival. The last two variants however are capacity planning procedures that rebalance vehicle flows periodically. The next three subsections will describe the variants in more detail.

5.1 Local coordination using dispatching rules (EVM1 and EVM2)

5.1.1 First-come-first-served (EVM1). The simplest variant of global empty vehicle management (EVM1) is to dispatch available vehicles to terminals on a First-Come-First-Served (FCFS) basis. Vehicle requests arrive at the global empty vehicle manager at times tA_n. At any point in time, the set of available vehicles is defined as:

- empty vehicles that are not dispatched to a terminal yet, i.e. vehicles that are waiting in or driving to a (local or central) parking,
- loaded vehicles that can be assigned a next order, to be processed when the current order is finished.

Hence every vehicle (loaded and empty) has as attribute the location where the next order should be picked up. Available vehicles are vehicles without next pickup location. Note that a vehicle becomes available for assignment if it starts loading at a dock in the terminal given by its next pickup location. The time at which this vehicle can pick up the next order can be estimated from the expected order handling time.

Now the procedure for *global empty vehicle management* is as follows. Each time when a load, say index n, arrives at a terminal i (at time tA_n), the global empty vehicle manager receives a vehicle request. If there are still vehicles available, the vehicle that is nearest to terminal i is dispatched. This is the vehicle with the earliest expected arrival time at terminal i. Computation of this time depends on the current status and location of the vehicle. If no vehicle is available, the request is added to a backorder list. This backorder list is sorted according to order arrival. As soon as a vehicle becomes available, the first request from the backorder list is satisfied.

As mentioned in the previous section, *local empty vehicle management* handles empty vehicles *within* the terminal. The following two events require a decision:

1) An empty vehicle arrives at the entrance of terminal i, having i as next pickup location. The vehicle may have already been assigned to a dock to pick up an order, but in the case of a local control concept, this decision has not been taken yet (see Sect. 4). In the latter case, the order scheduling has to decide upon the load dock to be assigned. If all load docks are occupied or no unassigned load orders are present, the empty vehicle is sent to a local parking. If the local parking is full, the empty vehicle waits in front of the terminal until room becomes available.
2) A vehicle becomes empty after it has unloaded at some dock in terminal i.
   a. If it has no next pickup location, it is directed to the nearest central parking. In periods of heavy traffic, this situation is not likely to occur because the backorder list will not be empty then.
   b. If it has a next pickup terminal $j \neq i$, the vehicle is dispatched to terminal j (this has already been decided by the global empty vehicle manager). If an order with the same destination is available at the terminal and a load dock is available too, the vehicle loads this order. In this way, empty vehicle dispatch from the terminal and order processing can be combined without significant delay.
   c. If it has terminal i as next pickup location, a load is assigned to it (via order scheduling), if possible. If the vehicle cannot be assigned to a load dock, it is sent to a local parking. If the local parkings are full, the vehicle is dispatched to the nearest central parking with notification to the global empty vehicle manager. As the vehicle was assigned to pick up an order, the global empty vehicle manager immediately dispatches the nearest available vehicle to terminal i as replacement. If no other vehicle is available, the request is inserted at the top of the backorder list. As another option, one may postpone the replacement until free space becomes available at the terminal. However, then significant time may pass until the free space is occupied, because of the relative long travel times for external transportation systems that we have in mind. Therefore we decided to implement an immediate reaction by the global empty car manager.

Note that a simple FCFS rule may not be as bad as it seems, because orders are released in each terminal according to a latest departure time priority rule (see Sect. 4) and because empty vehicle dispatch is combined with load orders if possible.

5.1.2 Look-ahead variant (EVM2). If orders are known some time in advance (for example $1/2$-1 hour), we may use this information to improve the planning procedure. When order n becomes known in the system (at tK_n), a vehicle request is sent to the global empty vehicle manager. Similarly to the first variant (EVM1), the vehicle requests are satisfied FCFS and if multiple vehicles are available, the vehicle m with the earliest expected arrival time at terminal i is selected. However, if vehicle m is dispatched immediately, it may arrive too early at the terminal and cause congestion. For vehicles from the central parking, the empty vehicle manager can prevent this by delaying its departure (which is never useful for EVM1, as vehicles cannot arrive too early). Therefore the global empty vehicle manager dispatches vehicle m to terminal i as follows:

- if vehicle m is loaded and has no next pickup location, assign terminal i as next pickup location.
- if vehicle m is empty and waiting in or driving to a central parking, reserve the vehicle and dispatch it at its earliest dispatch time which guarantees that the vehicle will not arrive too early and cause excessive queues at the terminal entrance; if dispatch is delayed, the vehicle waits in the central parking.

Note that the earliest dispatch time is introduced here, because early empty vehicle arrival is only possible in the case of prior information about orders. Under myopic planning, a load will always be present for every vehicle that arrives at a terminal, because the vehicle has been requested at $t_{A,n}$. Further, note that local empty vehicle management is equivalent to the procedure from the previous subsection.

5.2 Hierarchical coordination (EVM3)

In the first two options, the global empty vehicle manager does not take into account priorities when dispatching vehicles to terminals. Also, the vehicles cannot be fully locally controlled in a terminal, because some vehicles may already have a next pickup location. In this section, we discuss improvements on both issues. That is, we improve coordination between terminals. We neglect finite handling capacities (caused by a finite number of docks).

We assume that the global empty vehicle manager has full control of vehicles outside the terminals, i.e. the destination of each individual empty vehicle may be changed at any point in time. Within a terminal however, a global empty vehicle manager only controls the *number* of vehicles. This fits into the local logistics control concept, even though complete scheduling is possible (cf. Sect. 5.3). A global empty vehicle manager may request for empty vehicles to be sent to another location (central parking, terminal), but the local authority may decide *which* vehicles *when* to dispatch. The global empty vehicle manager sets priorities to empty vehicle requests using the latest dispatch time sL_n, i.e. the latest time that an empty vehicle has to be dispatched in order to pick up load n at terminal j before the latest departure time tL_n.

The *global empty vehicle manager* periodically plans empty vehicle redistribution between terminals and central parkings (e.g. every 10 minutes). Two lists are available for planning purposes:

- a list of all known load orders at all terminals (both present and underway) that have not been loaded yet;
- a list of all vehicles with status (loaded or empty) and approximate location.

The coordinated empty vehicle planning is as follows. First, the list of known load orders is sorted in increasing order of latest departure times tL_n. This list is processed sequentially. To each order, we assign the vehicle that can be available at the earliest point in time. The planning procedure continues until the order list is fully assigned or all vehicles are assigned.

A *local empty vehicle manager* assigns tasks to empty vehicles within a terminal. An empty vehicle becomes available in a terminal at the following events:

1) An empty vehicle enters the terminal.
2) A vehicle becomes empty after it has unloaded at some dock.

To assign a task to an empty vehicle that becomes available, the local empty vehicle manager uses the following information:

A. a list of orders to dispatch empty vehicles, sorted on latest dispatch time $\underline{sL}_m$, m=1,2,...
B. a list of known load orders, sorted on latest departure time $\underline{tL}_n$, n=1, 2 ...

The underlined variables refer to the respective sorted lists. If $\underline{sL}_1$¡$\underline{tL}_1$ (or if the list of known load orders is empty), the local empty vehicle manager dispatches the vehicle according to the first destination on list A. If $\underline{tL}_1 \leq \underline{sL}_1$ (or if the list of empty vehicles to be dispatched is empty)[1] , the local empty vehicle manager assigns the empty vehicle to the first load order on list B. If both lists A and B are empty, the local empty vehicle manager sends the vehicle to the local parking if it is needed in the near future and if the local parking is not full. Otherwise the vehicle is sent to the nearest central parking (for which a notification is send to the global empty vehicle manager).

The local empty vehicle manager tries to combine orders if possible. That is, if an empty vehicle dispatch order to terminal j has highest priority (sL_1¡tL_1), a load order for the same terminal j with smallest latest departure time tL_n is searched. If such an order is available and a load dock is available too, this order is scheduled, otherwise the vehicle is dispatched empty to terminal j. Vice versa, if a load order for some terminal j has highest priority ($tL_1 \leq sL_1$), the list of empty vehicle dispatch orders is searched for an order with smallest latest dispatch time having the same destination j. If such an order is found, it is removed from the list and the global empty vehicle manager is informed that the request is fulfilled. In response, the

[1] Note that the indices of $\underline{sL}_m$ and $\underline{tL}_n$ refer to *different* order lists, one for load orders at the terminal under consideration and one for load orders for which this terminal should supply an empty vehicle. Hence $\underline{tL}_1$ and $\underline{sL}_1$ refer to the first orders on each sorted list and thus *not* to the same order.

global empty vehicle manager may provide another empty vehicle dispatch order to the terminal, but this order will have (considerably) less priority.

This procedure has advantages above the simple rules from Section 5.1 in the sense that priorities are used by both the global and the local empty vehicle manager and the two decision layers with responsibilities are properly separated.

5.3 Integral planning (EVM4)

In the previous variant, the global and local empty vehicle managers seem to be well coordinated. A drawback however is the fact that only the next assignment of each vehicle is taken into account, without considering its impact on the distribution of vehicles through the transportation system later on. For example, the global empty vehicle manager may dispatch empty vehicles to terminal 1, knowing that they will be merely used to transport cargo to terminal 2. Hence these vehicles can be used to transport loads from terminal 2 to another destination later on. If the global empty vehicle manager ignores this information, as in EVM3, the consequence can be that empty vehicles will be dispatched to terminal 2 that appear not to be necessary later on. This may be overcome by making an integral planning in which all orders and empty vehicle trips are considered. This integral planning is the main difference between EVM4 and EVM3. A requirement for integral planning is that all information about orders, vehicles and docks is centrally available.

5.3.1 Relation between global and local empty vehicle manager. Based on information about the system status and all orders, the global empty vehicle manager optimises a transportation schedule for all orders and trips. This means that the global empty vehicle manager virtually assigns a sequence of jobs to each vehicle. When planning more jobs ahead, as in this integral approach, some jobs may seem inefficient locally, but lead to a better schedule globally. Therefore, based on the transportation schedule, the global empty vehicle manager provides the local empty vehicle managers with a sequence of load jobs for each terminal and a list of empty vehicle jobs with dispatch times. The local empty vehicle managers follow this schedule in the sense that load jobs are handled in the prescribed sequence and empty vehicles are dispatched at the prescribed time or as soon as possible afterwards. Note that the dispatch times of empty vehicles are required to avoid that an available empty vehicle, waiting for a load job, is used for another empty vehicle job. Recall that the global empty vehicle manager as described in the previous section (EVM3) just provides a list of empty vehicle dispatch jobs with *latest dispatch time* to the terminals.

The assignment of docks, the scheduling of unload jobs and the exact timing of load jobs, given the order prescribed by the global empty vehicle manager, is left to the local order scheduler. Also, the local empty vehicle manager is responsible for the assignment of particular empty vehicles to particular jobs. Although the latter seems to be a marginal authority, this element of the control structure is still important. It provides the local empty vehicle managers with the authority to deal with local circumstances, such as vehicle positioning related to terminal layout and reacting on local disturbances (dock and vehicle failures). Assigning all authority to the network level would lead to an unwieldy control system with excessive information exchange (e.g. all local disturbances should be known at network level).

5.3.2 The integral approach for the global empty vehicle manager. The idea for the integral planning approach is as follows. Periodically the global empty vehicle manager constructs an integral schedule. This can be considered as an off-line problem, since new orders that become known later on are neglected. New orders will only be incorporated when the global empty vehicle manager makes a new plan. Therefore, for such an approach to be fruitful, the planning period for the global empty vehicle manager should not be too large, since otherwise too many new orders are missed, but it should also not be too small since otherwise the efficiency of the global plan is not obtained. Of course, it may vary from case to case what too small and too large means. We will deal with this issue in Section 6 when we discuss the numerical experiments.

For the integral planning by the global empty vehicle manager EVM4 we apply a so called serial scheduling method which uses priorities of jobs to sequentially build a schedule (cf. Kolisch, 1996). The objective is to optimise due time performance, i.e. to minimise the maximum lateness of the jobs. Next, we describe this method to optimise a schedule, given a fixed system status. As mentioned, the information is continuously subject to changes due to the dynamics of the system. This can be handled by calling the serial scheduling method either periodically (with a fixed time interval) or event triggered (at each order arrival). As the second option would lead to excessive planning time because of the high order arrival rate, we chose in our application for scheduling with a fixed time interval.

5.3.3 The serial scheduling method. Input to the scheduling are the current status of the system and a list of orders. The status of the system is defined by the current time, the expected travel times for all origin-destination pairs in the system, the destination and the expected arrival time of every vehicle and the status of a vehicle (loaded or empty). The list of orders contains all orders n known to the system ($tK_n \leq t_0$), including orders in process. Recall that an order in process may either be on a load dock waiting to be handled, it may be in transportation on a vehicle, or it may be at a dock, being unloaded.

We apply a serial scheduling method, based on deterministic information. That is, we treat all expected handling, travel and arrival times as deterministic variables. Recall that these times are actually random variables, so the assumption of deterministic information serves as an approximation. This assumption is justified by the fact that the variations in the on-line, time dependent forecasts for these random variables are relatively small. Note however that the values of e.g. mean handling times may vary, depending on workload. The following steps are taken in our serial scheduling method:

1) *Initialisation.* From the status of the system, compute so called *initial empty vehicle profiles* for every central and local parking. An initial empty vehicle

profile describes the number of available empty vehicles for any $t \geq t_0$, given that no further jobs are started. Furthermore, we compose a sorted list of jobs to be scheduled, containing all jobs n that still have to be started, sorted by increasing latest departure time tL_n.

2) Let job n be the first job on the sorted list. Evaluate all possible assignments of empty vehicles to job n, and assign that empty vehicle that serves job n as early as possible. The resulting assignment leads to an *earliest serving moment* tE_n and a slack $tS_n = tL_n - tE_n$ for job n.
3) Try if any of the other jobs on the list can be *combined* with job n. By our definition, a job m can be combined with job n, if and only if the destination of job m equals the origin of job n. If such a combination is possible, it prevents unnecessary empty vehicle movements, but it also may cause a delay compared to the earliest serving moment tE_n. Combined transportation is controlled by two parameters, namely the *maximum delay* in serving time and the *minimum slack* that must remain for job n (i.e. the time gap between the scheduled starting time and the latest departure time of job n). To combine transportation, compute per terminal the job with the highest priority that can be combined with job n, such that the minimum slack and maximum delay are not violated. Amongst all these jobs, accept the job m that leads to the smallest delay for job n. Note that job m, and afterwards job n, may use an other empty vehicle than the one assigned to job n in step 2. In that case the empty vehicle assignment of step 2 is cancelled.
4) Remove job n and, in case of a combined transportation, remove job m from the list of jobs to be scheduled. Update the empty vehicle profiles based on these scheduled jobs.
5) Repeat steps 2 to 4 until the list of jobs to be scheduled is empty.

We notice that the orders are leading in the scheduling process for EVM4. That is, unlike the planning options as described in Sections 5.1 and 5.2, in the integral planning approach a vehicle may wait empty for an order to be processed instead of starting with the most urgent order amongst those orders that can start immediately.

Clearly, this sequential scheduling approach may improve the performance of empty vehicle management in terms of due time performance. The efficiency of the method may depend on two parameters, maximum delay and minimal slack, that may be worthwhile to be tuned to a specific case. Note that larger values for the maximum delay and smaller values for the minimum slack leave more room for combined transportation. On the other hand a larger maximum delay may postpone more jobs and a smaller minimum slack may result in more jobs that are scheduled close to their latest departure time. Such decisions may appear efficient for the integral off-line planning. However, the efficiency in a real-time, dynamic environment will depend on the length of the planning period and the dynamics and stochasticity of the environment.

Note that the scheduling method neglects two finite capacities, namely the capacity of the docks and the capacity of the parkings. Thus it assumes that (1) every job can be handled immediately at the dock and (2) that every empty vehicle can stay at the terminal where it becomes available. The first assumption may be critical at some moments. In our specific application, the availability of the vehicles is the bottleneck rather than the capacity of the docks. For other cases, we believe that it is possible to extend our method towards finite docking capacities. The second assumption is less critical if the terminal has reasonable parking space, as excess empty vehicles will be sent to another parking by the empty vehicle manager anyway. Besides, each terminal has queuing space in front of the entrance (see Sect. 2). As we shall see in Section 6, the simulation results suggest that these limiting assumptions are not too harmful in the case under consideration. This could change however if terminals hardly have parking space. Then refinements are possible by introducing dock and parking profiles.

The flexibility of the serial scheduling approach is a clear advantage. Several schedules can quickly be generated, based on different performance criteria, priority rules and parameter settings (maximum delay, minimum slack). We can also use a given maximum computation time as restriction for the number of schedules to generate.

6 Numerical investigation

We applied our logistics planning and control structure to the Dutch pilot project around Amsterdam Airport Schiphol as described in the introduction. An object-oriented simulation model based on eM-Plant has been constructed and all planning variants have been embedded. This discrete event simulation model reflects many details of the logistics system, such as individual load movements between buffers and docks and individual AGV movements between locations in the system along some well-defined track system (docks, parkings, junctions, etc.). To this end, we could use predefined objects from the eM-Plant simulation library. Detailed AGV-behaviour (acceleration, distance control) has not been included in our model, as we focused on network performance. Where required, we used a rough model of the AGV behaviour. For example, we used a statistical relation between the number of vehicles on a terminal and the effective AGV speed. This relation has been estimated using a separate, detailed simulation model that has been constructed by Verbraeck et al. (1998).

Although the eM-Plant object library provides a sound basis for our simulation model, the logistics control system almost had to be constructed from scratch. Most heuristics and rules were implemented in eM-Plants' own programming language. Only for empty vehicle manager EVM4, we have coded the serial scheduling method as an external routine in C++ that is periodically called by the eM-Plant simulation model.

In the remainder of this section, we first present the experimental setting and procedure (6.1). Next, we discuss some initial experiments on planning parameters (6.2). The numerical results of our simulation experiments are the subject of Subsection 6.3.

6.1 Experimental setting and procedure

For all our experiments, we use the layout as shown in Figure 1. The transportation system is further specified by the following parameters:

1) The time required for loading and unloading has mean 2 minutes and standard deviation 15 seconds. In the simulation experiments, these values were sampled from a gamma distribution with these characteristics.
2) At each dock only one AGV can load or unload, but there is a waiting position for a second AGV.
3) The number of locations to park empty AGVs temporarily between unloading and loading equals the number of docks.
4) The network contains one central parking that is located between the AAS-terminals and the junction to Hoofddorp and Aalsmeer.
5) The AGV speed is 6 m/s in the underground tubes, 2 m/s in the terminal and 3m/s on slopes between the underground tubes and the terminals at the surface.
6) The order pattern is based on projections for the year 2020 as generated by the Dutch Economical Institute and expressed as the number of standard load units. It is expected that the system will be in full operation in 2020. We used a peak day with heavily imbalanced transportation flows (Tuesday), a peak day with more balanced transportation flows (Monday), and an average day (Friday) to establish the capacities required (number of docks and AGVs) and to test our decision rules. In the simulation model, orders were sampled from a Poisson distribution at the VBA and AAS terminals. At the rail terminal (RTH), 6 batches per hour arrive where the number of orders in each batch were sampled from a Poisson distribution. Batch arrivals at the rail terminal are justified because trains containing a large number of loads arrive according to some fixed schedule.
7) We have various priority classes of orders, specified by their throughput time requirements. The throughput time requirements are as follows:
   - to and from Amsterdam Airport Schiphol (AAS): within 45 (10% of the orders), 60 (20%) 90 (20%) and 120 minutes (50%);
   - between Hoofddorp (RTH) and Aalsmeer (VBA): within 60 (10% of the orders), 75 (20%), 90 (20%) and 120 minutes (50%).
8) Energy provisioning of AGVs is arranged via electric wires in the tube system. While driving on the terminals, the AGVs use batteries that are recharged while driving in the tubes. As a consequence, battery recharging or changing is not required (cf. Ebben, 2001 for the analysis of alternative options of energy provisioning using batteries).
9) Equipment failures (AGVs, docks) are not taken into account in the numerical experiments, although the control frameworks and rules still apply (cf. Ebben, 2001 for the impact of equipment failures and failure management).

The logistics performance is measured using a large number of statistics, such as dock and AGV utilisation, throughput times, order waiting times, lateness and fill rates. The statistics are collected time dependent (e.g. for each day or even hour) and per route (terminal of origin and destination). In this section, we focus on the key performance characteristic, namely the fill rate. Recall that the fill rate is defined as the fraction of orders that is delivered before the due time (cf. Section 2). All numerical results presented in this paper are based on simulation runs of 30 consecutive days. The CPU requirements are in the range 1-2 hours on a Pentium II-266 MHz PC with 128 MB RAM.

As there are hardly any unserviced orders left around midnight, the statistics gathered per day are treated as independent observations. Formally the observations are dependent, but our experiments showed that the correlation between successive days is negligible. Therefore we can give an indication for the accuracy of the results corresponding to a run length of 30 days. We found that the confidence intervals of the fill rates are quite small in the range that is of interest to us, a fill rate of about 99% . Then the 0.95 confidence interval length for the fill rate is in the range 0.1–0.25% . When the fill rates are smaller than 98% , the confidence intervals become larger, even larger than 1% , but the required system performance should be around 99%.

In the numerical study we examined the following effects:

1) The performance of the four options EVM1-EVM4.
2) The impact of the period for which orders are known in advance (value of information).

Prior to the simulation study, a number of experiments were performed in order to determine the following model parameters:

a) Resource capacities
b) The choice of the planning frequency of task allocation to docks, see Section 4
c) Parameter settings for EVM4
d) The choice of the planning frequency when empty vehicle management is coordinated at network level (EVM3 and EVM4)

The results are presented in the following subsection.

6.2 Initial experiments

6.2.1 Resource capacities. Before examining the performance of the various logistics control procedures, we studied the key resource capacities. Our analysis revealed that 2 docks at each AAS terminal, 5 docks at each VBA terminal and 10 docks at the RTH terminal are sufficient to handle all orders. The system should contain at least 185 AGVs to handle all loads in peak hours on Tuesday. In this situation, vehicle availability is the bottleneck rather than dock capacity. For Monday, 120 AGVs appeared to be sufficient. Because the transportation flows are better balanced, less empty vehicle movements are needed and hence vehicles are better utilized. For Friday, transportation flows are quite imbalanced, yielding a number of 165 AGVs to handle the peak demand. We used these figures for our simulation experiments.

6.2.2 Planning frequency of task allocation to docks. For the period during which the task allocation is fixed, we found that the effect of fixing the tasks for some

Table 2. Effect of fixed dock task period (Tuesday, 185 AGVs, EVM3 planning each 30 minutes with orders being known 30 minutes in advance)

	Fill rate						
Fixed dock task period	AAS-RTH	AAS-VBA	RTH-AAS	RTH-VBA	VBA-AAS	VBA-RTH	Overall
15 min.	99.7	99.7	99.6	100.0	98.9	99.0	99.3
30 min.	99.3	99.3	99.5	100.0	97.8	98.8	99.1
60 min.	98.1	98.9	98.4	100.0	96.5	96.9	97.7

period of 15-30 minutes has only a slight impact on system performance, see Table 2. Once the period increases to one hour or more, the performance may decrease significantly. As a longer period has technical and organisational advantages (less task changes), we chose 30 minutes for the rest of our numerical experiments.

6.2.3 Parameter settings for EVM4. As mentioned in Section 5.3, the empty vehicle manager EVM4 periodically performs an integral planning run. We may call the serial scheduling method once at each run, but we can also call it several times for different parameter settings. Then, the best schedule obtained is selected. Possible parameters to be varied are the maximum delay and the minimum slack, the control parameters for combined transportation. We may also use other priority rules than the latest departure time. However, using the latest departure time as priority rule allows EVM4 to be evaluated against other empty vehicle managers. We found that EVM4 generally requires about half a second of CPU time per schedule for our application. Because we needed lengthy simulation runs to compare the various empty vehicle managers, we evaluated only four schedules per integral planning run to keep the run time acceptable. Preliminary analysis shows that the schedules are more sensitive to the minimum slack than to the maximum delay. Therefore, we decided to use only one value for the maximum delay, namely 8 minutes. Since the sum of the (expected) loading and unloading time is 4 minutes, a maximum delay of 8 minutes allows to wait for an empty vehicle or for the release time of a job for at most 4 minutes. A small minimum slack generally increases the number of combined transportation jobs and reduces empty vehicle travel. However, this may lead to a larger maximum lateness. We found that a minimum slack between 5 and 20 minutes yields good results. A minimum slack smaller than 5 minutes reduces the on-line fill rate significantly because too many jobs are scheduled close to their latest departure time. On the other hand, a minimum slack larger than 20 generally did not result in significantly different schedules. Among the four schedules generated, we selected the schedule with the smallest maximum lateness. In case of ties, the schedule with the smallest total travel time for empty vehicles was selected.

6.2.4 Planning frequency for coordinated empty vehicle management. For coordinated empty vehicle management (EVM3 and EVM4), the planning frequency has to be determined. As the system situation changes quickly, infrequent rescheduling may decrease the system performance. Table 3 shows some corresponding simula-

Table 3. Effect of empty vehicle planning frequency (Tuesday, 185 AGVs, orders being known 30 minutes in advance)

	Fill rate							
EVM period	AAS-RTH	AAS-VBA	RTH-AAS	RTH-VBA	VBA-AAS	VBA-RTH	EVM3 Overall	EVM4 Overall
10 min.	99.3	99.3	99.5	100.0	97.8	98.8	**99.1**	**98.9**
20 min.	99.0	99.4	99.1	100.0	96.6	97.7	**98.4**	**98.4**
30 min.	98.9	99.3	99.2	100.0	97.0	97.4	**98.2**	**96.8**
40 min.	98.2	98.8	99.0	100.0	95.7	95.6	**97.0**	X
50 min.	96.8	97.6	99.5	100.0	90.9	91.9	**94.7**	X
60 min.	93.3	96.0	99.0	99.9	82.1	84.6	**89.8**	X

Table 4. Effect of empty vehicle management (orders being known 30 minutes in advance)

		Overall fill rate			
Day	AGVs	EVM1	EVM2	EVM3	EVM4
Monday	120	94.7	97.1	99.8	100
Tuesday	185	80.3	89.9	99.1	98.9
Friday	165	85.5	93.2	99.7	99.6

tion results when orders are known 30 minutes before cargo arrival at the terminal of departure. Note that a prerequisite for proper functioning of EVM4 is that the planning period should be less than or equal to the time for which orders are known in advance. For example, if the planning period is 40 minutes and orders are known 30 minutes in advance, the orders arriving after 30 minutes will not be scheduled until the next EVM planning. Hence these loads will wait until the next schedule is generated, even if sufficient vehicle and dock capacity is available. Obviously, this is not a practical option. Therefore, Table 3 only shows the performance of EVM4 if the planning frequency is at least once per half hour.

Surprisingly, Table 3 shows that the fill rate is not extremely sensitive to the empty vehicle manager planning frequency. As long as the frequency is at least once every half hour, the fill rate remains high. For the time being we choose to reschedule every 10 minutes, yielding high customer service.

6.3 Numerical results

6.3.1 Comparison of empty vehicle management strategies: the value of coordination. An important research question is the effectiveness of the various options for empty vehicle management. We tested these options for the three weekdays as mentioned in Section 6.1. The overall fill rates are shown in Table 4 for all cases.

We see that planning coordination (included in EVM3 and EVM4) has a significant impact on the customer service. It is remarkable that the effects are considerably smaller for Monday than for the other two days. A possible explanation for this phenomenon is the fact that transportation flows are more balanced on Monday.

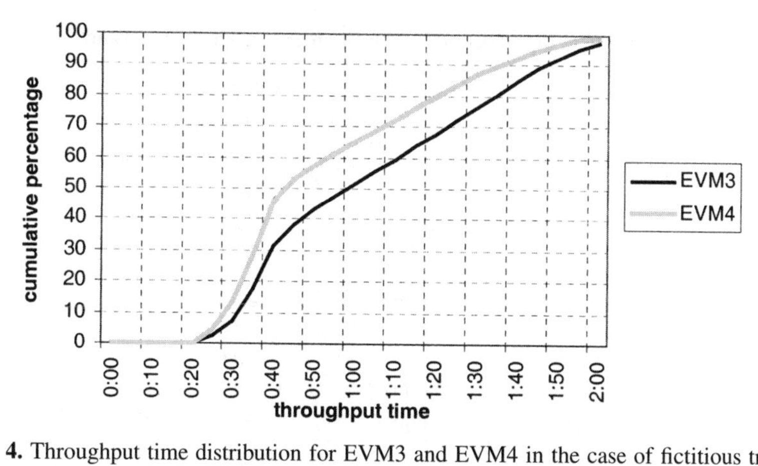

Fig. 4. Throughput time distribution for EVM3 and EVM4 in the case of fictitious transportation flows with quickly moving peak levels between routes

To give an indication, on Monday the flows from VBA to RTH are similar to or at most double the return flows, whereas this ratio is 5-10 on the other days. Hence, vehicles tend to be positioned at the right location without much coordination on Monday. Especially when transportation flows are heavily imbalanced (Friday and particularly Tuesday), proper empty vehicle management may increase the system performance considerably.

For the cases mentioned in Table 4, EVM3 and EVM4 have more or less the same overall fill rate. To judge whether this is a general result, we constructed an additional artificial case. This case is based on the same network, but with a different transportation pattern. To test the empty vehicle managers under difficult conditions, we defined transportation patterns with peak levels moving quickly from one route to the other. This means that anticipation of heavily fluctuating transportation demand is essential to attain high customer service levels. For this case, we found that EVM4 performs significantly better than EVM3 (fill rate 98.5% versus 95%). Also, the throughput time distribution under EVM4 appears to be significantly better, see Figure 4. Hence we conclude that EVM4 is preferable in the sense that it performs at least equal to EVM3, and under difficult conditions considerably better than EVM3. Moreover, as stated in Section 6.2.3., we could run EVM4 with more parameter settings or priority rules to improve its performance.

For the two best empty vehicle managers, EVM and EVM4, we compared the average empty travel distance per day. In this respect, we found that EVM4 requires 3-10% less empty travel than EVM3, depending on the specific case. Hence EVM4 is also more efficient from an environmental point of view. Besides, empty travel reduction is relevant if AGVs use batteries for energy provisioning, because the recharging frequency is reduced. The choice of energy provisioning (electric or combustion engine) is beyond the scope of this paper.

6.3.2 Impact of the information horizon: the value of information. As mentioned before, pre-information about orders may facilitate planning and improve the sys-

Table 5. Effect of the period for which orders are known in advance

Day	AGVs	orders known	Overall fill rate EVM2	EVM3	EVM4
Monday	120	0 min.	94.7	99.4	99.0
		30 min.	97.1	99.8	100
		60 min.	90.3	99.9	100
Tuesday	185	0 min.	80.3	84.7	92.4
		30 min.	89.9	99.1	98.9
		60 min.	93.0	98.5	99.2
Friday	165	0 min.	85.5	90.0	94.8
		30 min.	93.2	99.7	99.6
		60 min.	95.6	99.9	99.6

tem performance. However, this requires additional communication and hence more sophisticated organisation and information systems. Therefore we analyse the effect of the period for which orders are known in advance. As pre-information is only used for the empty vehicle manager options 2, 3 and 4, we omit option 1 (FCFS myopic). Obviously, EVM2 (FCFS-look ahead) is identical to EVM1 if pre-information is not available (horizon = 0). The main results are shown in Table 5.

We notice a similar phenomenon as in the previous subsection, namely that the impact of pre-information is highest if the transportation flows are heavily imbalanced (Tuesday and Friday). Pre-information appears to have even a stronger effect on customer service than coordination. Of course, proper coordination is facilitated by the availability of sufficient information. The marginal value of additional pre-information clearly decreases.

It is remarkable that for EVM2 the overall fill rate on Monday is even considerably less when orders are known 60 minutes in advance compared to only 30 minutes pre-information. This phenomenon can be explained as follows. Vehicles are dispatched to the terminals FCFS once the orders are known. If orders are known 60 minutes in advance, they arrive too early at the terminal and have to wait for a long time before they can be loaded, causing a significant capacity loss on a bottleneck resource (AGVs).

In the pilot project one hour of pre-information will be sufficient for practical purposes because of the maximum travel times between the various locations in the network.

7 Conclusions and directions for further research

In this paper, we developed several options for empty vehicle management in automated transportation networks and embedded these planning methods in a logistics planning and control framework. The various methods have been implemented and tested in a simulation environment for the case of an underground transportation system using AGVs around Amsterdam Airport Schiphol.

We found that both information about future orders and planning coordination between terminals gives considerable advantage in terms of customer service (fill rates). A relatively simple method to balance empty vehicle flows in the system (EVM3), taking into account some future order information and communication with local (terminal) levels, already appears to provide considerable benefits in many cases. The advantage of EVM3 is the relatively low level of complexity and information exchange between local (terminal) and central (network) level.

A more advanced serial scheduling method (EVM4) can improve the performance further, especially in difficult cases when peak demand quickly moves from one route to another. Then it is worthwhile to plan a sequence of orders in an integral way for a longer time horizon. Also, integral planning has the advantage of empty travel reduction. Moreover, the serial scheduling method offers a broad range of possibilities for further refinements. In this respect, inclusion of finite terminal capacities, both docking capacities and parking space, is an interesting subject for further research.

Regarding further research, another interesting topic is the consideration of additional resources in the integral planning aproach. For example, the pilot project on underground transportation around Schiphol Airport has evolved to a network containing bi-directional tubes (cf. van der Heijden et al., 2000). That is, the system contains single tubes for traffic in two directions in order to save investment. As a consequence, travel times between locations may fluctuate heavily, because vehicles may have to wait before entering a bi-directional tube. The simple coordinated method EVM3 can easily incorporate these fluctuating travel times, but the performance of this method may decrease. In general, the combination with finite terminal capacities is especially hard, as scheduling on finite capacities requires predictable travel times. So modelling the bi-directional tubes simply as random variables with considerable variance does not seem to be appropriate. A challenging alternative is the extension of EVM4 to incorporate both bi-directional tube scheduling and finite terminal capacities. We expect that the flexible approach should allow this global scheduling method. This issue is the main focus for our further research.

References

Akturk MS, Yilmaz H (1996) Scheduling of automatically guided vehicles in a decision making hierarchy. International Journal of Production Research 34(2): 577–591

Cheung RK, Chen C-Y (1998) A two-stage stochastic network model and solution methods for the dynamic empty container allocation problem. Transportation Science 32(2): 142–162

Cheung RK, Powell WB (1994) An algorithm for multistage dynamic networks with random arc capacities, with an application to dynamic fleet management. Operations Research 44(6): 951–963

Dejax PJ, Crainic TG (1987) A review of empty flows and fleet management models in freight transportation Transportation Science 21(4): 227–247

Ebben MJR (2001) Logistics control of automated transportation networks. PhD thesis, University of Twente, Faculty of Technology and Management (forthcoming)

Egbelu PJ, Tanchoco JMA (1984) Characterization of automated guided vehicle dispatching rules. International Journal of Production Research 22: 359–374

Gademann AJRM, van de Velde SL (2000) Positioning automated guided vehicles in loop layout. European Journal of Operational Research 127: 565–573

van der Heijden MC, van Harten A, Ebben MJR, Saanen Y, Verbraeck A, Valentin E (2000) Safeguarding Schiphol Airport accessibility for freight transport: The design of a fully automated underground transportation system and the role of simulation. Working paper, University of Twente and TRAIL Research School (submitted for publication)

Holmberg K, Joborn M, Lundgren JT (1998) Improved empty freight car distribution. Transportation Science 32(2): 163–173

Hu CH, Egbelu PJ (2000) A framework for the selection of idle vehicle home locations in an automated guided vehicle system. International Journal of Production Research 38: 543–562

Jordan WC, Turnquist MA (1983) A stochastic, dynamic network model for railroad car distribution. Transportation Science 17(2): 117–145

Kim KH, Kim JY (1997) Estimating mean response time and positioning idle vehicles of automated guided vehicle systems in loop layout. Computers and Industrial Engineering 33: 669–672

Klein CM, Kim J (1996) AGV dispatching. International Journal of Production Research 34: 95–110

Kolisch R (1996) Serial and parallel resource-constrained project scheduling methods revisited: Theory and computation. European Journal of Operational Research 90(2): 320–333

Powell WB (1996) A stochastic formulation of the dynamic assignment problem, with an application to truckload motor carriers. Transportation Science 30(3): 195–219

Powell WB, Carvalho TA (1998) Dynamic control of logistics queuing networks for large–scale fleet management. Transportation Science 32(2): 90–109

Powell WB, Sheffi Y, Nickerson KS, Butterbaugh K, Atherton S (1988) Maximizing profits for North American Van Lines' truckload division: A new framework for pricing and operations. Interfaces 18: 21–41

Silver EA, Pyke DF, Peterson R (1998) Inventory management and production planning and scheduling. Wiley, New York

Ulusoy G, Bilge U (1993) Simultaneous scheduling of machines and automated guided vehicles, International Journal of Production Research 31: 2857–2873

Verbraeck A, Saanen YA, Valentin E (1998) Logistic modeling and simulation of automated guided vehicles. In: Bargiela A, Kerckhoffs E (eds) Simulation technology: science and Art, pp. 514–519. SCS, San Diego CA

White W, Bomberault A (1969) A network algorithm for empty freight car allocation. IBM Systems Technical Journal 8: 147–171

van der Zee DJ (1997) Simulation as a tool for logistics management. PhD Thesis, University of Twente

Modeling of capacitated transportation systems for integral scheduling*

Mark Ebben[1], **Matthieu van der Heijden**[1], **Johann Hurink**[2], **and Marco Schutten**[1]

[1] University of Twente, School of Business, Public Administration and Technology, P.O. Box 217, 7500 AE Enschede, The Netherlands
[2] University of Twente, Faculty of Electrical Engineering, Mathematics and Computer Science, P.O. Box 217, 7500 AE Enschede, The Netherlands

Abstract. Motivated by a planned automated cargo transportation network, we consider transportation problems in which the finite capacity of resources (such as vehicles, docks, parking places) has to be taken into account. For such problems, it is often even difficult to construct a good feasible solution. We present a flexible modeling methodology which allows to construct, evaluate, and improve feasible solutions. This new modeling approach is evaluated on instances stemming from a simulation model of the planned cargo transportation system.

Keywords: Transportation – Scheduling – Modeling – Heuristic

1 Introduction

In this paper, we consider transportation scheduling problems in which the finite capacity of resources (such as vehicle parking places and docks for loading and unloading) has to be taken into account. The motivation of our research is a planned automated cargo transportation network using automatic guided vehicles (AGVs) around Schiphol Airport, the Netherlands (see van der Heijden et al. [12]). To run such a network, a wide range of coherent decisions has to be taken, such as prioritizing transportation jobs, load consolidation, assignment of loads to AGVs and docks (for loading), assignment of loaded AGVs to docks (for unloading), and redistribution of empty AGVs to cope with imbalanced transportation flows. A suitable logistic planning and control system should be fast and flexible, that is, it should be able to take decisions in real time, taking into account rapidly changing circumstances (such as the arrival of rush jobs and AGV failures), and the limited

* The authors gratefully acknowledge the constructive comments of the anonymous referees, which helped to improve the presentation of the paper a lot.
Correspondence to: M. Schutten (e-mail: m.schutten@utwente.nl)

resource capacities. Van der Heijden et al. [12] developed a local control concept, in which they use relatively simple heuristics and algorithms for each decision. Although such an approach is fast and flexible indeed, efficient resource usage is not guaranteed. Integration of some key decisions may improve the system performance, see Van der Heijden et al. [11] and Ebben et al. [10]. Therefore, it is interesting to determine to which extend and under which circumstances integral optimization improves the system performance in terms of reaching the same service levels using less resources (or reaching higher service levels using the same resources). Obviously, such an integral approach should still be fast and flexible.

Real-life situations often incorporate a lot of side constraints, which are often not included in the available approaches. In our approach, we are able to include resource capacity constraints, such as limited loading/unloading capacity and limited parking space on the terminals. In this paper, we will present a generic approach to model integral network scheduling problems that can meet these requirements. The emphasis is on developing a flexible modeling methodology, facilitating construction, evaluation, and improvement of feasible solutions. Although our approach facilitates optimization, we do not consider such optimization algorithms with their performance in this paper. As an example, we will show how our approach can be implemented for the automated transportation network mentioned above. However, our approach is suitable to deal with other heavily automated transportation and transshipment systems as well, such as scheduling of container terminals, where loading and unloading operations have to be scheduled using cranes and vehicles.

The literature on transportation scheduling under multiple resource constraints is limited. Kozan [14] investigates the minimization of handling and traveling times of containers from the time the ship arrives until all containers from that ship leave the port. He performs a sensitivity analysis with respect to equipment type, trucks, highstackers and shore cranes. Interactions between vehicles and, for example, limited parking space for vehicles are not taken into account. Kozan and Preston [13] present a genetic algorithm for the related problem of optimizing container transfers at multimodal terminals.

Bostel and Dejax [4] address the problem of rail-rail transshipment shunting yards. They propose optimal and heuristic methods to minimize the number of container moves in a terminal. In their models they take intermediate container storage capacity into account.

Meersmans [16] considers the optimization of container handling systems, where he designs models for integrated scheduling of the handling equipment. To solve the models, he develops exact and heuristic algorithms and evaluates the performance of these algorithms. In his models he does not consider finite parking capacity, i.e. the queuing of AGVs in front of quay cranes and stacking lanes.

Alicke [2] investigated an intermodal terminal concept called Mega Hub. He models the terminal as a multi-stage transshipment problem where sequence-dependent duration of empty moves, alternative assignments of containers to cranes and a sequence-dependent number of operations have to be handled. An optimization model based on Constraint Satisfaction is formulated and heuristics are developed.

The stated requirements for the transportation problem indicate that the considered transportation systems are quite complex and general, and that the resulting problems form a combination of scheduling and vehicle routing aspects. This combination is hardly considered in the literature. Within the scheduling area, transports or routing aspects are often neglected or only simple variants of these aspects are treated:

- within robotic cells only simple movements by a robotic arm are possible (for a survey cf. Crama et al. [8])
- shop problems with transportation times incorporated consider basicly the vehicles as additional machines, i.e. they assume fixed transportation times and neglect the parking aspects within terminals and the handling of empty vehicles (see e.g. Bilge and Ulusoy [3] or Hurink and Knust [13])
- shop problems with buffers consider the parking storage aspects between machines but do not combine these aspects with transportation issues or vehicles (see e.g. Nowicki [17] or Brucker et al. [5]).

On the other hand, in the routing area capacities within locations or timing restrictions are often neglected or treated very simplified:

- in the vehicle routing problem with time windows only timing restrictions on the load/unload operation are given (see e.g. Cordeau et al. [7])
- the capacitated vehicle routing problem considers only capacities for the vehicles but not for the locations (see e.g. Ralphs et al. [18])

Thus, the considered problem forms a new challenging problem where two areas with different approaches are combined. For this problem it is already difficult to develop a reasonable constructive heuristic which takes into account all the given constraints.

The remainder of the paper is organized as follows. In Section 2, we will give a more detailed description of the problems that we consider. Then, in Section 3, we present our approach to model these problems. Section 4 describes how we construct a feasible solution and we apply this approach in Section 5 to a real-life application. In Section 6, we discuss some model extensions, stemming from practical cases. Section 7 ends this paper by giving a number of conclusions and suggestions for further research.

2 Problem description

Transportation systems in general have two main layers: the physical layer and the logistic layer. The physical layer of the system is considered to be fixed. It specifies the locations, where decisions with respect to the transportations have to be made, the possible connections between these locations, and the equipment (vehicles) which is used for the transportations. The logistic layer uses the physical part as input and specifies the orders or tasks which have to be performed by the system. In the following three subsections we specify the transportation system considered in this paper by describing the components of the two layers in more detail. Based on the specifications given in these three subsections, in Section 2.4 we discuss

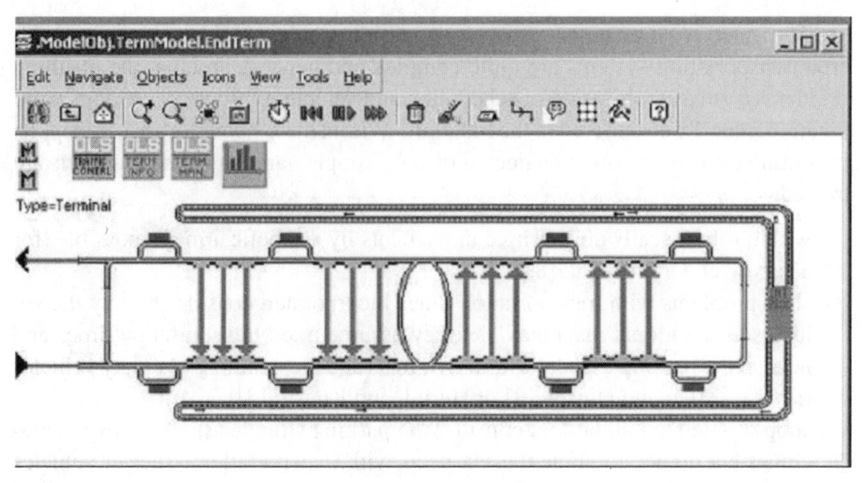

Fig. 1. Layout of a terminal

the type of decisions which have to be taken within the considered transportation model.

2.1 The system network

The system network is the part of the physical layer which specifies the 'decision' locations (called *terminals* in this paper) and connections between them. In our model, terminals are locations where orders may depart and arrive and, thus, loading and/or unloading operations take place. These terminals may have a rather complex substructure. In the considered transportation system, a terminal consists of a terminal parking to store vehicles and a set of docks to handle load and unload operations. Within the terminal, connections between the parking and the docks are given, i.e. the terminal contains a small 'system network' itself. We assume that the structure within a terminal can be expressed by travel times necessary to travel between the parking and the docks. In Figure 1 a layout of one of the planned terminals for the cargo transportation system around Schiphol Airport (see also Sect. 5) is given. Although, in general, terminals may have a rather complex structure, also special terminals which contain only a parking or only docks may exist.

Between the terminals connections exist. We assume that each existing direct connection between two terminals consists of a unidirectional single track without any intermediate terminals on this track. For each connection a length is given. This length is used to estimate the travel time for the connection.

It remains to describe the ingredients of a terminal - docks and parkings - in more detail.

Docks. Docks are the places where the vehicles are physically loaded and unloaded. They consist of a dock parking (often very small - place for one or two vehicles or even no parking) and a set of parallel servers (often only one). A vehicle enters a dock via the parking and will be directed to one of the servers, where the load or unload operation takes place. Within a dock the times for loading or unloading a vehicle on a server and the setup time between two consecutive vehicles on a server (minimal time between the departure of a vehicle and the arrival of the next vehicle on the server) are given.

Parkings. A parking (terminal parking as well as dock parking) is specified by the number of vehicles which jointly may occupy the parking and by the operating mode of the parking. As operating modes we consider FIFO (the vehicles have to leave the parking in the same order in which they entered it) or an arbitrary mode.

2.2 The transportation equipment

In general there are three important characteristics of vehicles. One main characteristic is given by the number of items a vehicle can transport. The simplest case occurs if each vehicle can transport only one *transportation unit* (e.g. a pallet or a container) at a time. A transportation unit, however, can consist of several items. AGV systems usually use transportation units to speed up loading and unloading operations. In this single-unit case, the route of a vehicle is characterized by a sequence of orders assigned to the vehicle. The vehicle has to travel from the origin to the destination of an order and, afterwards, to the origin of the next order. In our model we assume that there is exactly one route between each pair of locations. The more general case where each vehicle can transport multiple transportation units at a time leads to a more complex routing problem: for the vehicle a sequence in which the orders assigned to the vehicle are picked up and delivered has to be determined. This sequence must be chosen such that the capacity of the vehicle is respected.

A second important characteristic of the transportation equipment is given by the speed of the vehicles. Again, the homogeneous case, where all vehicles have the same speed, is the easiest to deal with. For the heterogeneous case (different speeds for the vehicles) congestion may have a large influence on the behavior of the system.

A third characteristic of the transportation equipment is the size of the vehicles. If vehicles of different size are present, some of the terminals/docks may not be reachable by all vehicles and, thus, side constraints on the assignment of orders to vehicles have to be taken into account.

In this paper we consider the simplest case of homogeneous vehicles with respect to speed and size which can transport only one order/item a time. As a consequence, in our model no congestion on tracks is taken into account and the arrival time of a transportation is equal to the departure time plus a fixed travel time. Furthermore, the only specification for this part of the system is the number of available vehicles. However, the presented approaches can be adapted to different vehicle sizes without much effort (see Section 6). An adaptation to vehicles with different speed and to vehicles that can transport several orders a time is not straightforward and asks for new techniques.

2.3 The orders

Following the choice for the vehicles, the orders are considered to be one-item orders. The origin and destination of an order are specified on terminal level. Furthermore, an earliest departure time (release date) from the origin terminal and a latest arrival time (due date) at the destination terminal are given.

2.4 Decisions

The basic decision within the presented model is when and how to transport the given orders from their origin to their destination terminal.

These decisions lead to

- an assignment of orders to vehicles,
- for each vehicle a sequence in which this vehicle transports the orders assigned to it,
- an assignment of vehicles to parkings and/or docks and servers, and
- a timing of all resulting transportations.

All these decisions have to be done in such a way that the resulting solution is feasible. To achieve feasibility, one has to ensure that

- the capacities of the parkings are respected,
- the parking type (FIFO, arbitrary) is respected,
- servers handle only one vehicle at a time,
- the 'service' times (load, unload, setup, ...) and transportation times are respected, and
- the release dates are respected.

We do not consider due dates as a restriction, but we incorporate them in a performance measure.

3 Representation of solutions

In the previous section we have seen that solutions consist of assignments, sequences, and timing of orders, vehicles, and transportations taking into account the capacities within the terminals and the timing constraints. Such a solution may be represented by a set of transportations and a corresponding timing of them. However, for the considered problem it is already difficult to develop a reasonable constructive heuristic which takes into account all the given constraints. This difficulty arises from the fixed travel times between locations and the finite capacities of the parkings. Furthermore, in the area of scheduling it is not very handy to use the concrete timing of activities within the representation of solutions since iterative as well as enumerative methods have problems with such representations. As a consequence, mainly assignments and sequences are used to represent solutions. These representations are chosen such that a corresponding timing can be calculated efficiently.

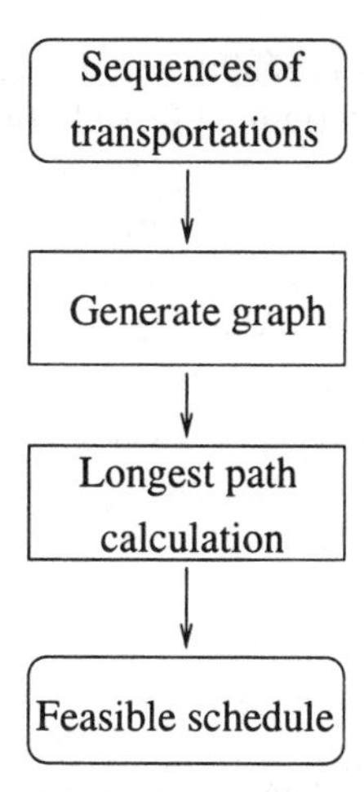

Fig. 2. Solution representation

In the following, we give such a representation of solutions for the considered transportation model (Section 3.1) and a corresponding method for the calculation of the timing (Section 3.2). The basic idea is to represent solutions by sequences of transportations. For a given set of sequences a graph is constructed and via a longest path calculation in this graph a timing of the transportations (i.e. a feasible schedule) is achieved (see Fig. 2). To construct a feasible schedule using this representation, it remains to give a concrete set of sequences of transportations (first element in Fig. 2). In Section 4, we give a corresponding method for constructing such a set of sequences. The basic idea is to use a heuristic to construct a schedule for a relaxation of the considered problem (a relaxation for which a reasonable constructive heuristic is easy to achieve). Then, we extract from this schedule the used transportations and their sequences within the locations. Finally, we calculate for these sequences, with the method developed in this section, a feasible schedule for the whole problem (i.e. the non-relaxed problem).

The basic elements of the representation are transportations. Each transportation is characterized by its origin and destination location and the order which is transported (also 'empty' orders may be assigned). The basic structures of the representation are sequences of these transportations. The sequences are introduced for certain types of locations (docks, parkings) and for vehicles. For a vehicle, the sequence gives all transportations to be carried out by the vehicle (including empty transportations). For the locations, the sequences are chosen such that they enable us to determine a timing of the transportations within the locations. Whereas for pure scheduling problems or scheduling problems with simple transportation models one sequence per location (machines, robots, etc.) is sufficient (see e.g. [13]), it will turn out that for docks and parkings one insequence and one outsequence are needed. Given a set of transportation sequences for the locations, in a straightforward manner, the assignments of vehicles to parkings/docks can be achieved by just looking in which dock- or parking-sequence a certain vehicle occurs. The assignment of vehicles to servers within a dock and the timing of all the transportations needs some more effort. For each location, the corresponding sequences are checked on consistency (local feasibility). This consistency is necessary but not sufficient for being able to determine a feasible timing on the base of the given sequences. Together with the consistency, for a dock also the assignment of the vehicles to the servers is determined. Furthermore, based on the transportation sequences for the locations and vehicles, timing relations for the transportations are derived. Together with some additional timing constraints resulting from the orders, these relations are represented by a directed graph and the timing of the transportations results from longest path calculations in this graph (during this calculation, also global infeasibility can be detected). In the remainder of this section, we present this approach in more detail.

3.1 Required transportation sequences

As mentioned above, the basic elements of the representation are transportations. A transportation consists of:

- an origin and a destination location: These locations are either a terminal parking or a dock. Between these two locations, a direct transportation (without passing any other terminal) has to be possible;
- a vehicle: It indicates which vehicle handles the transportations;
- an order: It indicates which order is transported (the order may be an 'empty' order);
- a travel time.

For such a transportation the departure time at the origin location and the arrival time at the destination location have to be determined. These times have to differ exactly by the travel time.

By specifying a set of transportations (without timing!), already some decisions are made: For each order, the subset of transportations corresponding to this order specifies the route the order will take through the network and whether or not the order has to go via a terminal parking. Furthermore, the transportation starting at the origin terminal specifies the dock at which the order is loaded and the transportation ending at the destination terminal specifies the dock at which the order is unloaded. Finally, the assignment of vehicles to orders has been made.

To achieve a timing of the transportations and, thus, a complete schedule, we have to couple the transportations to avoid conflicts at the different resources (vehicles, parkings, servers, etc.). This is realized by sequences of transportations.

- Each vehicle gets one sequence. This sequence indicates which transportations are carried out by the vehicle and the corresponding ordering.
- Each dock and each terminal parking get an insequence and an outsequence. They indicate the ordering in which the transportations (i.e. the vehicles which carry out the transportations) enter and leave the dock or parking.

Each transportation is inserted in the outsequence of the origin location, in the insequence of the destination location, and in the sequence of the vehicle which handles the transportation.

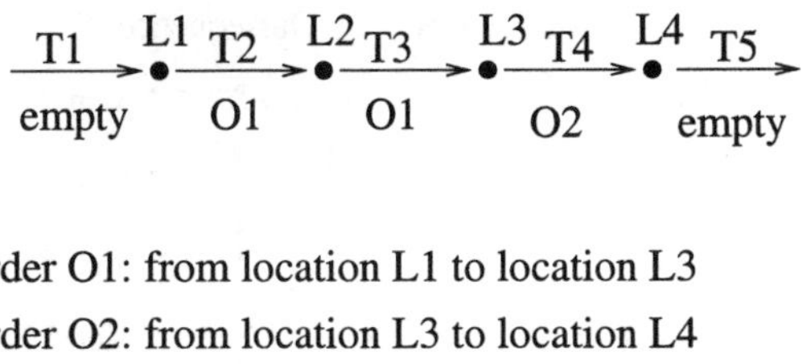

T1 L1 T2 L2 T3 L3 T4 L4 T5

empty O1 O1 O2 empty

Order O1: from location L1 to location L3

Order O2: from location L3 to location L4

Fig. 3. A sequence of transportations for a vehicle

The sequences indicate the ordering in which the transportations will be handled. However, they do not directly lead to a schedule (start times for the transportations). In the remainder of this section, we present necessary conditions which have to be fulfilled by the sequences belonging to a location or to a vehicle to enable a feasible schedule respecting a given set of transportation sequences and a method to calculate a timing of the transportations.

For consistency, the following conditions have to be satisfied:

- **Vehicle sequence**:
  Two consecutive transportations in a vehicle sequence must 'fit' together; i.e.
  - the destination location of the first transportation must be equal to the origin destination of the second,
  - if the orders of the two transportations are different (see transportations $(T1, T2)$, $(T3, T4)$, $(T4, T5)$ in Fig. 3),
    - the first transportation either must be an empty transportation (transportation $T1$ in Fig. 3) or its destination location must be equal to the destination location of the first order (where it will be unloaded) (transportations $T3$ or $T4$ in Fig. 3) and
    - the second transportation either must be an empty transportation (transportation $T5$ in Fig. 3) or its origin location must be equal to the origin location of the second order (where it will be loaded) (transportations $T2$ or $T4$ in Fig. 3),
  - all transportations related to an order must be sequenced consecutively.
- **Parking:**
  If the operating mode of the parking is FIFO, the in- and outsequence of the parking must be equal in the sense that the vehicles corresponding to the ith transportations in both sequences are the same. If we have a parking with arbitrary operating mode and capacity c, we have to ensure that if a vehicle arrives at the parking, not more than $c - 1$ vehicles are allowed to be in the parking and, thus, the vehicle can be no more than $c - 1$ positions earlier in the outsequence than in the insequence, i.e. if a transportation occurs within the outsequence at position i, its vehicle has to correspond to a transportation which occurs in the insequence not later than at position $i + c$. Figure 4 gives an example with $c = 3$. If in this example Vehicle 5 would be at position 2 of the outsequence (i.e. change its position with Vehicle 2), then at the time Vehicle 5 arrives at the parking, Vehicles 1, 2, and 4 have to be waiting in the parking. But since the parking has only capacity 3, Vehicle 5 has no free track to pass these vehicles.

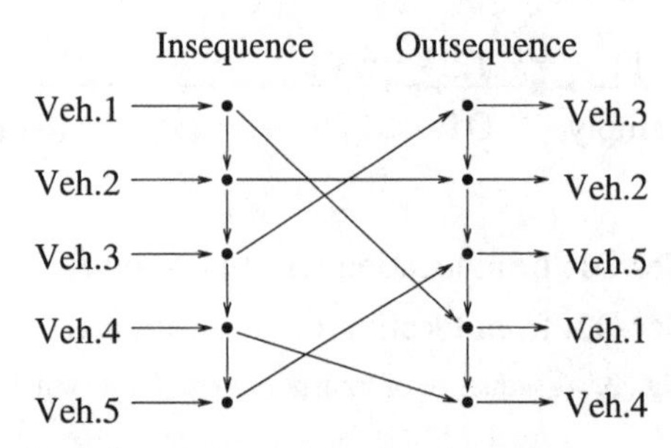

Fig. 4. Arbitrary Parking with capacity $c = 3$

- **Docks:**
  A dock consists of a dock parking (denote its capacity by c) and a set of servers (denote the number of servers by s). Within the dock a vehicle may pass only a limited number of other vehicles. This number depends on s and, if the operating mode of the dock parking is not FIFO, also on c: if the parking is a FIFO parking, a vehicle may pass $s - 1$ vehicles and, otherwise, it may pass $s + c - 1$ vehicles. Thus, the restrictions on the sequences can be derived in the same way as for parkings.

3.2 Timing relations

After checking the consistency of the sequences, it remains to determine a timing of the transportations. The sequences at one location or for one vehicle lead to 'local' timing relations between transportations. These relations may be represented by a directed graph. Using this graph, a schedule respecting the given sequences can be calculated by longest path calculations (similar to critical path calculations for project scheduling, cf. [6]) and feasibility checking is reduced to checking whether the graph contains a positive cycle or not. In the following we will describe the construction of the graph, representing the timing relations resulting from a given set of sequences, in more detail.

The timing relations will be represented by a graph $G = (V, A, l)$, where V denotes the set of vertices, A the set of directed arcs, and $l : A \to \mathbb{R}$ is the length function for the arcs. Each vertex represents either a departure or an arrival of a transportation and a solution is given by an assignment of times $s(i)$ to all vertices $i \in V$. An arc $(i, j) \in A$ with length l_{ij} expresses a timing relation of the form

$$s(i) + l_{ij} \leq s(j),$$

i.e. event j cannot start before the start of the event i plus l_{ij}. Since the above timing relations correspond to longest path conditions, a schedule respecting all timing relations from the arc set A can be achieved by longest path calculations.

It remains to explain how the graph corresponding to a fixed set of transportations and sequences is constructed (we assume that the given sequences are consistent; see Sect. 3.1).

- **Vertices**:
  For each transportation t two vertices t_d and t_a are introduced. Vertex t_d corresponds to the departure and vertex t_a to the arrival of t.
- **Travel times**:
  Since the travel time (let Δ be this time) is fixed for a given transportation t, the times $s(t_d)$ and $s(t_a)$ must differ exactly Δ; i.e.

$$s(t_d) + \Delta = s(t_a).$$

  This relation can be expressed by two arcs (t_d, t_a) and (t_a, t_d) of length $l_{t_d,t_a} = \Delta$ and $l_{t_a,t_d} = -\Delta$. The arc (t_d, t_a) of length Δ ensures that $s(t_a) \geq s(t_d) + \Delta$, whereas the arc (t_a, t_d) of length $-\Delta$ ensures that $s(t_d) \geq s(t_a) - \Delta$. Combining these two inequalities results in the required equality.
- **Order**:
  We have to ensure that the departure of the first transportation belonging to an order does not depart before the release date. This can be ensured by an arc from a dummy node representing the start of the schedule and the node corresponding to the departure of the first transportation with a length equal to the release date.
- **Parking**:
  If the parking has FIFO mode, the in- and outsequence of the parking must be the same. First, we have to ensure that the transportations arrive in the order given by the insequence, depart in the order of the outsequence, and that between two consecutive arrivals (departures) a minimal time lag s_{in} (s_{out}) is taken into account. These time lags represent, e.g., safety distances. More precisely, if $\tilde{t}$ and $\hat{t}$ are two consecutive transportations in the insequence (outsequence), the timing relation

$$s(\tilde{t}_a) + s_{in} \leq s(\hat{t}_a) \quad (s(\tilde{t}_d) + s_{out} \leq s(\hat{t}_d)) \tag{1}$$

  has to be fulfilled. Furthermore, we have to ensure that a vehicle does not leave the parking with a transportation $\hat{t}$ before the vehicle has arrived with its previous transportation $\tilde{t}$ i.e.

$$s(\tilde{t}_a) + s_{park} \leq s(\hat{t}_d), \tag{2}$$

  where s_{park} denotes a minimal time needed between the vehicle entering and leaving the parking. Finally, we have to ensure that the parking capacity is respected, i.e. for a parking with capacity c the $(i+c)$th transportation $\tilde{t}$ cannot enter the parking before the ith transportation $\hat{t}$ has left the parking; i.e.

$$s(\tilde{t}_a) \geq s(\hat{t}_d). \tag{3}$$

- **Dock**:
  In the same way as for parkings, we ensure that the transportations respect the in- and outsequence and corresponding minimal time lags, the vehicle sequence, as well as the capacity of the dock (see (1), (2) and (3)). It remains to take into account the minimal time distance between two transportations using

the same server. These minimal time distances represent, e.g., setup times between two consecutive loading or unloading operations on a server. If in the dock only one server is available, these timing relations can be incorporated in the relation (1) for the outsequence. If more than one server is available, first an assignment of vehicles to servers is calculated and then the minimal time lags for two transportations processed consecutively on a server are added. A best assignment respecting the given in- and outsequence can be calculated as follows: when a server gets free (the order in which they get free is determined by the outsequence) we try to put directly onto the free server that vehicle (transportation) which comes first in the outsequence and which has not been assigned to a server yet. If this is not possible (the capacity of the dock parking does not allow this vehicle to enter the dock) we first move vehicles from the dock parking to servers until the specified vehicle can enter the dock and move to the server. The choice, which vehicles (transportation) leave the dock parking is again determined by the order in which the transportations occur in the outsequence. In Figure 5 an example of a dock with 2 servers and a parking of capacity 2 is given. Using the given in- and outsequences as input for the above procedure, we get that Vehicle 1 and 2 enter the parking, that Vehicle 3, 5 and 2 are handled by Server 1, and that Vehicle 4 and 1 are handled by Server 2. The resulting arcs together with their corresponding length now can be added to the graph G. Note that the given construction is not based on a concrete timing of the transportations, but considers only sequences in which tasks are executed.

The (untill now unknown) values $s(i)$ express the starting times of the corresponding events and have to fulfill all the introduced timing relations. Therefore, these values can be achieved by calculating the longest path from the dummy node representing the start of the schedule to all vertices. If this longest path calculation detects no global infeasibility (positive cycles), the resulting schedule has earliest possible departure and arrival times within the class of all schedules respecting the given vehicle-, dock-, and parking sequences. Note that due to the fixed travel times also negative arc lengths l_{ij} occur and, thus, more general longest path methods like the Floyd-Warshall algorithm (cf. Ahuja et al. [1]) have to be used to calculate a best schedule respecting a given set of sequences.

3.3 Possible objective functions

As indicated above, the longest path calculation ensures that for the achieved schedule the starting times of the transportations cannot be decreased without changing in at least one location one of the sequences of the transportations. Therefore, our approach is able to cope with any objective function for which it is optimal that, *given the sets of sequences for the transportations*, all transportations are performed as early as possible (i.e. for regular objectives). Such objective functions include minimizing the number of late orders, minimizing the maximum or mean lateness, minimizing the time to complete all orders ('makespan'), and minimizing the total length of empty vehicle movements. For the latter objective function, the timing of the transportation is not important, given the sets of sequences for the

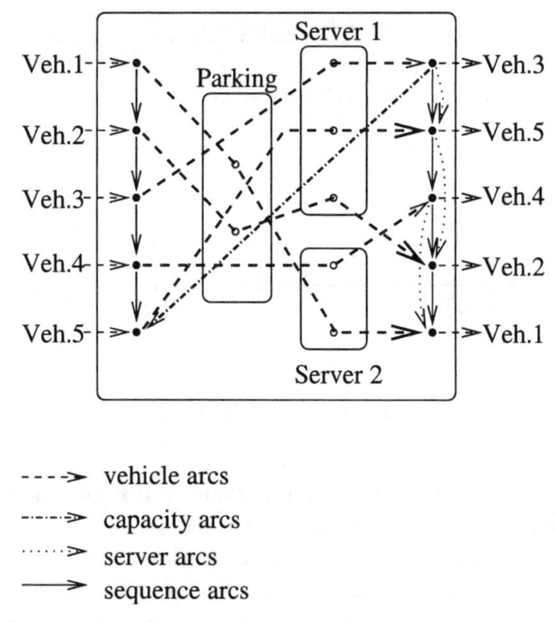

Fig. 5. A dock with 2 servers and a parking of capacity 2

transportations. Therefore, it is also optimal to perform the transportations as early as possible. An example of an objective function that cannot be handled by our model is minimizing earliness costs. In this case, it might be better to postpone a transportation, although it would be feasible to perform it now.

4 Heuristic approach

In this section, we show how the representation given in the previous section can be used to develop in a very simple way a heuristic solution approach for the considered transportation problem. The basic idea is to first treat a relaxation of the problem and solve this relaxation with a constructive heuristic. From the resulting schedule (which in general is infeasible for the original problem) now only the relevant information for a solution representation (vehicle and in- and outsequences) is extracted and used as input for the method presented in the previous section. This results in a feasible schedule for the original non-relaxed transportation problem (see Fig. 6). In the following we describe this method a bit more detailed.

Due to the fixed travel times between locations and the finite capacities of parkings, it is not trivial to construct a good feasible schedule for the considered transportation problem. Therefore, we chose for a relaxation where the capacity constraints of a terminal parking and minimal time lags for vehicles arriving at a terminal may be violated. The resulting problem now can be solved using a standard 'event based' heuristic using priority rules for making the decisions. The outcome of this priority driven heuristic is a schedule which consists of a set of transportations

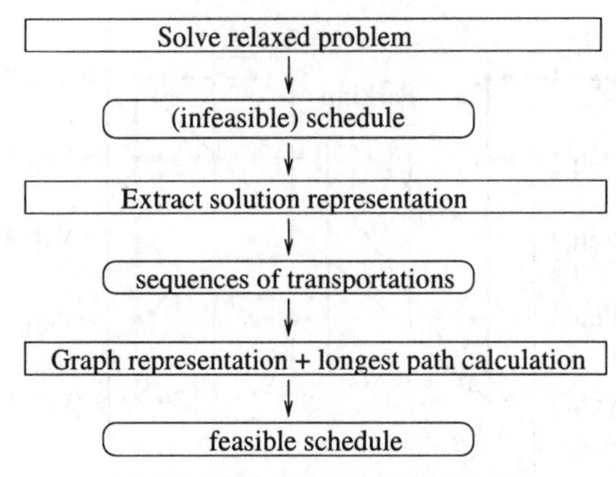

Fig. 6. Structure of the heuristic

and a corresponding timing of these transportations. However, this schedule may violate some of the constraints mentioned in Sections 2 and 3.

To achieve a feasible schedule we neglect the determined timing of the transportations and use only the structural information of the 'infeasible' schedule (the introduced transportations, the vehicle sequences, and the in- and outsequences of the locations). This structural information gives a representation of a solution, and based on the method presented in Section 3 a new schedule is calculated. Although this new solution has the same structure as the previous solution, it may differ from it:

- transportations from different locations may arrive at a parking at the same time in the previous schedule but must have a safety distance in the new schedule
- in the previous schedule, a vehicle may arrive at a terminal at a time that no parking space is available, whereas in the new schedule this vehicle has to leave the origin terminal later to be able to insert it directly in a parking
- in the previous schedule, a vehicle is sent to its next location/order (parking, server, next order) at the time the next location gets free or the next order is assigned to it, whereas in the new schedule the vehicle can anticipate on the next task already earlier.

Summarizing, the new schedule now takes into account all restrictions and the transportations may anticipate on the 'knowledge' that servers/parkings get free or orders arrive. Consequently, the makespan of the new schedule may differ from that of the previous schedule.

5 An application

In this section, we apply our model and heuristic to instances derived from a practical case. It is one of the planned layouts for an automated cargo transportation system around Schiphol Airport (The Netherlands), which we discussed in the introduction. We will refer to this application as the OLS case, in which 'OLS' is a Dutch abbreviation for underground logistic system. Section 5.1 gives some details

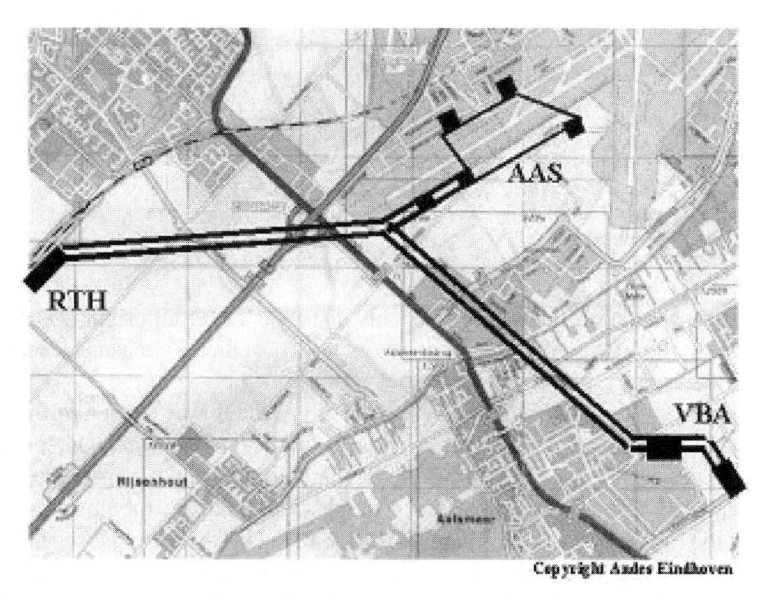

Fig. 7. Considered layout at Schiphol

on this case. Afterwards, in Section 5.2, we give some results of computational experiments.

5.1 The OLS case

In this section, we give a rough sketch of the application that we used to test our modeling framework. For details, we refer to Ebben [9].

To avoid road congestion, a highly automated underground transportation system using automatic guided vehicles (AGVs) is being developed around Schiphol Airport, connecting the airport with the world's largest flower auction market in Aalsmeer and a planned rail terminal (near Hoofddorp or Schiphol Airport). The system should be able to handle different order deadlines and should guarantee reliable throughput times. It is unique in its scale, incorporating 16-25 km tubes connecting five to 20 terminals, and it includes up to 200 AGVs to transport an estimated 3.5 million tons of cargo in 2020. To operate this system in a reliable and efficient way, it would need an innovative planning and control system for logistics, which takes the order deadlines and limited resource capacities at the terminals into account.

In the layout, we consider in the experiments in the next subsection, the system consists of 2 terminals with 6 docks at Aalsmeer (VBA), 5 terminals with 2 docks at Schiphol Airport (AAS), which are located on a one-directional loop, and 1 rail terminal with 10 docks at Hoofddorp (RTH) (see Fig. 7). There is a central vehicle parking at Schiphol Airport and each terminal has a limited parking capacity. We investigate three different patterns of transportation flows, with the following characteristics:

Table 1. Average results for the scheduling instances

Case	Nodes	Arcs	GC Time	Path time	Makespan		Late jobs	
					Heur	Path	Heur	Path
Case 1	3,338.3	9,742.1	0.03	0.01	232.3	230.1	23.3	22.6
Case 2	5,311.6	15,438.4	0.08	0.01	274.3	258.1	168.1	151.7
Case 3	5,236.9	15,271.7	0.05	0.01	231.4	230.3	25.9	24.0

1. The transportation flows are reasonably distributed over all origin-destination pairs. In this case, the system requires about 120 vehicles to ensure acceptable service levels.
2. Most transportations are between two locations: the flower auction (VBA) and the rail terminal (RTH). Therefore, most traffic is on one route, which should make the vehicle planning rather easy. The total number of transportations is larger than in the previous case and the distance between the flower auction and the rail terminal is the longest distance in the system; this causes that 200 vehicles are needed.
3. Peak levels in transportation flows move quickly from one route to another, which means that it is essential to anticipate heavily fluctuating transportation demands. To guarantee acceptable service levels 165 vehicles are used.

5.2 Results

For the OLS case mentioned above, a simulation model in eM-plant is available (cf. Ebben [9]). Using this model, we simulated 30 days for each of the 3 mentioned cases. Each 30 minutes of simulation time, the simulation model saves the current status information to a database. This information primarily concerns the current vehicle activities (current load, position, and destination) and the available orders. This database is then used as input for making a schedule. We start with applying the event-based heuristic, which may result in a non-feasible schedule. From this schedule, we use the found sequences to construct a graph in which we perform a longest path calculation to determine the timing of each transportation (see Section 4). The schedule is then saved to the same database and read by the simulation model which then follows this schedule for the next 30 minutes. After that, new status information is saved to the database and a new schedule is constructed. In this way, we generate 1441 scheduling instances for each of the three cases.

Table 1 shows scheduling results for the generated instances. The column 'Nodes' shows the average number of nodes in the constructed graph, whereas the column 'Arcs' specifies the average number of arcs in this graph. The column 'GC Time' gives the average time (in seconds) to construct the graph (using a computer with an AMD Athlon 1800+ processor and 512 MB of memory). Column 'Path Time' shows the average time (in seconds) to do the longest path calculations (using the same computer). The columns under 'Makespan' give the average makespan in minutes for the event-based heuristic (column 'Heur') and for the schedule based on the longest path calculations (column 'Path'). Analogously, the

Table 2. Maximum result values for the scheduling instances

					Makespan		Late jobs	
Case	Nodes	Arcs	GC time	Path time	Heur	Path	Heur	Path
Case 1	5,066	15,340	0.08	0.02	300	289	199	199
Case 2	9,944	29,877	0.22	0.05	482	421	609	571
Case 3	6,636	19,852	0.08	0.02	273	272	121	116

next two columns present this data for the average number of late jobs. Table 2 shows the same results, but now for the maximum values instead of averages.

From Tables 1 and 2, we conclude that most of the required computation time is spent on the construction of the graph and that the time for the longest path calculations is small. Therefore, we see good possibilities to use our model in local search methods. In these methods, the local changes imply small changes in the graph (which can be realised by updates instead of complete new constructions) after which the consequences of the changes can be evaluated by (fast) longest path calculations.

The final schedule is on average and on the maximum always better than the heuristic solution (on makespan and on the number of late jobs). A detailed analysis of the results shows that only for 8 (4) instances out of 4,323 the event-based heuristic had a slightly better makespan (number of late jobs), whereas for 3,413 (2,772) instances the final schedule is better. Therefore, apart from correcting infeasibilities, the final schedule improves the solution quality.

6 Model extensions

The graph representation of a transportation system as presented in this paper has proven to be flexible with respect to incorporating practical extensions. In this section, we will discuss some possible extensions.

In Section 2, we discussed that one of the characteristics of a parking is whether vehicles have to leave the parking in a FIFO order or not. In the OLS case, parkings have an additional characteristic. For each parking it is given what type of vehicles can use this parking: Empty vehicles, loaded vehicles, or both. This feature is easily incorporated in the model by taking care of this in the heuristic. The heuristic produces then in- and outsequences for the parkings taking into account the load status of the vehicles. These sequences are used to perform the longest path calculations.

The second extension in the OLS case is more difficult. Due to the large costs of building underground tubes, in the current plans for the track layout, the connection between Schiphol and Aalsmeer will be a single track to be used in both directions. This means that when a vehicle traveling from Schiphol to Aalsmeer enters the tube, only upon arrival of this vehicle in Aalsmeer another vehicle can enter the tube traveling from Aalsmeer to Schiphol. Therefore, we have to decide when to open a track for vehicles traveling in a certain direction. We modeled this bidirectional track as a terminal (in which important decisions have to be made). A terminal can consist of docks and parkings, but now also of a bidirectional track. Associated with each bidirectional track, there are two FIFO parkings (one for each direction). For the bidirectional track, we consider now *two insequences* and *one outsequence*. Each insequence represents the vehicle arrivals in one direction. The outsequence represents the order in which the vehicles use the bidirectional track. If two consecutive vehicles in the outsequence travel in the same direction, then there should be a small safety time between them. If two consecutive vehicles travel in opposite directions, then there should be a large 'safety time' between them: The first vehicle has to travel the complete track before the second vehicle can enter it. As done in Section 3 for the other terminals, we can derive for this new type of terminal a procedure, which checks the local feasibility of the sequences and a procedure which creates the timing restrictions resulting from the sequence. Furthermore, the event-based heuristic has to be enlarged with the event of an arrival at a 'bidirectional terminal' and a procedure that determines when the track can be used for transportations in a certain direction.

Load and unload times at the servers of a dock may depend on the goods that are loaded or unloaded. For example, large loads may require more load time than small loads. This can easily be incorporated by varying the length of the arcs between the node for the arrival of a vehicle at a dock and the node for the next departure of this vehicle at this dock.

Sometimes, (parts of) loads are consolidated at a terminal to a 'new' load. This means that there are precedence relations between the arrival of the loads that will be consolidated and the departure of the consolidated load. This can easily be incorporated in the model by adding arcs between the nodes associated with the arrival of the loads that will be consolidated and the node associated with the departure of the consolidated load. Furthermore, if the consolidation decisions have to be made within the model, the solution representation does not change; only the heuristic has to be adapted.

In some systems, vehicles may have different sizes, which implies that the 'large' vehicles may not be able to enter all terminals. In this case, only the procedure that assigns vehicles to orders has to be adapted.

The above discussion shows that the introduced concept of dealing with transportation systems with restricted capacities is quite general and can be adapted to more general systems: If in a transportation system a decision structure comes up, where the local restrictions on feasibility and timing can be expressed by in- and outsequences, these structures can be integrated in the presented approach.

7 Conclusions and further research

We presented a model that is able to deal with a broad class of capacitated transportation systems for integral scheduling. These type of problems require a combination of scheduling and transportation approaches. We have chosen to adapt the graph representation used in scheduling theory to handle combined scheduling and vehicle routing problems. The base of our methodology is a representation of solutions using sequences for the arrivals at and the departures from a location. Based on these sequences, a timing of the transports can be determined by using longest path calculations. We demonstrated the capabilities of our representation by showing that the sequences resulting from an (infeasible) solution generated by a simple event-based heuristic can be transformed to better and feasible solutions.

Experiments for a real-life case have shown that the computational effort for constructing a graph that represents a solution and the longest path calculation in this graph are very small. Moreover, in almost all cases the final schedule improves the solution of the event-based heuristic. In addition, we demonstrated that our approach can be exted to various practical extensions of the real-life case. Therefore, we may state that our structural solution representation has high value in modeling capacitated transportation systems.

Furthermore, our representation facilitates the use of concepts which have led to powerful local search approaches for various scheduling problems. These concepts are based on neighborhoods changing relevant sequences in the graph representation (see, e.g., Vaessens et al. [19]).

Future research may include a local search approach in which the neighborhood structure is based on changes in the sequences. The effect of a change in a sequence can then be evaluated easily by performing a longest path calculation on the resulting graph. For example for the OLS case, it is also interesting how this approach performs in a rolling horizon environment where new orders arrive and all kinds of disturbances may occur. An important question is then what a good strategy is for rescheduling the system.

References

1. Ahuja RK, Magnanti TL, Orlin JB (1993) Network flows: theory, algorithms and applications. Prentice-Hall, New Jersey
2. Alicke K (2002) Modeling and optimization of the intermodal terminal Mega Hub. OR Spectrum 24: 1–18
3. Bilge Ü, Ulusoy G (1995) A time window approach to simultaneous scheduling of machines and material handling system in an FMS. Operations Research 43: 1058–1070
4. Bostel N, Dejax P (1998) Models and algorithms for container allocation problems on trains in a rapid transshipment shunting yard. Transportation Science 32(4): 370–379
5. Brucker P, Heitmann S, Hurink JL (2003) Flow-shop problems with intermediate buffers. OR Spectrum 25: 549–574
6. Chao X, Pinedo M (1998) Operations scheduling with applications in manufacturing and services. McGraw Hill, New York
7. Cordeau J, Desaulniers G, Desrosiers J, Solomon M, Soumis F (2002) The vehicle routing problem with time windows. In: Toth P, Vigo D (eds) The vehicle routing problem. SIAM monographs on discrete mathematics and applications, vol. 9, pp 157–193. Philadelphia, PA
8. Crama Y, Kats V, Van de Klundert J, Levner E (2000) Cyclic scheduling in robotic flowshops. Annals of Operations Research 96: 97–124
9. Ebben MJR (2001) Logistics control of automated transportation networks. PhD thesis, University of Twente, Twente University Press, Enschede, The Netherlands
10. Ebben MJR, van der Heijden MC, van Harten A (2002) Dynamic resource-constrained vehicle scheduling. Working paper, University of Twente, Faculty of Technology and Management, The Netherlands

11. Van der Heijden MC, Ebben MJR, Gademann N, van Harten A (2002) Scheduling vehicles in transportation networks. OR Spectrum 24(1): 31–58
12. Van der Heijden MC, van Harten A, Ebben MJR, Saanen YS, Valentin E, Verbraeck A (2002) Using simulation to design an automated underground system for transporting freight around Schiphol Airport. Interfaces 32(4): 1–19
13. Hurink J, Knust S (2001) Tabu search algorithms for job-shop problems with a single transport robot. Memorandum No. 1579, University of Twente, Faculty of Mathematical Sciences, The Netherlands. EJOR (to appear)
14. Kozan E (2000) Optimising container transfers at multimodal terminals. Mathematical and Computer Modelling 21: 235–243
15. Kozan E, Preston P (1999) Genetic algorithms to schedule container transfers at multi-modal terminals. International Transactions in Operational Research 6, 311–329
16. Meersmans P (2002) Optimization of container handling systems. PhD thesis, Erasmus University Rotterdam, The Netherlands
17. Nowicki E (1999) The permutation flow shop with buffers: a tabu search approach. European Journal of Operational Research 116: 205–219
18. Ralphs TK, Kopman L, Pulleyblank WR, Trotter Jr LE (2003) On the capacitated vehicle routing problem. Mathematical Programming 94: 343–359
19. Vaessens RJM, Aarts EHL, Lenstra JK (1996) Job shop scheduling by local search. INFORMS Journal on Computing 8: 302–317

Modeling and optimization of the intermodal terminal *Mega Hub*

Knut Alicke

Institut für Fördertechnik und Logistiksysteme, Universität Karlsruhe, Kaiserstraße 12, 76128 Karlsruhe, Germany (e-mail: Knut.Alicke@mach.uni-karlsruhe.de)

Received: May 2, 2000 / Accepted: July 4, 2001

Abstract. The convergence of European states can be expected to lead to an increase in the trading of goods within the next few years and thus to a growing demand for transport. Overland intermodal transport is an important development, because it combines the advantages of rail for long distance transportation with the effective area cover offered by road. Different terminal concepts and production forms have been developed to increase the flexibility of intermodal transport and to make it more attractive for the customer.

The intermodal terminal concept investigated in this paper is called Mega Hub. The configuration and the control of the terminal is a complex and challenging task. Here, the terminal is modeled as a multi-stage transshipment problem. In this approach, sequence-dependent duration of empty moves, alternative assignments (of containers to cranes) and a sequence-dependent number of operations have to be handled. An optimization model based on Constraint Satisfaction is formulated and heuristics for the search procedure, especially value and variable ordering are developed.

Key words: Intermodal transport – Multistage sequencing – Constraint satisfaction

1 Introduction

Hub and spoke systems are used to connect transshipment terminals in a corridor-like network so that load units for different destinations can be loaded onto one train. Trains arriving from different origins (spokes or other hubs) are loaded with units for different destinations. At the hub, the transshipment between the trains takes place and trains now having only one destination leave the hub. New terminal concepts replace the former used large shunting yards which are very inflexible and time-consuming. A study recently published determined the average speed for container transportation to $v = 7$ [km/h]. For a competitive intermodal transport concept, new terminals and production forms were developed.

In the concept of the Mega Hub, the connection of containers to wagons is not fixed, therefore no time consuming shunting is necessary. Load units change trains like in passenger transport. Up to 12 cranes span six railway tracks and operate in parallel to transship the containers from train to train in an efficient way. The stop over time of the trains in the terminal is reduced significantly compared to the former used shunting yards. The trains operate according to time-tables, therefore delays have the same effect like in passenger transport and should be avoided.

In Section 2 the terminal is introduced and modeled in Section 3. In Section 4 the model is transformed into a Constraint Optimization Problem. Furthermore, the demand-profiles to solve the problem are introduced. The model is used to determine the optimal transshipment-sequence with the results of the practical application presented in Section 5. Finally, a short summary is given.

2 Terminal Mega Hub

A possible layout of the terminal is shown in Figure 1 with up to 12 cranes spanning six rails for loading and unloading of trains, a lorry loading lane and the fully automatic sorting system. The operating areas of the cranes can overlap, so that a low utilized crane can support a highly utilized one. Theoretically, the cranes can serve the entire length of the terminal.

The *cranes* are mainly portal cranes with a cantilever on both sides, similar to cranes used today for the transshipment of load units from road to rail. These cranes

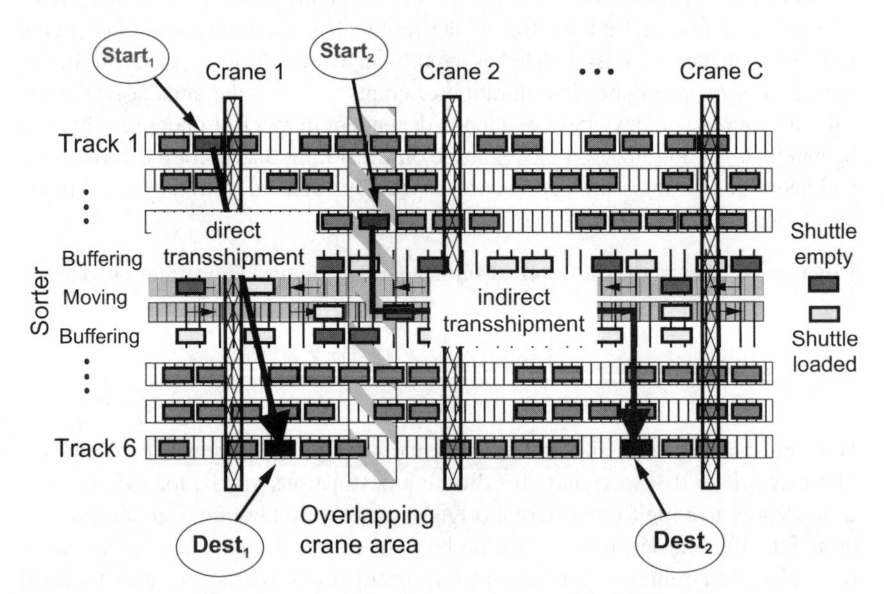

Fig. 1. Layout of the terminal; shown are the tracks, sorting system, cranes and the difference between a direct and indirect transshipment

use flexible spreaders to transfer ISO containers, swap bodies and semi-trailers with weights of up to 41 t, in the following summarized as containers for ease of reading.

The *sorter* consists of a system of runways running parallel and orthogonal to each other. Along the runways, cars are conducted fully automatically both lengthwise and crosswise. The shuttle cars are rail-mounted and operate bi-directional (straight and sideways). The units for drive and position detection are integrated into the runway. The shuttle cars are driven by contact-free linear synchronous motors, which are distributed over the unit according to the requirement of driving force. The shuttle cars can be positioned exactly with an accuracy of +/- 3 mm [4].

The transshipment of containers requires cranes, shuttle cars and personnel. The departure time stated in the time-table should be met and the terminal should operate at minimal cost. The trains arrive and depart in so called *bundles* of 6 trains in which the transshipment takes place. The goal can be defined as to minimize the maximum lateness L_{max}. To achieve this, the following problems have to be solved:

1. In which sequence should the containers be transshiped?
2. Should the operating areas of the cranes be fixed or variable?
3. Of which size should the operating areas be?
4. Which crane should tranship a container within an overlapping area?
5. Which track should be used by the incoming trains?

A framework has been developed to model problems (1) to (4) as Constraint Satisfaction Problem. The model was implemented based on ILOG Solver and Scheduler [9] with some own extensions. Problem (5) can be formulated and solved as a quadratic assignment problem (QAP) as presented in [2].

As the terminal concepts are a new development and the control task is very special, not many researchers have worked in this area. One simulation based approached was developed by [10]. He developed different priority rules for the transshipment of containers and tested the performance using a simulation model of the Mega Hub based on petri nets. The rules are very complex and show the difficulty of determining an optimal transshipment-sequence for the terminal.

The problem is related to cyclic multi hoist scheduling problems originating in the chemical treatment process for printed circuit boards. The hoists correspond to cranes (transport-devices) in the transshipment process and the circuit boards can be seen as containers (transport-objects). The application of Constraint Logic Programming (CLP) approaches together with a formulation as Mixed Integer Problem (MIP), to solve the cyclic multi hoist scheduling problem, is presented by [12]. Here, no alternative assignments are considered. An approach for the cyclic multi-hoist scheduling problem with overlapping partitions is presented by [16].

The container allocation problem is investigated by [6]. They developed an approach to find a sequence in which the containers should be transshiped. The terminal under consideration is less complex than the Mega Hub, because the cranes can only be moved in one direction over one row of containers.

A lot of research has been conducted in the area of scheduling, but most of the approaches do not consider sequence-dependent issues, alternative assignment and arbitrary routings. For an overview see [5].

3 Model of the terminal

To transship a container from the source position to the destination position, one or two cranes and in many cases a shuttle car of the sorter is needed. A transshipment is possible between every position of the trains, containers can even be sorted within one train, i.e. by positioning a container from the back of a train to the front. The transshipment matrix is known in advance and contains the pick-up (source) and drop (destination) position for every container. Depending on the source and destination position of the container, the transshipment can be classified as (see Fig. 1):

- *Direct transshipment:* The crane moves empty to the source-position of the container, picks it up, moves loaded to the destination position and drops it off. The container is handled once by one crane.
- *Indirect transshipment:* The container is handled more than once, by the same or different cranes and/or a shuttle car. Due to constraints of the time-table or the pick-up/drop position it can be necessary to buffer the container. The container is picked up, moved to the sorter and placed on an empty shuttle. The shuttle moves to the destination area. There, a crane picks up the container, moves to the destination position and drops the container. The container is therefore handled twice by a crane (same or different) and once by a shuttle.

3.1 Duration of transshipment

The duration of a transshipment can be calculated based on the kinematic data of the resources. We approximate the acceleration a_s and deceleration a_b according to [3] with $a = \frac{2 \cdot |a_s \cdot a_b|}{a_s + |a_b|}$. Thus, the duration t to move a resource a distance l with the velocity v is given by

$$t = \begin{cases} \frac{l}{v} + \frac{v}{a} & \text{for } l \geq \frac{v^2}{a} \\ 2 \cdot \sqrt{\frac{l}{a}} & \text{for } l < \frac{v^2}{a} \end{cases} \tag{1}$$

The crane can move simultaneously (l)ong- and (s)idewise. If the durations of the single moves are given as t_l and t_s, the duration t of the entire move is $t = \text{ MAX } [t_l, t_s]$.

The transshipment of a container is composed of different steps. The crane moves *empty* to the source position of the container, positions the spreader, *picks up* the container, moves *loaded* to the destination position and *drops* the container after positioning again. Then, the crane waits until the next container has to be transshiped. The time needed for a transshipment can be stated as

$$t = t^{empty} + t^{pick} + t^{load} + t^{drop} \tag{2}$$

The time to position the spreader is included in t^{pick} and t^{drop}, respectively. With a given transshipment matrix, the duration of the loading step, the duration to pick up and drop a container can be calculated in advance. The duration of the empty move is sequence-dependent.

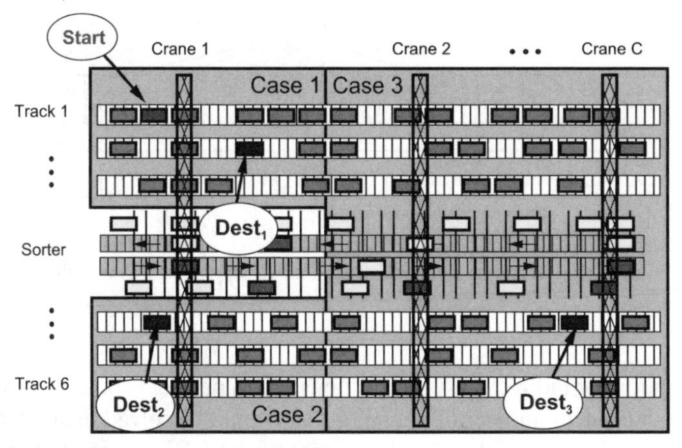

Fig. 2. Different cases of transshipment. Shown are three different destination positions of a container $Dest_1$, $Dest_2$ and $Dest_3$ for the same start position $Start$

3.2 Classification of transshipments

We define the transshipment of a container as a *job*, which is build by one to three *operations*, depending on the source and destination position of the containers and the size of the crane-areas. The operations are assigned to *resources* which are the cranes and the shuttles. The three cases are (see Fig. 2):

- *Case 1* (F_1): The source and destination positions of the container are located in the same crane area and on the same side of the sorter. The stop over time of the source and destination train overlaps, thus a *direct* transshipment is possible. A job like this is modeled by one operation: source $\rightarrow$ destination.
- *Case 2* (F_2): The source and destination positions of the container are located in the same crane area, on different sides of the sorter though. Due to constraints of the time-table it can be necessary to buffer the container on the sorter and complete the transshipment later. It is assumed that the containers can be buffered on the way from the source- to the destination position. The job of this case is modeled with two operations: source $\rightarrow$ sorter and sorter $\rightarrow$ destination. This applies also for transshipments of case 1 if the stop over time of the trains does not overlap.
- *Case 3* (F_3): Source and destination positions of the container are located in different crane areas, thus the usage of the sorter is required leading to an *indirect* transshipment. This is independent of the stop over time of the source- and destination train. Jobs falling into this case are modeled with three operations: source $\rightarrow$ sorter (source-area), sorter (source-area) $\rightarrow$ sorter (destination-area) and sorter (destination-area) $\rightarrow$ destination.

3.3 Direct versus indirect transshipment

As many jobs of case 2 as possible should be transshiped directly. It has to be decided during the sequence-generation whether a container should be transshiped directly

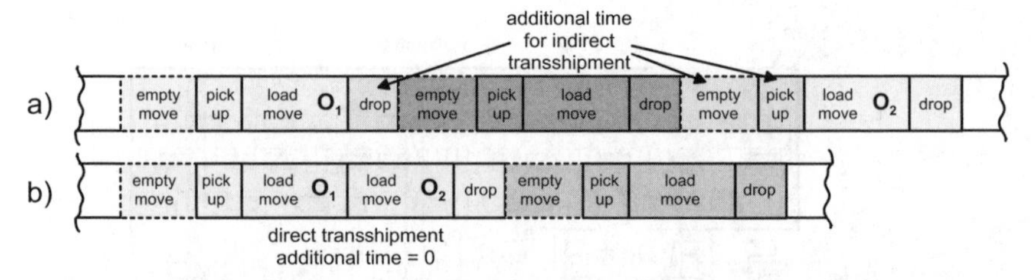

Fig. 3. Modeling the sequence-dependent number of operations as **a** indirect or **b** direct transshipment

(one operation) or indirectly (two operations). This problem could be described as sequence-dependent number of operations and could be solved using the *mode* of transshipment, increasing the complexity of the model. We use an elegant, efficient way to capture the problem, namely sequence-dependent setup-times to distinguish between direct and indirect transshipment. The total duration of the transshipment of case 2 is given by

$$t = t_1^{empty} + t_1^{pick} + t_1^{load} + t_1^{drop} + t_2^{empty} + t_2^{pick} + t_2^{load} + t_2^{drop} \tag{3}$$

If the operations are performed consecutively, $t_1^{drop} + t_2^{empty} + t_2^{pick}$ can be saved or set to 0 (see Fig. 3). The time for dropping and picking up the container is independent of the container, thus $t^{drop} + t^{empty} + t^{pick}$ can be defined as setup time between two operations and the sum of the load moves $t_1^{load} + t_2^{load}$ as the processing time. The setup-time is 0 if two consecutive operations are refering to the same job (container) and are handled by the same crane.

3.4 Overlapping crane areas - alternative cranes

The cranes can operate in disjunct or overlapping areas. Using disjunct areas can lead to unfavorable states as shown in Figure 4 (left). The border between two crane areas is indicated by the bold line. The two jobs are classified as case 3, if $Start$ indicates the source-position and $Dest_1$, respectively $Dest_2$ the destination-position. If a transshipment performed by resource c (crane) or s (shuttle) is described by $\xrightarrow{c}$ and s(Start) refers to the start position of the shuttle used for transport, the following operations are necessary, c and $c+1$ refer to adjacent cranes:

Dest 1: $Start \xrightarrow{c} s(Start) \xrightarrow{s} s(Dest_1) \xrightarrow{c+1} Dest_1$

Dest 2: $Start \xrightarrow{c} s(Start) \xrightarrow{s} s(Dest_2) \xrightarrow{c+1} Dest_2$

Savings are achieved through classifying jobs as case 1 or 2, which formerly were classified as case 3. This is achieved in the example if the transshipment is performed by only one crane c or $c+1$. We define the number of rows as r where the containers can be transshiped alternatively by the adjacent cranes. The containers of the alternative area are not assigned to one resource, therefore the classification has to be reconsidered.

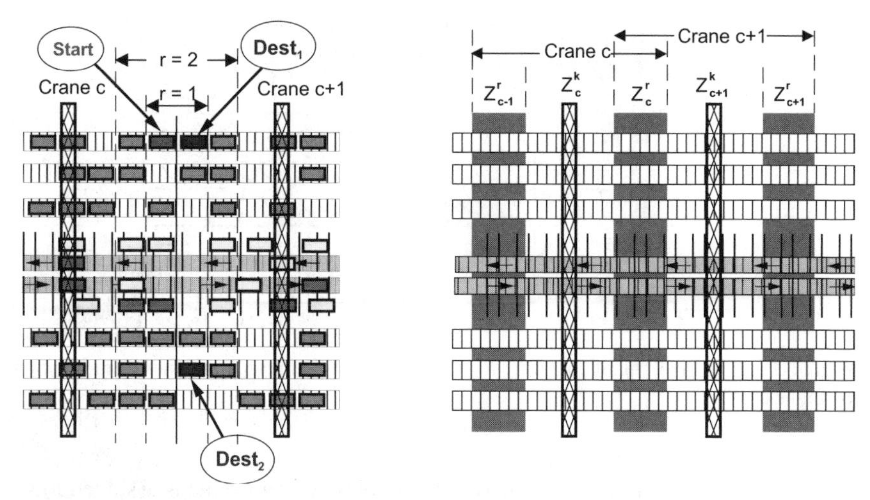

Fig. 4. Alternative cranes and implications for the assignment of containers

Table 1. Assignment of jobs to cranes considering the overlapping area, this serves as basis for the classification. Z states the whole area operated by all cranes

$P_s(o)$	Z_c^k	Z_c^r	Z_{c-1}^r	Z_c^r
$P_z(o)$	$Z_c^k \vee Z_{c-1}^r \vee Z_c^r$	$Z_c^k \vee Z_{c-1}^r$	$Z_c^k \vee Z_c^r$	$Z \backslash (Z_c^k \vee Z_{c+1}^k \vee Z_c^r)$
Crane		c		$c, c+1$

The core area of a crane c is stated as Z_c^k. In this area only crane c is operating. The overlapping area where crane c and $c+1$ are able to operate is defined by Z_c^r. $P_s(o)$ refers to the source-position of an operation, $P_z(o)$ to the destination respectively. With that the classification can be updated (see Table 1).

The assignment operation to a resource should be performed dynamically depending on the utilization of the cranes and is therefore integrated in the problem-formulation. Furthermore, it has to be ensured that the cranes do not collide. This is done by blocking the overlapping area when a crane enters it. In case of large overlapping areas this would lead to high idle times of the crane which is actually not operating in the overlapping area. Thus, a dynamic blocking is appropriate. Depending on the actual position of the moving crane, rows of containers are dynamically blocked, as shown in Figure 5.

4 Formulation as a constraint optimization problem

A Constraint Satisfaction Problem (CSP) is defined as a set of constraint variables $x \in X$, the corresponding domains D_x and a set of constraints which are defined on the variables $c(x_1, \ldots, x_n)$. A feasible solution to the CSP is obtained by assigning a value to every variable so that no constraint is violated. A lot of work has been done in the area of Constraint Satisfaction Problems and Constraint Propagation, see [11] for an overview.

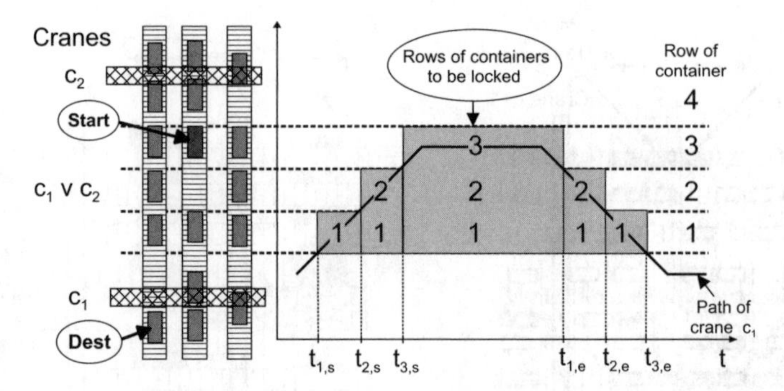

Fig. 5. Dynamic blocking of container rows, in this example the red container is transshiped by crane c_1

The solution approach uses the set of variables X^{na} no value is assigned to so far and the set of variables X^a a value is already assigned to, where $X = X^{na} \cup X^a$. The CSP is solved in three steps:

1. *Propagation*: Values which do not take part in any combination of feasible values are removed from a variables (x) domain D_x.
2. *Assignment of values*: If there is no variable left in X^{na}, a feasible solution is found. Otherwise a variable x is chosen from X^{na} and a value out of its domain D_x is assigned that satisfies all constraints defined on the variables in X^a. As a consequence x is moved from X^{na} to X^a and the algorithm proceeds with step 1. If there is no such value, the assignment made so far does not comply with the constraints and backtracking (step 3) takes place.
3. *Backtracking*: The last (or the last n) assignment(s) is (are) deleted. Step 2 follows.

The main problem is to choose a variable and value in step 2. Critical variables should be chosen first. Such conflicts (failures) occur early and the search tree remains flat. On the other hand, values that are likely to be part of a feasible solution should be assigned first, in order to find this solution as fast as possible.

In the work of [13], [14] and [15] heuristics for variable and value sorting are developed to minimize backtracking. The underlying idea is to identify the most critical variable which has a high probability to cause backtracking during the search and to assign the best possible value which would be content in a large number of different solutions. They introduce the concept of *demand-profiles* to select the most critical variables first. The concept is illustrated in Figure 6.

A CSP is transformed into a Constraint Optimization Problem (COP) by introducing an additional constraint representing the objective function. To determine the optimal solution, the (objective) constraint is set to the upper bound, which is known to give a feasible solution to the problem. Now we can obtain the optimal solution with different iterative techniques, based on the idea of tightening the bound of the objective function and solving the resulting CSP. This gives a high flexibility in problem-solving.

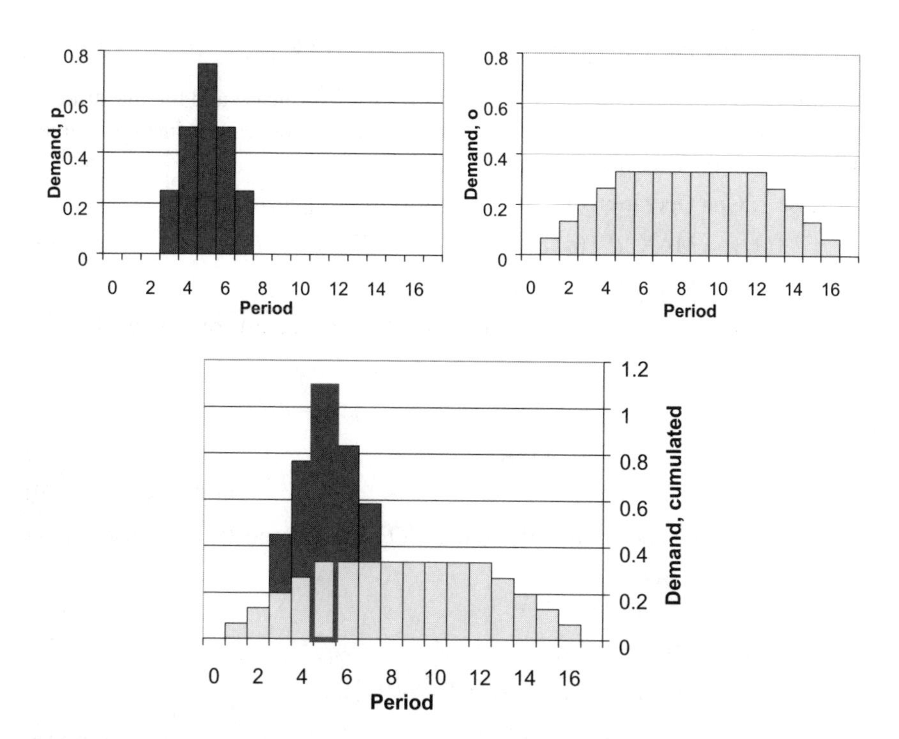

Fig. 6. Demand profiles for operation o with $pt(o) = 5$, $D_o = \{1, \dots, 11\}$ (top), operation p with $pt(p) = 3$, $D_p = \{3, 5\}$ (middle) which are both assigned to resource r. The cumulated demand profile of the resource r (bottom) is shown with the critical period and the critical operation p

To formulate the problem as a constraint optimization problem, we define the transshipment of a container as a job $j \in \mathcal{J}$, which is built by one to three operations $O_i^j \in \mathcal{O}$, depending on the classification as introduced in Section 3.2. In the following, we refer to the operations as $o, p \in \mathcal{O}$ for ease of reading. The cranes $c \in \mathcal{C}$ have a capacity of one. Two adjacent cranes build an area where a crane can be used alternatively. With $fr(o)$ for each operation the feasible resources are given. The capacity of a resource is stated by $cp(r)$, the demand (for capacity) of an operation is $sz(o)$. The sorter is modeled as one resource r_s where the capacity is equal to the number of shuttles $cp(r_s) = f$. The processing time of an operation on a resource is given with $pt(o, r)$; for each operation the release time $t^{rel}(o)$ and the due time $t^{due}(o)$ are given. The binary relation $\succ$ defines a sequence between two operations. The setup time between two operations is given by $t^{set}(o, p)$ and can be sequence-dependent, the horizon of the model is defined from $0, \dots, H$.

The matrix of transshipments and therefore the positions on the trains are known in advance. The set of trains is denoted as $\mathcal{Z}$, the assignment of a job to a train is stated by $az_s(j)$ for the arriving and $az_d(j)$ for the departing train. The time-table defines the arrival $t^{rel}(z)$ and departure times $t^{due}(z)$ for all trains. The goal is to determine a schedule minimizing the maximum lateness L_{max} of all trains.

4.1 Constraints due to due dates

The time-table of the trains defines constraints for the release and the due-time of the operations, restricting the corresponding domains. This constraints are defined depending on the classification as

- Case 1: One Operation O_1^j (source/destination) is generated.

$$t^{rel}(O_1^j) = \text{MAX}\ [t^{rel}(az_s(j)); t^{rel}(az_d(j))] \tag{4}$$

$$t^{due}(O_1^j) = \text{MIN}\ [t^{due}(az_s(j)); t^{due}(az_d(j))] \tag{5}$$

- Case 2: Two operations O_1^j (source) and O_2^j (destination) are generated.

$$t^{rel}(O_1^j) = t^{rel}(az_s(j)) \tag{6}$$

$$t^{due}(O_1^j) = \text{MIN}\ [t^{due}(az_s(j)); \quad t^{due}(az_d(j)) - pt(O_2^j)] \tag{7}$$

$$t^{rel}(O_2^j) = t^{rel}(az_d(j)) - pt(O_2^j) \tag{8}$$

$$t^{due}(O_2^j) = t^{due}(az_d(j)) \tag{9}$$

- Case 3: Three operations O_1^j (source), O_2^j (sorter) and O_3^j (destination) are generated.

$$t^{rel}(O_1^j) = t^{rel}(az_s(j)) \tag{10}$$

$$t^{due}(O_1^j) = \text{MIN}\ [t^{due}(az_s(j)); \quad t^{due}(az_d(j)) - (pt(O_2^j) + pt(O_3^j)] \tag{11}$$

$$t^{rel}(O_2^j) = t^{rel}(az_s(j)) + pt(O_1^j) \tag{12}$$

$$t^{due}(O_2^j) = t^{due}(az_d(j)) - pt(O_3^j) \tag{13}$$

$$t^{rel}(O_3^j) = \text{MAX}\ [t^{rel}(az_d(j)); t^{rel}(az_s(j))] \tag{14}$$

$$t^{due}(O_3^j) = t^{due}(az_d(j)) \tag{15}$$

4.2 Constraints due to occupation of the destination position

To transship a container, the destination-position has to be free, otherwise the containers would collide. This is ensured by stating a sequence constraint using $P_s(O_i^j)$ and $P_z(O_i^j)$ as introduced in Section 3.4:

$$O_i^j \succ O_l^k \quad \forall j, i, k, l \mid P_z(O_i^j) = P_s(O_l^k) \tag{16}$$

4.3 Constraints to model direct and indirect transshipment

The sequence-dependent setup-time $t^{set}(o, p)$ to model the distinction between direct and indirect transshipments (Section 3.3) is defined as follows:

$$t^{set}(o,p) = \begin{cases} 0 & \alpha \\ t(P_z(o), P_s(p)) + t^{pick} + t^{drop} & \text{otherwise} \end{cases}$$

$$\alpha \Leftrightarrow o \gg p \wedge j = k \ \forall \quad ra(o) = ra(p) \tag{17}$$

The relation $\gg$ states that the operations are performed *directly* after each other, $t_s(p) = t_e(o)$.

4.4 Capacity constraints

The following capacity constraints ensure that the given capacity is never exceeded.

Capacity constraints of the sorter
The capacity of the sorter is demanded at the time when the shuttle is positioned at the buffer in the outer track, which equals the moment the crane reaches the destination-position (above the shuttle). The processing time is equal to the time needed to transport the container to the destination crane area. Due to the fact that the processing does not necessarily take place immediately after reaching the destination area, the capacity is released when the crane of the destination area picked up the container and the spreader is in the upright position.

Let O_1^j be the operation of the crane of the source-position, O_2^j the operation of the sorter and O_3^j the operation of the destination crane. The capacity constraint can be stated as:

$$t_e(O_1^j) + pt(O_2^j) \leq t_s(O_3^j) - t^{drop}$$
$$\sum_{O_2^j | (O_1^j \succ O_2^j \succ O_3^j) \wedge (t_e(O_1^j) < t < t_s(O_3^j))} sz(O_2^j) \leq cp(r_s) \; \forall \, j \in F_3 \tag{18}$$

The capacity demand of O_2^j is $sz(O_2^j) = 1$. The constraint is also necessary for transshipments falling into case 2 where a shuttle is used to buffer the container. For the operations $O_1^j \succ O_2^j$ it has to be ensured that:

$$\sum_{(O_1^l \succ O_2^l) \wedge (t_e(O_1^l) < t < t_s(O_2^l))} sz(O_1^l) \leq cp(r_s) \quad \forall \, l \in F_2 \tag{19}$$

Both constraints consider the sorter, thus they have to be combined to:

$$\sum_{O_2^j | (O_1^j \succ O_2^j \succ O_3^j) \wedge (t_e(O_1^j) < t < t_s(O_3^j))} sz(O_2^j) + \sum_{(O_1^l \succ O_2^l) \wedge (t_e(O_1^l) < t < t_s(O_2^l))} sz(O_1^l) \leq cp(r_s)$$
$$\forall \, j \in F_3 \; \wedge \; l \in F_2 \tag{20}$$

Constraints due to the alternative resources
If the operations can be assigned to alternative cranes, it has to be ensured that the cranes are not in conflict during the transshipment. Avoiding conflicts can be achieved by stating disjunctive constraints between operations where the source- or destination-position is located in the overlapping area. During the transshipment of a container of the area Z_c^k (see Table 1) no other container of this area is allowed to be transshiped:

$$(o \succ p) \vee (p \succ o)$$
$$\forall \, o, p, c \mid P_s(o) \in Z_c^r \vee P_z(o) \in Z_c^r \vee P_s(p) \in Z_c^r \vee P_z(p) \in Z_c^r \tag{21}$$

Constraints to dynamically lock the overlapping area
The alternative rows in the overlapping crane area should be locked dynamically as shown in Figure 5. Depending on $P_z(o)$ the rows of containers are locked. $P_z(o)$ is reached at $t_e(o)$; the whole row containing $P_z(o)$ is locked when the corresponding crane enters the row, which happens at $t_{z,s}(o) = t_e(o) - t^{move}(l_{ct})$. Here, l_{ct} represents the length of a container and $t^{move}(l_{ct})$ the duration for this move. The crane exits the row and therefore the unlocking can take place at $t_{z,e}(o) = t_{z,s}(o) + t^{drop} + t^{pick} + t^{move}(l_{ct})$. The time entering and exiting the rows before the destination row (in moving direction of the crane) can be formulated as

$$\begin{aligned}
t_{z-1,s}(o) &= t_{z,s}(o) - t^{move}(l_{ct}) \\
t_{z-1,e}(o) &= t_{z,e}(o) + t^{move}(l_{ct}) \\
&\vdots \quad \vdots \\
t_{z-n,s}(o) &= t_{z,s}(o) - t^{move}(n \cdot l_{ct}) \\
t_{z-n,e}(o) &= t_{z,e}(o) + t^{move}(n \cdot l_{ct})
\end{aligned} \tag{22}$$

Based on these equations, the time for locking and unlocking rows can be computed. All containers with start and destination position in the alternative area may not be transshiped in the given time interval. This is realized by enhancing the sequencing constraint (21) with the row $r_z(o) \in [1, \ldots, r]$ corresponding to an operation o which is dependent on the crane the operation is assigned to. The row can contain the start or destination position and the alternative rows $1, 2, \ldots, r_z(o)$ can be locked with the following constraint

$$\begin{aligned}
&t_s(p) > t_{z-\Delta r,e}(o) \;\vee\; t_e(p) < t_{z-\Delta r,s}(o) \\
&\forall\, o, p \mid r_z(o) > r_z(p), \text{with } \Delta r = r_z(o) - r_z(p)
\end{aligned} \tag{23}$$

4.5 Assignment for alternative cranes

To formulate the constraints given in equations (21) to (23), the assignment of operations to cranes must be defined. If the constraints are stated (posted) during the search, the assignment can be done during the search.

A simple assignment heuristic for a balanced utilization of the cranes may be formulated as follows. The utilization V_c of a crane c is estimated with the already assigned operations as

$$\begin{aligned}
V_c = &\sum_{O_i^j \mid ra(O_i^j)=c} \Big[pt(O_i^j) + (t^{pick} + t^{drop})_{j \in F_1 \cup F_3} \\
&+ \frac{1}{2}(t^{pick} + t^{drop})_{j \in F_2}\Big]
\end{aligned} \tag{24}$$

The setup-times are sequence-dependent, therefore only the durations of load moves, pick up and drop off are included in the calculation. For jobs classified as cases 2 $j \in F_2$ the term $\frac{1}{2}(t^{pick} + t^{drop})$ is used to determine the lower bound as the decision of direct or indirect transshipment depends on the sequence, too.

The operations where source and/or destination positions are near at a crane should be assigned to this crane. The distance of the source- $x_{s,o}$ and destination-position $x_{d,o}$ to the middle of the crane area $x_{m,c}$ is defined as $\Delta x_{o,c}$:

$$\Delta x_{o,c} = |x_{s,o} - x_{m,c}| + |x_{z,o} - x_{m,c}| \tag{25}$$

The assignment-algorithm chooses the crane with minimum V_c and assigns the operation with minimum $\Delta x_{o,c}$. Then, V_c is updated and the procedure is repeated until every operation is assigned to a crane.

5 Results

To prove the practical relevance of the model, we applied the algorithm to a real life problem instance based on data sets provided by Noell GmbH and the Deutsche Bahn AG. The goal was to determine a solution within 5 [min], being the duration of a train from passing the last signal until entering the terminal. We measured the performance and quality of the algorithm using the CPU time t_{cpu} and the value of the objective function L_{max}. The experiments were performed on a Pentium Pro 450Mhz (128 MB RAM). The class libraries ILOG Solver and Scheduler with Microsoft Visual C++(5.0) were used to implement the algorithms.

We tested two different time-tables provided by [10] where trains arrive every 6 (8) [min]. The trains remain together in the terminal for 12 (16) [min] before they depart again in the sequence of arrival every 6 (8) minutes. Here, only the time-table based on 8 min is used (see Appendix). In [1] the results for both time-tables are provided, also the experimental design is described in more detail.

It is assumed that the containers are equally distributed over the arriving and departing trains. Moreover, the containers are transferred only to trains in the same bundle. In case of a delay of a train, it is not possible to wait for the next bundle to decrease L_{max}.

The calculation of the processing time was based on kinematic data and the real distances provided by Noell GmbH, which can be found in the Appendix. To validate the COP approach, a simulation model of the terminal was build using SiMPLE++.

We compared the following heuristics for the variable/value ordering:

- **MinSizeInt**: Standard heuristic where the variable with the smallest number of values left in its domain is chosen for the next assignment.
- **MinEndMax**: Standard heuristic, which selects the operation with the minimum earliest start time and - among these - the one with the latest finish time.
- **MinEndMin**: Standard heuristic, which chooses the operation with the earliest start time and - among these - the one with the earliest finish time.
- **ORRstatic**: Identifies critical operations by calculating the probability that an operation is processed on a resource at a certain point of time (*individual* demand profile). This calculation takes place before solving the problem, during initialization. Adding all operations' individual demand profiles that require the same resource results in a *cumulated* demand profile of that resource. The operation that contributes most to the most critical (busiest) resource is chosen.

Table 2. Comparison of the ordering heuristics; for MinEndMin and MinEndMax it is further distinguished if the assignment heuristic was used or not used (n.u.). The first 7 rows correspond to scenario 1, the last 7 rows refer to scenario 2

Sorting heuristic	Assign. Heur.	First solution t_{cpu}	L_{max}	C_{max}	Best solution t_{cpu}	L_{max}	C_{max}	# Sol.
MinEndMin	used	5.87	-214	3332	30.37	-656	3580	6
MinEndMax	used	7.69	-173	3403	28.02	-656	3522	5
MinEndMin	n. u.	25.26	-410	3136	105.24	-656	3542	3
MinEndMax	n. u.	25.37	-368	3208	142.2	-656	3484	5
MinSizeInt	(n. u.)	$t_{cpu} > 3600$ [sec]						
ORRstatic	(n. u.)	27.3	-410	3595	100.63	-656	4246	4
ORRdynamic	(n. u.)	27.52	-404	3602	94.03	-656	3743	2
MinEndMin	used	$t_{cpu} > 3600$ [sec]						
MinEndMax	used	$t_{cpu} > 3600$ [sec]						
MinEndMin	n. u.	69.97	-538	3079	266.11	-656	3267	3
MinEndMax	n. u.	70.08	-521	3164	282.04	-656	3267	4
MinSizeInt	(n. u.)	$t_{cpu} > 3600$ [sec]						
ORRstatic	(n. u.)	73.38	-396	3386	276.5	-656	3401	3
ORRdynamic	(n. u.)	74.37	-566	3244	259.31	-656	3294	2

- **ORRdynamic**: In contrast to ORRstatic, the calculation of individual and aggregate demand profiles is performed several times during the solving process.

The sorting heuristics MinEndMin and MinEndMax were tested in two variants. In the first, assignment and sequencing are calculated in two steps using the heuristic introduced in Section 4.5. In the second assignment and sequencing are done simultaneously. The other heuristics are tested for the simultaneous assignment and sequencing.

The terminal configuration of scenario 1 consists of 90 shuttle cars, 10 cranes and one overlapping row. Scenario 2 was built by 90 shuttle cars, 12 cranes and two rows for the overlapping area. This configuration are based on the results of [1]. The performance results are shown in Table 2. Next to L_{max}, the makespan C_{max} is shown, but not considered in the objective function. The last column provides the number of solutions found during the search. The best solution($L_{max} = -656$ [sec]) is determined in all cases where the time restriction of 3600 [sec] is met. When L_{max} becomes negative, all trains could leave the terminal ahead of the time-table.

The performance of the simple heuristics is quite good for scenario 1 where ORRdynamic outperforms these in scenario 2. It is interesting that no solution was found within the time bound applying MinEndMax and MinEndMin using the assignment heuristic. However the performance of the assignment heuristic could probably be improved. If no assignment heuristic is used, the complexity increases dramatically, here ORRdynamic outperforms the other heuristics.

Table 3. Comparison of the measures relevant for effective serach, $|\mathcal{C}|$ states the number of constraints, $|\mathcal{O}|$ the number of operations and $|\mathcal{R}|$ the number of resources. The first 7 rows correspond to scenario 1, the last 7 to scenario 2

Sorting heuristic	Assignm. Heur.	#Var.	$\|\mathcal{C}\|$	$\|\mathcal{O}\|$	$\|\mathcal{R}\|$	#Choice points	Memory [KB]
MinEndMin	used	1387	2282	462	29	3230	1815
MinEndMax	used	1387	2282	462	29	2769	1819
MinEndMin	n. u.	1684	2579	462	29	2239	1945
MinEndMax	n. u.	1684	2579	462	29	5603	1945
MinSizeInt	(n. u.)	$t_{cpu} > 3600$ [sec]					
ORRstatic	(n. u.)	1684	2579	462	29	4323	2338
ORRdynamic	(n. u.)	1684	2579	462	29	2998	2432
MinEndMin	used	$t_{cpu} > 3600$ [sec]					
MinEndMax	used	$t_{cpu} > 3600$ [sec]					
MinEndMin	n. u.	2161	3454	455	35	2880	2324
MinEndMax	n. u.	2161	3454	455	35	3598	2332
MinSizeInt	(n. u.)	$t_{cpu} > 3600$ [sec]					
ORRstatic	(n. u.)	2161	3454	455	35	4173	2795
ORRdynamic	(n. u.)	2161	3454	455	35	2920	2716

In Table 3 the number of choice points (of the search tree) is given. Comparing ORRstatic and ORRdynamic, the higher effort to calculate the demand-profiles becomes clear, where the memory utilization is still very low. The choice points indicate the number of branching decisions during the search procedure. A very low number of choice points refers to a good pruning of the search space or a sophisticated search, which holds for ORRdynamic. Comparing the number of choice points, it becomes clear that the additional computational effort for dynamically calculating the demand profiles pays back.

The time needed to determine the optimal solution is sufficient in all cases where the time bound was met. The trains were able to leave the terminal in time, an important constraint for the practical application of the model. Here a schedule could also be calculated dynamically when a train is delayed to determine which containers could be transshiped meeting the time-table or which delay would be necessary.

6 Summary

The intermodal terminal Mega Hub was modeled as a CSP and transformed into a COP. Practical relevant constraints like the distinction between direct and indirect transshipment as well as the overlapping crane areas are included in the model. Real-world problem instances can be solved very fast on state-of-the-art PCs.

Table 4. Distance matrix of the real layout

d_{ij} $[m]$	Tr. 1	Tr. 2	Tr. 3	Sor. 1	Sor. 2	Tr. 4	Tr. 5	Tr. 6
Track 1	0	8,1	13	19,8	31,3	39,05	43,95	52,35
Track 2		0	4,9	11,7	23,2	30,95	35,85	44,25
Track 3			0	6,8	18,3	26,05	30,95	39,35
Sorter 1				0	11,5	19,25	24,15	32,55
Sorter 2					0	7,75	12,65	21,05
Track 4						0	4,9	13,3
Track 5							0	8,4
Track 6								0

Table 5. Time-table of the trains, the arrival sequence of the trains is equal to the departure sequence. The stop over time of the trains overlaps for 16 [min]. The durations are given in [min] and ([sec])

Train	t^{rel}	t^{due}
1	0 (0)	56 (3360)
2	8 (480)	64 (3840)
3	16 (960)	72 (4320)
4	24 (1440)	80 (4800)
5	32 (1920)	88 (5280)
6	40 (2400)	96 (5760)

This performance allows the application of the model in the real terminal. A recalculation of the transshipment sequence in case of lateness of the trains is also possible. The model can both be used for the configuration (long-term) and the operation (short-term) of the terminal.

Modeling the problem as a CSP allows to include constraints on a detailed and aggregate level as well as stating the objective in a clear way.

The results show the practical relevance of the algorithms. The new heuristics for variable ordering outperform standard heuristics. The Constraint Satisfaction approach makes it possible to model the system on a very detailed and aggregated basis with a high flexibility.

The performance results show the applicability of the model to a practical setup both for calculating an initial schedule or to reschedule in case of a delay of a train.

Appendix

The kinematic data are given to

$$v_{long} = v_{cross} = 3\ [\frac{m}{s}], \quad a_{long} = a_{cross} = 0,54\ [\frac{m}{s^2}]$$

$$v_{up/down,loaded} = 0,75\ [\frac{m}{s}], \quad v_{up/down,empty} = 1,5\ [\frac{m}{s}]$$

$$a_{up/down,loaded} = a_{up/down,empty} = 1\ [\frac{m}{s^2}]$$

$$v_{shuttle} = 3\ [\frac{m}{s}], \quad a_{shuttle} = 0,3\ [\frac{m}{s^2}]$$

References

1. Alicke K (1999) Modellierung und Optimierung mehrstufiger Umschlagsysteme. PhD thesis, Universität Karlsruhe, Institut für Fördertechnik und Logistiksysteme
2. Alicke K, Arnold D (1998) Optimierung von mehrstufigen Umschlagsystemen. Fördern und Heben 8(10): 769–772
3. Arnold D (1995) Materialflußlehre. Vieweg, Wiesbaden
4. Bauer R (1998) Innovative linear motor-based transfer technology allows intelligent container handling. In: Proceedings of EURNAV 98, Hannover
5. Blazewicz J, Domschke W, Pesch E (1996) The job shop scheduling problem: Conventional and new solution techniques. European Journal of Operational Research 93: 1–33
6. Bostel N, Dejax P (1998) Models and algorithms for the container allocation problem on trains in a rapid transshipment yard. Transportation Science 32(4): 370–379
7. Franke K-P (1997) Mega-Drehscheibe für den kombinierten Verkehr. EIA Symposium Europe Towards Intermodal Transport
8. Franke K-P, Häffner G (1996) Automatisierung der Umschlagsknoten als Beitrag der Industrie, Transportketten wirtschaftlicher zu machen. VDI-Berichte 1274: 181–195
9. ILOG (1998) ILOG Optimization Suite - Discovering a competitive advantage. White Paper
10. Meyer P (1999) Entwicklung eines Simulationsprogramms für Umschlagterminals des kombinierten Verkehrs. Universität Hannover
11. Prosser P (1993) Hybrid algorithms for the constraint satisfaction problem. Computational Intelligence 9: 268–299
12. Rodosek R, Wallace M (1998) One model and different solvers for hoist scheduling problems. Unpublished working paper, Presentation at INFORMS 1998, Montreal
13. Sadeh N (1991) Look-ahead techniques for micro-opportunistic job shop scheduling. PhD thesis, Carnegie Mellon University
14. Sadeh N (1994) Micro-opportunistic scheduling: The micro-boss factory scheduler, ch 4, pp 99–135. Morgan Kaufmann, San Francisco
15. Sadeh N, Fox MS (1996) Variable and value ordering heuristics for the job shop scheduling constraint satisfaction problem. Artificial Intelligence 86: 1–41
16. Varnier C, Bachelu A, Baptiste P (1997) Resolution of the cyclic multi-hoists scheduling problem with overlapping partitions. Information Systems and Operational Research 35(4): 1–16
17. Zwieben M, Fox MS (1994) Intelligent scheduling. Morgan Kaufman, San Francisco

A dispatching method for automated guided vehicles by using a bidding concept*

Jae Kook Lim[1], **Kap Hwan Kim**[2], **Kazuho Yoshimoto**[3], **Jun Ho Lee**[2], **and Teruo Takahashi**[1]

[1] Institute of Asia-Pacific Studies, Waseda University, Sodai-Nishiwaseda bldg. 6F, 1-21-1 Nishiwaseda Shinjuku-ku Tokyo, 169-0051, Japan
(e-mail: {jklim; taka12}@mn.waseda.ac.jp)

[2] Department of Industrial Engineering, Pusan National University, Jangjeon-dong, Kumjeong-ku, Pusan 609-735, Korea
(e-mail: kapkim@pusan.ac.kr; chiwoo@sktelecom.com)

[3] Department of Industrial & Management Systems Engineering, Waseda University, 3-4-1 Okubo Shinjuku-ku Tokyo 169-8555 Japan
(e-mail: kazuho@yoshi.mgmt.waseda.ac.jp)

Abstract. A dispatching method is suggested for automated guided vehicles by using an auction algorithm. The dispatching method in this study is different from traditional dispatching rules in that it looks into the future for an efficient assignment of delivery tasks to vehicles and also in that multiple tasks are matched with multiple vehicles. The dispatching method in this study is distributed in the sense that the dispatching decisions are made through communication among related vehicles and machines. The theoretical rationale behind the distributed dispatching method is also discussed. Through a simulation study, the performance of the method is compared with that of a popular dispatching rule.

Key words: Automated guided vehicle – Distributed dispatching method – Simulation

1 Introduction

The dispatch of automated guided vehicles (AGVs) can be defined as the assignment of vehicles to a delivery task so that some performance objectives of a shop are achieved. The most popular strategy for dispatching vehicles is to match a delivery

* The research was financially supported by the Sasakawa Scientific Research Grant from The Japan Science Society. The original version of the simulation program is provided by Professor Jae Yeon Kim at Dong Yang University, Korea.

Correspondence to: J. K. Lim

request with an idle vehicle whenever a delivery request is issued or whenever a vehicle becomes idle. When the dispatching decision is triggered by an occurrence of a delivery task, one of vehicles idle at that time is selected for the new delivery task. Also, when a vehicle becomes idle, one of the waiting delivery-tasks is chosen for the new idle vehicle. Traditional dispatching rules for the former case and the latter case are called "task-initiated dispatching rules" and "vehicle-initiated dispatching rules," respectively (Egbelu and Tanchoco, 1984).

Although the traditional dispatching rules are simple to use, one drawback of these rules is that they are myopic in a sense that only one vehicle is considered in case of the vehicle-initiated rules and only one delivery task is considered in case of the task-initiated rules. For example, suppose that a new idle vehicle was assigned to a delivery task at a workstation (let it be workstation A) because workstation A was located nearest to the vehicle among all the workstations with delivery tasks. However, suppose that, just after the (first) vehicle was assigned, another (second) vehicle became idle at a location nearer to workstation A than the first vehicle did. In this case, if the dispatching decision had been made considering where and when the second vehicle would have become idle, the second vehicle must have been assigned to workstation A instead of the first vehicle.

To overcome this problem, when a vehicle becomes idle and needs a delivery task to be assigned, all the vehicles must be considered simultaneously as well as all the delivery tasks. That is, the dispatching decision must be made in many-to-many basis instead of one-to-many basis. However, the optimal decision-making for all the vehicles is computationally impractical especially when the number of vehicles involved is large. Also, dispatching decisions must be made by a central processor that has complete information on states of all the vehicles and workstations. However, when the size of a material handling system is large, it is risky to depend on only a central controller because of possible breakdowns or overloading of the central controller. Also, it is not economical for the central controller to make the dispatching decision again for all the vehicles whenever a small change in the system state occurs.

Thus, the dispatching algorithm in this study attempts to satisfy the following desirable conditions:

(1) A high level of system performance must be obtained. For the high performance, the dispatching decision must be near optimal form the perspective of the empty travel distance of vehicles or the response time of vehicles to calls from workstations.
(2) The decision process for dispatching must be distributed so that the central controller is not overloaded and the entire material handling system becomes robust to various failure and breakdowns of some components and the central controller.
(3) The effect of small changes in the system's state must be localized. For example, when a new delivery task arrives, the changes in the dispatching decision are usually confined to several related vehicles. Thus, it is necessary to develop a method for identifying the related vehicles and revising the dispatching decisions only for the related vehicles without having to solve the entire assignment problem again.

Bartholdi and Platzman (1989) proposed the first-encountered-first-served (FEFS) rule, which, with a closed single-loop guide path, attempts to deliver tasks as quickly as possible for AGVs. Egbelu (1987) suggested the demand-driven rule in which AGVs are first dispatched to delivery tasks that are bound for input buffers whose lengths are below a threshold value. Kim et al. (1999) suggested an AGV dispatching method in which balancing workloads among different workstations is the first criterion for selecting the next delivery task. For selecting a vehicle, although the dispatching method proposed by Kim et al. (1999) looks beyond the times when vehicles become idle, their method basically selects in sequence one task among several tasks and then one vehicle among many vehicles sequentially. Thus, their method does not attempt to optimally match multiple tasks with multiple vehicles simultaneously. Sabuncuoglu, I. and Hommertzheim (1992) proposed a dynamic scheduling method for both vehicles and machines.

Klein and Kim (1996) compared multi-attribute dispatching rules with single-attribute dispatching rules and showed that the former rules are superior to the latter rules. Lee et al. (1996) and Bilge and Tanchoco (1997) treated the dispatching problem for AGVs with multi-load capacities. Taghaboni-Dutta (1997) suggested a dispatching rule based on an index of values added during the operations of a job in a shop.

Bilge and Ulusoy (1995) provided a simultaneous scheduling method for operation of machines and transfer of materials by AGVs. The scheduling problem was decomposed into two subproblems: a machine scheduling subproblem and a vehicle scheduling subproblem. An iterative procedure was suggested for each subproblem. Their paper is related to this study in that future delivery tasks and idle vehicles are considered for the vehicle scheduling. However, the problem solved in their study was limited in size because of the complexity of the suggested algorithm. Co and Tanchoco (1991) provided a good review about the various aspects of the operation of AGVs.

This study suggests a dispatching method based on a bidding concept. The bidding-based dispatching method (BDM) assumes a market system in which vehicles attempt to earn money as much as possible by performing delivery tasks with the highest possible price at the lowest possible costs, and delivery tasks attempt to pay charges as less as possible by hiring vehicle with the lowest possible opportunity cost.

It is shown in property 2 that BDM results in the optimal solution of an assignment problem in which cost coefficients correspond to empty travel times or response times. BDM considers currently busy vehicles as well as currently idle vehicles by looking ahead to the change of states of vehicles. Thus, BDM makes dispatching decisions on a many-to-many basis instead of a one-to-many basis, as was in the case of workstation-initiated rules or vehicle-initiated rules in previous studies. This property of BDM coincides with the first desirable condition of the dispatching algorithm.

However, when a decision is made about the assignment of multiple vehicles to multiple delivery tasks, it is time-consuming to reconsider all the dispatching decisions already made whenever a new delivery task arrives at the shop or whenever a vehicle becomes idle. In BDM, a newly idle vehicle submits a bid to the task giving the maximum margin to the vehicle, and then the task sends a cancellation notice to a previously assigned vehicle (let it be the second vehicle) so that the second vehicle looks for another task, and so on. When a new delivery task arrives at the shop, similar things happen. Thus, the dispatching procedure of BDM is distributed in that no central controller is necessary. Also, the effect of small changes in the system's state is localized because the changes in the dispatching decision are confined to several tasks and vehicles. Thus, second and the third desirable conditions of the dispatching algorithm are satisfied.

In the next section, an auction-based assignment algorithm and the concept of economic equilibrium are introduced. Section 3 suggests a process of bidding-based dispatching and a method of constructing bids. Section 4 describes results of a simulation study to evaluate the performance of the dispatching procedure described in this paper. Conclusions are provided in the final section.

2 An assignment problem and the economic equilibrium

In order to explain the rationale of the bidding-based dispatching method, the concept of economic equilibrium is first introduced. In this section, the dispatching problem is considered as an assignment problem in which multiple delivery tasks are matched with multiple vehicles and objective function is minimizing the total travel distance or the total response time of vehicles. Assuming that there are n vehicles and m delivery tasks to be matched with each other at a specific point in time, it is attempted to solve the assignment problem through a market mechanism (Bertsekas, 1990), viewing each task or a vehicle as an economic agent acting for its own best interest.

Let a_{ij} be the cost of vehicle i to perform task j. The cost may include only that for empty travel or for both empty and loaded travel. And, let x_{ij}= 1 if vehicle i is assigned to task j; 0, otherwise. Note that the number of vehicles and the number of delivery tasks are not usually the same. When the number of vehicles is larger than the number of tasks, dummy tasks will be added to the list of vehicles. On the other hand, when the number of tasks is larger than the number of vehicles, dummy vehicles will be added. The cost parameters of each dummy vehicle (task), a_{ij}, are set to be the same for all tasks (vehicles). In the following discussion, without the loss of generality, it is assumed that the number delivery tasks waiting for the assignment of a vehicle is larger than or equal to the number of assignable vehicles. Thus, it is assumed that dummy vehicles may be added for the formulation of the assignment problem.

Then, the minimum cost assignment problem can be formulated as follows:
(P1)

$$\text{Minimize} \sum_{i=1}^{m} \sum_{j=1}^{m} a_{ij}\, x_{ij}$$

subject to

$$\sum_{j=1}^{m} x_{ij} = 1, \text{ for all } i,$$

$$\sum_{i=1}^{m} x_{ij} = 1, \text{ for all } j,$$

$$x_{ij} \geq 0, \text{for all } i \text{ and } j.$$

The dual of (P1) becomes
(D)

$$\text{Minimize} \sum_{i=1}^{m} v_i - \sum_{j=1}^{m} p_j$$

subject to

$$v_i - p_j \quad \geq -a_{ij} \ \text{ for all } i \text{ and } j. \tag{1}$$

The concept of the complementary slackness is useful to introduce the relationship between the optimality of an assignment and the condition that values of prices and margins must satisfy as follows:

(**An optimality condition**) A solution x_{ij} and (v_i, p_j) is optimal if x_{ij} and (v_i, p_j) are feasible for (P1) and (D), respectively, and satisfy the complementary slackness conditions of linear programming, which can be stated as follows (Hillier and Lieberman, 1986):

$$(v_i - p_j + a_{ij})x_{ij} = 0 \text{ for all } i \text{ and } j.$$

The scalar p_j will be referred to as the price of delivery task j, which task j must pay to the corresponding vehicle performing the delivery task. Also, v_i can be interpreted as the profit margin of vehicle i that performed task j.

In the following, a concept of "equilibrium," – which is useful in devising a bidding-based dispatching method – is introduced:

(**Definition of the equilibrium**) Suppose that task j is assigned to vehicle i. Vehicle i will be *happy* if $v_i = p_j - a_{ij} = \max_{k=1,\ldots,m} \{p_k - a_{ik}\}$. We will say that a feasible assignment and a set of prices and margins are at *equilibrium* when all the vehicles are *happy*.

The following property guarantees the optimality of prices and margins that are at equilibrium:

Property 1: (The optimality of the equilibrium condition) For a feasible assignment, if the a set of prices and margins are at equilibrium, the assignment is optimal.

Proof. The fact that $v_i = p_j - a_{ij} = \max_{k=1,\ldots,m} \{p_k - a_{ik}\}$ for all i implies that $v_i \geq p_k - a_{ik}$ for all i and k, which in turn means that the set of v_i and p_k is feasible for constraint (1). Also, the fact that $v_i = p_j - a_{ij}$ for all i and j with positive x_{ij} (based on the definition of the equilibrium) implies that the complementary slackness conditions are satisfied. Thus, the conclusion holds. □

For devising a bidding procedure which guarantees to be stopped within a finite number iterations, the concepts of "almost equilibrium" and "almost happy" will be introduced which are almost equivalent to those of "equilibrium" and "happy" in the following (Bertsekas, 1990):

(Definition of almost equilibrium) Suppose that task j is assigned to vehicle i. Vehicle i will be *almost happy* if $v_i = p_j - a_{ij} = \max_{k=1,\ldots,m} \{p_k - a_{ik}\} - \varepsilon$ for a small positive real number ε. Task j will be *almost happy* if $p_j = v_i + a_{ij} = \min_{k=1,\ldots,m} \{v_k + a_{kj}\} + \varepsilon$ for a small positive real number ε . We will say that a feasible assignment and a set of prices and margins are at *almost equilibrium* when all vehicles or all tasks are *almost happy*.

The following section describes a dispatching algorithm for obtaining an assignment of candidate trucks to candidate tasks that satisfies the conditions of the almost equilibrium.

3 A bidding-based procedure for dispatching vehicles

BDM assumes that the price of each delivery is determined through a bidding process. During the bidding process, each vehicle selects the delivery task that maximizes its own margin which is the price of the task minus the cost of performing the task, while each task chooses the vehicle that requests the least compensation which is the transportation cost required by the vehicle for performing the task plus the minimum margin requested by the vehicle.

For a given assignment of vehicles to delivery tasks, a set of prices and margins is said to be at "equilibrium" if a vehicle cannot increase its margin by changing its currently assigned task and a delivery task cannot decrease the compensation by changing its currently assigned vehicle (see Sect. 2 for more formal definition of the equilibrium).

In the dynamic situation assumed in this study, all loaded or idle vehicles are candidates for dispatching. Note that empty vehicles traveling to pick up a load is excluded from the set of candidates. Tasks in the output buffer space at each workstation are considered to be candidates for dispatching. A dispatching decision process is initiated whenever a vehicle is loaded or a new delivery order is issued. When a vehicle is loaded, the ASSIGN-TASK-TO-A-NEW-VEHICLE procedure is initiated to secure the next delivery task. In case a delivery order is issued, the ASSIGN-VEHICLE-TO-A-NEW-TASK procedure is initiated. Once either of the two procedures is initiated, a feasible assignment and prices almost at equilibrium – which is defined in Section 2 – are obtained. A dispatching decision on a vehicle or a delivery task is fixed and implemented in either of two following cases: when a vehicle becomes idle and it has an assigned delivery task or when a task is assigned an idle vehicle. Thus, once a vehicle starts an empty travel for implementing a delivery task, both the vehicle and the task will be excluded from the candidate list for dispatching.

During the ASSIGN-TASK-TO-A-NEW-VEHICLE procedure, prices of tasks and margins of vehicles decrease. However, during the ASSIGN-VEHICLE-TO-A-NEW-TASK procedure, prices of tasks and margins of vehicles increase. The prices of tasks are limited by a pre-specified upper bound of the price from the above, while, above zero, margins of vehicles change.

In the following, two procedures (ASSIGN-TASK-TO-A-NEW-VEHICLE, ASSIGN-VEHICLE-TO-A-NEW-TASK) will be introduced for optimally matching vehicles with delivery tasks. The optimality of the resulting assignment will be proved In Property 2. Note that the number of available vehicles (n) may not be the same as the number of available delivery tasks (m). Assume that $m > n$ without the loss of generality. Then, even after an assignment is determined, one or more tasks may not be assigned ("unassigned and inactive": UI) to a vehicle. Both the number of vehicles and the number of tasks in the "assigned (A)" state are n. However, when a vehicle (task) becomes a new candidate for an assignment, it is "unassigned" but has a potential to be assigned. The vehicle (task) is said to be "unassigned but activated (UA)." Also, during the assignment procedure, a less competitive vehicle (task) may have its assigned task (vehicle) taken away by another more competitive vehicle (task). Then, the former vehicle (task) becomes "unassigned but activated (UA)," while the latter vehicle (task) becomes "assigned (A)."

The bidding process (ASSIGN-TASK-TO-A-NEW-VEHICLE) for the case of the vehicle initiation (when a new vehicle becomes idle) is illustrated in Figure 1. Before the new vehicle becomes idle, three vehicles are assigned to three tasks. Thus, they are in state "A" except one task that is not assigned to any vehicle and so is in state "UI" (see Fig. 1a). When a vehicle (vehicle D) becomes idle, its initial state is "UA", and it seeks a delivery task to perform (see Fig. 1b). Among waiting tasks, vehicle D selects a task (task 3) giving the maximum margin at the current price. Then, vehicle D submits a bid with a price lower than the current price of task 3. Then, task 3 accepts the bid because the suggested price is lower than the current price, and informs the cancellation of assignment to the vehicle (vehicle C) that task 3 was previously assigned to. The state of vehicle D is changed from "UA" to "A" and the state of vehicle C is changed from "A" to "UA" (see Fig. 1c). Then, vehicle C whose state became "UA" searches for the task that gives the highest margin at the current price. In Figures 1c, it is task 4. Because task 4 was in state of "UI," no vehicle turns to "UA" and so the bidding process is terminated (see Fig. 1d). Note that tasks 1 and 2 and vehicles A and B were not involved in the entire dispatching process. That is, the effect of changes was confined only to related tasks and vehicles. And also note that no central controller was involved in the bidding process.

A similar process (ASSIGN-VEHICLE-TO-A-NEW-TASK) will be followed when a new delivery task appears. Figure 2 illustrates the bidding process. The new task selects the vehicle (vehicle 1) with the lowest price that is the sum of the travel cost for vehicle to perform the new task and the current margin of vehicle 1 for performing the currently assigned task (task 2 in this example). The entering task submits a bid – which suggests a margin higher than the current margin of vehicle 1 – to vehicle 1. Then, vehicle 1 sends a cancellation notice to the currently assigned task (task 2). Then, task 2 begins the same procedure as what the new entering did.

The following describes how to obtain a feasible assignment and a set of prices and margins that are at *almost equilibrium* through two bidding processes: ASSIGN-TASK-TO-A-NEW-VEHICLE and ASSIGN-VEHICLE-TO-A-NEW-TASK.

First, the procedure of ASSIGN-TASK-TO-A-NEW-VEHICLE is described in the following:

This process is triggered when a new vehicle becomes idle. Let A(i) be the set of tasks that can be assigned to vehicle i. Also, let the initial price of task j, p_j, be

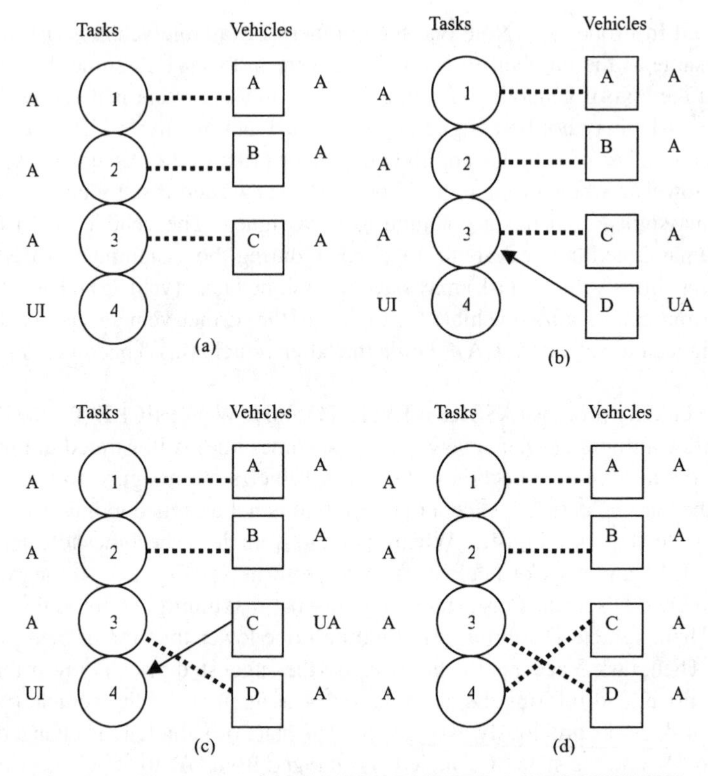

Fig. 1. An illustration of the ASSIGN-TASK-TO-A-NEW-VEHICLE

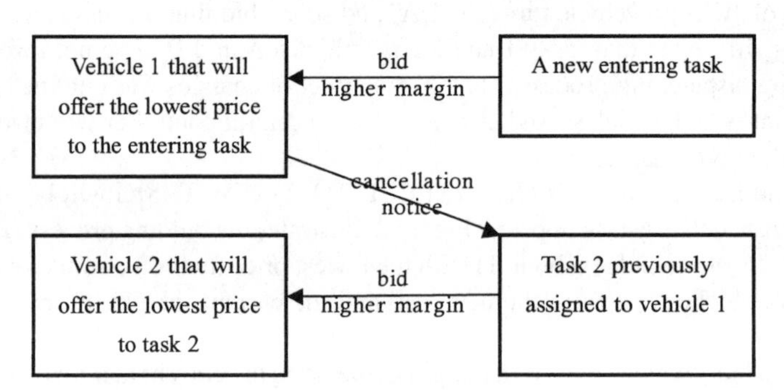

Fig. 2. Procedure of ASSIGN-VEHICLE-TO-A-NEW-TASK

a large number (p_0) for all j and the initial margin of vehicle i, v_i, be zero. The initial state of the new idle vehicle is set as UA.

Procedure for an unassigned vehicle: ASSIGN -TASK-TO-A-NEW-VEHICLE
The following procedure is repeated until no UA vehicle is found:

Preparing a bid by a UA vehicle
Let the UA vehicle be vehicle i.
Compute the current margin (profit) that vehicle i can earn by performing task $j \in A(i)$, which is given as

$$v_{ij} = \max\{p_j - a_{ij}, 0\}. \tag{2}$$

Find the best task j^* having the maximum value of

$$v_{ij^*} = \max_{j \in A(i)} v_{ij}.$$

(Task j^* will give vehicle i the maximum margin if it is assigned to vehicle i.)
If $v_{ij^*} = 0$, then it implies that there is no profitable task for vehicle i. (This happens when prices of tasks in A(i) are too low for vehicle i to obtain a positive margin by performing any task in A(i). In this case, vehicle i should remain idle.) Let $v_i = 0$ and the state of vehicle i be UI, and then stop.
Otherwise, find the highest margin offered by tasks other than j^*, which is given as

$$w_{ij^*} = \max_{j \in A(i), j \neq j^*} v_{ij} \tag{3}$$

If task j^* is the only task in A(i), then w_{ij^*} is set to be 0. (w_{ij^*} means the margin that vehicle i can earn when it is assigned to the second best task.)
Let the state of vehicle i be A. Compute the bid of vehicle i for task j^*. The price of the bid is given by

$$b_{ij^*} = p_{j^*} - v_{ij^*} + w_{ij^*} - \varepsilon. \tag{4}$$

(b_{ij^*} is the level of the price of task j^* that gives the same margin to vehicle i as the second best task of (3) does. Note that ε is related to the definition of "almost equilibrium.")
Submit the bid to task j^*.

Accepting the bid submitted to task j^*
If there is a vehicle that is currently assigned to task j^* (let it be vehicle i^*), make the state of vehicle i^* be UA. Assign task j^* to vehicle i and make the state of vehicle i be A. Let $p_{j*} = \min\{p_0, b_{ij^*}\}$ and $v_i = p_{j^*} - a_{ij}$. Inform the new assignment to vehicle i^* and i.
The following describes the ASSIGN-VEHICLE-TO-A-NEW-TASK, which is the procedure triggered when a new delivery task becomes available:

Procedure for unassigned tasks: ASSIGN-VEHICLE-TO-A-NEW-TASK
Let $V(k)$ be the set of vehicles that can be assigned to task k.
The following procedure is repeated until no UA task is found.

Preparing a bid by a UA task
Let the UA task be task k.

Compute the current minimum compensation required to induce each vehicle $i \in V(k)$, which is given by

$$c_{ik} = \min \{v_i + a_{ik}, p_0\}. \quad (5)$$

(Note that v_i represents the margin that vehicle i already secured.)

Find the best vehicle i^* having the minimum value

$$c_{i^*k} = \min{}_{i \in V(k)} c_{ik}.$$

If $c_{i^*k} = p_0$, then it implies that there is no vehicle that task k with the maximum price p_0 can afford. Then, let $p_k = p_0$ and the state of task k be UI, and then stop. (Task k will remain unassigned.)

Otherwise, find the minimum compensation for vehicles other than vehicle i^* by

$$d_{i^*k} = \min{}_{i \in V(k), i \neq i^*} c_{ik}. \quad (6)$$

(d_{i^*k} means the compensation that task k must pay so that it is to be assigned to the second best vehicle.)

If vehicle i^* is the only vehicle in $V(k)$, $d_{i^*k} = p_o$.

Let the state of task k be A.

Compute the bid of task k for vehicle i^*. The margin of the bid is given by

$$e_{i^*k} = v_{i*} - c_{i^*k} + d_{i^*k} + \varepsilon. \quad (7)$$

(e_{i^*k} is the level of the margin that vehicle i^* can earn when task k pays the same compensation to vehicle i^* as task k needs to do for inducing the second best vehicle of 7.)

Submit the bid to the manager of vehicle i^*.

Accepting the bid submitted to vehicle i^*
If there is a task that is currently assigned to vehicle i^* (let it be task j^*), make the state of task j^* be UA. Assign vehicle i^* to task k. Let $v_{i^*} = \min \{e_{i^*k}, p_0 - a_{i^*k}\}$ and $p_k = v_{i^*} + a_{iT^*k}$.

Inform the new assignment to tasks j^* and k.

Property 2: For a given set of candidate vehicles and tasks, procedures of ASSIGN-TASK-TO-A-NEW-VEHICLE and ASSIGN-VEHICLE-TO-A-NEW-TASK enable a feasible assignment to (P1), and a set of prices and margins that are almost at equilibrium in a finite number of iterations.

Proof. See Appendix.

The following provides a numerical example to illustrate step-by-step the distributed assignment procedure:

A scenario of the dynamic arrival of vehicles and delivery tasks is presented here. At the beginning, tasks 1 and 2 are waiting for vehicles. Next, AGV 1 and AGV 2 become available for assignment, one by one. The following shows how the assignment procedure is performed at each moment of the event. Table 1 shows

Table 1. Travel distance from locations of vehicles to pickup location of delivery tasks (unit: ft)

	Task 1	Task 2
AGV 1	360	260
AGV 2	700	200

Table 2. Assignment, prices, and margins for the example with two tasks and two AGVs

	Cost (assignment)		v_i
	Task 1	Task 2	
AGV 1	360 (1)	260 (0)	$240 - 2\varepsilon$
AGV 2	700 (0)	200 (1)	$300 - \varepsilon$
p_j	$600 - 2\varepsilon$	$500 - \varepsilon$	

the travel distance from the initial locations of vehicles to the pickup locations of the delivery tasks.

Let both the initial p_1 and p_2 be 1,000 ($= P_0$).

1) When AGV 1 becomes available,
   $v_{11} = 1,000 - 360 = 640. v_{12} = 1,000 - 260 = 740$. Thus, j^*=2, and $w_{12^*} = 640. b_{12} = p_2 - v_{12} + w_{12} - \varepsilon = 900 - \varepsilon$. State of AGV 1, which was assigned to task 2, changes from UA to A. $p_2 = 900 - \varepsilon$ and $v_1 = 640 - \varepsilon$.
2) When AGV 2 becomes newly available,
   $v_{21} = 1,000 - 700 = 300$, and $v_{22} = 900 - \varepsilon - 200 = 700 - \varepsilon$. Thus, $j^* = 2$ and $w_{22} = 300$. Also, $b_{22} = p_2 - v_{22} + w_{22} - \varepsilon = 500 - \varepsilon$. Task 2 is newly assigned to AGV 2. The state of AGV 1 changes from A to UA, while the state of AGV 2 becomes A. $p_2 = 500 - \varepsilon$, and $v_2 = 300 - \varepsilon$.
   AGV 1 whose state became UA constructs a new bid, as follows: $v_{11} = 1000 - 360 = 640$, and $v_{12} = 500 - \varepsilon - 260 = 240 - \varepsilon$. Thus, $j^* = 1. w_{11} = 240 - \varepsilon$, and $b_{11} = p_1 - v_{11} + w_{11} - \varepsilon = 600 - 2\varepsilon$. AGV 1 is assigned to task 1. The state of AGV 1 changes from UA to A. $p_1 = 600 - 2\varepsilon, v_1 = 240 - 2\varepsilon$. Table 2 shows the final assignment, margins, and prices.

4 A simulation experiment

A simulation was conducted to evaluate the performance of the bidding-based dispatching method (BDM). Figure 3 and Table 3 show respectively the guide path layout and the flow requirement used for the experiment. Each department has the area of 100×100 ft^2, and stations on each path segment are located 30 ft away from the nearest intersection. Table 3 shows the sequence of workstations that each product must be processed on. It is assumed that all the processing times follow uniform distributions, with the parameters as shown in Table 4.

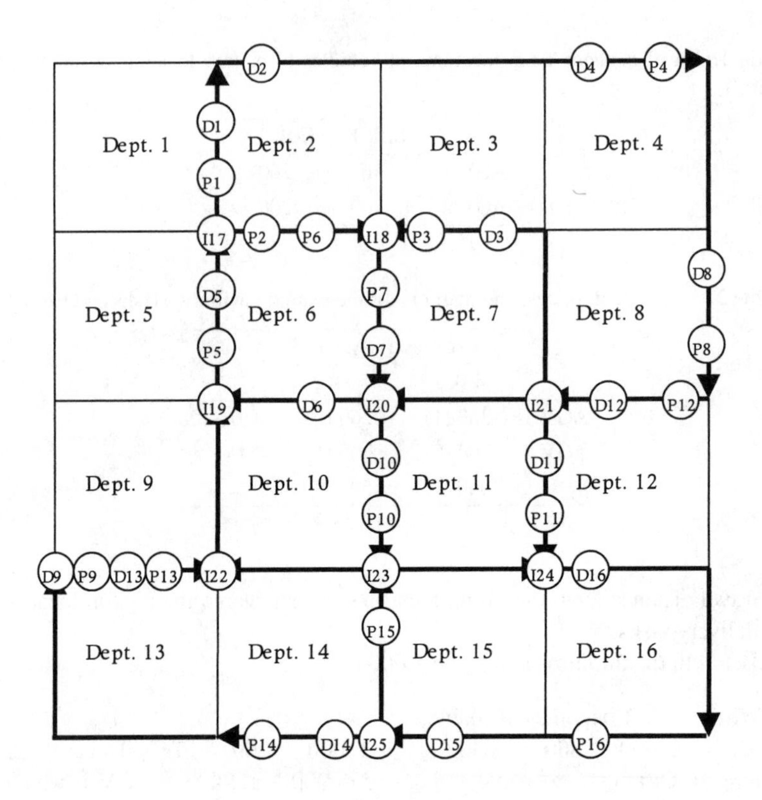

*P,D: pickup, delivery station of department

(I#) : intersections

* ➔ : guide paths

Fig. 3. Guide path and pickup and drop-off stations for vehicles

Table 3. Product mix and processing route

Products	Production mix (%)	Route (work station number)
Product 1	0.609	5-1-4-8-12-3-7-10-16
Product 2	0.229	6-7-10-16-14-9-13
Product 3	0.114	10-16-14-13-5-2-6
Product 4	0.048	5-8-11-16-15-1

The probability distributions of operation times shown in Table 4 were used in one of situations (Ex-3 in Table 5) assumed for the simulation. The average of the operation times shown in Table 4 is 4.5 minutes. However, the operation times were increased or decreased by multiplying every value in Table 4 by the same ratio for the simulation (See Table 5).

Table 4. Lower and uppper bounds of uniform distribution for the operation times at workstateion (unit: minutes)

Workstation	1	2	3	4	5	6	7	8
Bounds	[4,6]	[5,7]	[3,5]	[2,4]	[3,5]	[5,7]	[4,6]	[5,7]
Workstation	[9]	[10]	[11]	[12]	[13]	[14]	[15]	[16]
Bounds	[4,6]	[4,6]	[1,3]	[3,5]	[6,8]	[3,5]	[2,4]	[1,3]

Table 5. Data for the first experiment

Situation number	The number of vehicles	Inter-arrival time of orders (min.)	Average processing time at workstations
Ex-3	3	5.00	4.50
Ex-4	4	3.75	3.38
Ex-5	5	3.00	2.70
Ex-6	6	2.50	2.25
Ex-7	7	2.14	1.93

The following assumptions are introduced for the simulation:

1) The capacities of input and output buffers are infinite.
2) The speed of vehicles is constant, and the transfer time for pallets between vehicles and workstations is zero. However, the result of the simulation would be the same even if a strict positive transfer time is assumed.
3) A vehicle can only move one unit-load at a time.
4) Vehicles travel on the shortest distance route.
5) The inter-arrival time of production orders follows an exponential distribution.
6) The empty travel time of a vehicle from the delivery location of task i to the pickup location of task j is used as the value of a_{ij}.

Idle vehicles and loaded vehicles are candidates for task assignment. That is, empty but assigned vehicles are excluded from candidates for task assignment. Delivery tasks become candidates for vehicle assignment only after delivery requests for them are issued. Even if the qualifying range for candidates of assignment is changed, BDM in this study remains valid.

The number of runs in the simulation was 10 for each problem. The simulation time was 20,000 minutes. The first 1,000 minutes were considered as a warm-up period and so the data collected during the warm-up period was discarded when various statistics were calculated. The simulation study was performed by using ARENA 3.5.

Although there exist many different types of weights (a_{ij}) of assignments for modeling the assignment problem, the deadhead travel time is a good candidate for the weight. Minimizing the total travel distance is expected to maximize the efficiency of the manufacturing system in the long run and also to provide a robust rule under various different situations.

As the reference rule for comparisons with BDM, the nearest-workstation rule was used as the vehicle-initiated rule, while the nearest-vehicle rule was used as the workstation-initiated rule (Egbelu, 1984), which was denoted as SDR (the shortest distance rule). That is, under the SDR, when a vehicle becomes idle, the workstation with a delivery task nearest to the new idle vehicle is selected for the next service, while, when a workstation calls for a vehicle, it chooses the idle vehicle nearest to the location of the workstation. Because BDM makes dispatching decisions based on the state of the system at the moment of the decision, it can be called a dynamic decision rule.

The following statistics were collected for comparing the two dispatching methods:

NOA: The Number of Orders Arrived at the shop
NOP: The Number of Orders Produced during the simulation period
RPA: The Ratio of the number of orders Produced to the number of orders Arrived (NOP/NOA). The higher value of RPA implies the higher throughput rate of the production system that is one of the ultimate performance measures of the material handling system.
RT: The Response Time, which is the time between a call for a vehicle and the arrival of a vehicle for the call. RT consists of the waiting time of a task until the assignment of a vehicle and the empty travel time of the assigned vehicle.
ETT: The Empty Travel Time from the delivery of the previous task to the arrival time at the corresponding pickup station
TS: The Time that an order stays in the production System. This performance measure is related to the work-in-process inventory.
AU: The average AGV Utilization that is the ratio of the sum of loaded travel time, "empty but assigned" travel time, and load transfer time to the total time spent by vehicles
AGVQ: The average number of delivery orders waiting for pickup by an AGV
NC: The Number of Communications occurred for one event that needs a dispatching decision in the case of BDM. Communication is necessary for sending bids, cancellation notice, new price, and new margin. Thus, it is desirable the dispatching process is completed with fewest possible communications.

Two experiments were conducted to evaluate the performance of BDM. In the first experiment, the number of vehicles was varied between 3 and 7. For Ex-3, data for the processing time, shown in Table 4, were used. However, processing times used for other situations were values in Table 4 multiplied by the ratio of the number of vehicles for Ex-3 to that of the corresponding situation. The inter-arrival time was also adjusted in the same way. The basic idea for the adjustment was to make the work-load of deliveries be proportional to the number of vehicles so that the work-load of deliveries per vehicle can be maintained at the same level, even in different situations. Also, the work-load on each workstation was maintained at

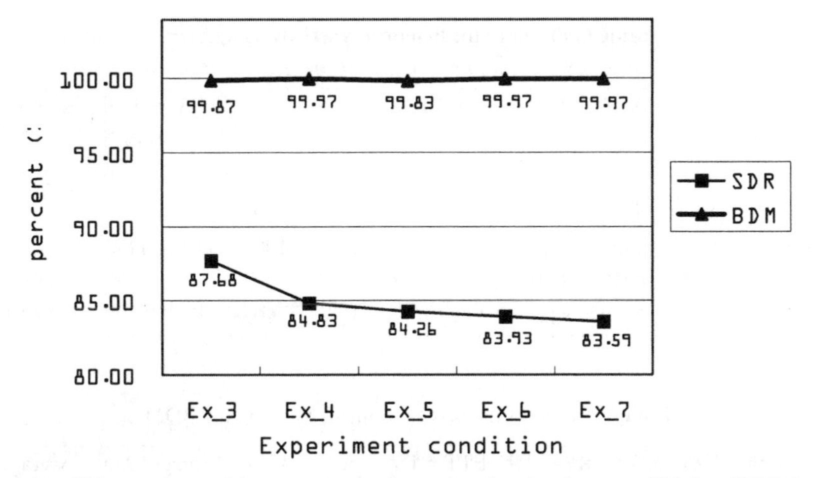

Fig. 4. Comparison of the percentage of completed orders between SDM and BDM

the same level so that the number of the works-in-process is maintained at a similar level. It was attempted to test the changes in performance measures for different numbers of vehicles and delivery tasks that participate to the auction process (the density of participants).

Table 6 shows the results of the experiment and each numeric value in Table 6 represents the average of results of the ten simulation runs. According to the results of the first experiment in Table 6, the production system using the SDR rule completed only 83–88 percent of all production orders issued during the planning horizon. In comparison, by using BDM (see Table 7), almost all production orders (over 99%) were completed during the same period. It was found that, when using SDR, the number of works-in-process waiting for vehicles (AGVQ) became higher and so the response time (RT) and the flow time (TS) were much longer than in the case using BDM. The poor performance of SDR is attributable to the longer empty travel time (ETT) in case of SDR than in the case of BDM. The last two columns of Table 7 show the maximum and the average numbers of communications, respectively. It is interesting to note that the average number of communications is less than 3, even when the number of vehicles is seven. Although the maximum number of communications went up to 120 when the number of vehicles became seven, it depends on the value of ε which was introduced to make the assignment algorithm terminate in a finite number of iterations and thus can be reduced by adjusting the value of ε. Also, note that in the result by BDM, the values of RT, ETT, and TS became smaller as the density of participants became higher. However, the results by SDR showed the trends opposite to those by BDM. The contrasting results comes from that BDM utilizes vehicles more efficiently than SDR does. Figure 4 shows the difference in the percentage of completed orders between two heuristic methods.

In the second experiment, the processing time and the inter-arrival time were maintained at the same level as Ex-5 in the first experiment. However, the number

Table 6. Result of the first experiment by using SDR

Situation	NOA	NOP	RPA	RT	ETT	TS	AU	Maximum AGVQ	Average AGVQ
Ex_3	3961	3473	87.7	203.0	1.63	1427	99.99	530	270.4
Ex_4	5350	4538	84.8	257.9	1.64	1761	100.00	859	457.5
Ex_5	6674	5623	84.3	257.5	1.65	1744	100.00	1104	568.4
Ex_6	7997	6712	83.9	263.7	1.66	1777	100.00	1359	696.7
Ex_7	9342	7809	83.6	277.3	1.66	1861	99.99	1623	853.5

Table 7. Result of the first experiment by using the BDM

Situation	NOA	NOP	RPA	RT	ETT	TS	AU	Max. AGVQ	Average AGVQ	Max NC	Average NC
Ex_3	4018	4013	99.9	2.0	1.10	119	93.84	10	1.3	38	1.6
Ex_4	5326	5325	100.0	1.7	1.05	56	92.40	11	1.3	57	1.8
Ex_5	6651	6640	99.8	1.5	1.01	70	91.35	10	1.2	75	1.9
Ex_6	7984	7982	100.0	1.4	0.97	66	90.26	11	1.3	91	2.0
Ex_7	9318	9315	100.0	1.4	0.95	58	89.56	12	1.5	120	2.2

Table 8. Result of the second experiment by using SDR

No. of AGVs	NOA	NOP	RPA	RT	ETT	TS	AU	Max AGVQ	Average AGVQ
3	6607	2285	34.6	1520.0	1.60	7036	100.00	4520	2362
4	6621	3857	58.3	794.5	1.63	4522	100.00	2891	1501
5	6674	5623	84.3	257.5	1.65	1744	100.00	1104	568
6	6652	6641	99.8	3.9	1.57	84	96.95	47	6
7	6686	6678	99.9	1.6	1.45	72.85	91.26	13	0.4

Table 9. Result of the second experiment by using BDM

No. of AGVs	NOA	NOP	RPA	RT	ETT	TS	AU AGVQ	Max AGVQ	Average Max	Max NC	Average NC
3	6684	5137	76.9	186.8	0.72	1517	100.00	1598	824	12	1.4
4	6686	6672	99.8	2.5	0.94	80	98.30	19	4	65	2.0
5	6651	6640	99.8	1.5	1.01	70	83.24	9	1	75	1.9
6	6681	6675	99.9	1.3	0.94	71	76.19	7	1	72	1.4
7	6590	6586	100.0	1.3	0.93	68	79.23	7	1	88	1.4

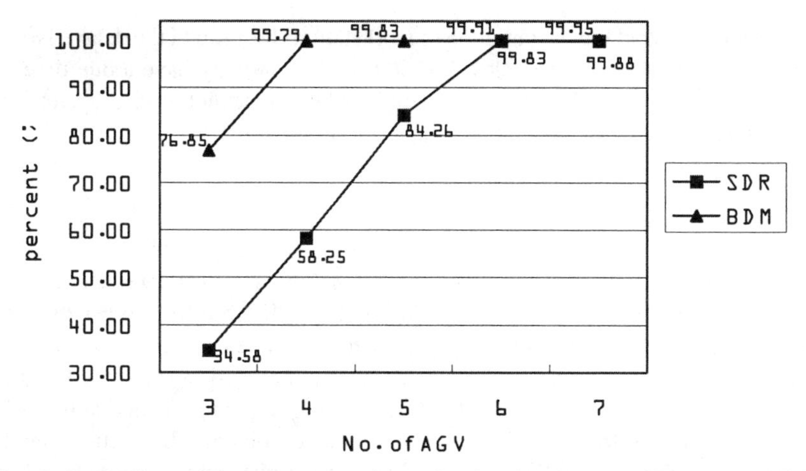

Fig. 5. Comparison of the percentage of completed orders between SDR and BDM for different number of vehicles

of vehicles was varied between 3 and 7. When the number of vehicles was high, the percentage of completed production orders was similar between the two dispatching methods. However, as shown in Tables 8 and 9, when the number of vehicles was low, the AGVs became the bottleneck of the production system, and the difference between two methods became significant. In Figure 5, the percentage of completed orders is compared between SDR and BDM for different number of vehicles

In certain layouts where some of the stations have favorable locations as opposed to others on account of reachability, it is known that SDR performs rather poorly, because calls from the unfavorable stations are usually not responded for very long durations. A common modification to the SDR for making up the drawback is to force a task to have the highest priority when the task had no response during a period longer than a specified time, which can be called "modified SDR." A dynamic algorithm as BDM would alleviate this problem, because multiple vehicles become candidates for assignment and thus there is a high chance for a vehicle to be assigned even to a distant station. However, it is still possible for a distant station to wait for a response for a longer period of time than other stations, because even in those rare occasions when a vehicle is assigned to a distant station, this decision will be quickly discarded as soon as a new task appears at a preferable station. For making up the possible drawback of BDM, it would be possible to modify BDM in a similar way to the modified SDR.

In this paper, a_{ij} was evaluated by using the empty travel time to pick up task j after performing task i. One of other alternatives for evaluating a_{ij} is to use the response time – which is the time from the current location of the vehicle performing task i to the pick-up location of task j – for minimizing the total response time of vehicles which is a possible objective of material handling systems. Also, there may be cases where delivery tasks have due times for pick-ups or deliveries. Typical examples are when workstations have a finite input or output buffer space for incoming or outgoing materials. In the case, the most important objective can be to minimize blocking or starvation of workstations because of the delayed transportation of materials by vehicles. Thus, each material may have a due time for pickup or delivery and the amount of delay – when the vehicle that is performing task i is assigned to task j – can be used as the value of a_{ij}.

5 Conclusion

A bidding-based method is suggested for dispatching automated guided vehicles. The dispatching method described in this study is different from other dynamic dispatching rules in that it looks into the future for an efficient assignment of delivery tasks to vehicles and that so it essentially assigns multiple tasks to multiple vehicles. The theoretical background of the bidding-based dispatching method is basically the auction method for the well-known assignment problem. The auction method developed by Bertsekas (1981) was modified to the dispatching method, and the rationale behind the bidding-based dispatching method is also provided.

A simulation was conducted to test the performance of the bidding-based dispatching method. It was shown that the bidding-based dispatching method significantly outperformed a popular dispatching rule (shortest distance rule) in throughput rate, flow time, response time, empty travel time, and average and maximum queue length of calls for vehicles. In addition, it was found that the number of communications among workstations and vehicles, which occurred during the dispatching decision process, was as low as three, even when the number of vehicles was seven. The low number of communications implies that the bidding-based dispatching process can be applied without a high load on the communication links among processors.

Because this paper introduces a new concept for distributed dispatching, only a limited number of practical factors were considered. However, in real systems, there may be more practical factors, such as types of guide paths, the size of buffer spaces, starvation, blocking, and deadlocks, which must be considered in the dispatching processing. Thus, further research is necessary on the method before it can be applied in practice. Also, empty vehicles traveling to pick up a load as well as the loaded vehicles and idle vehicles may be considered in the dispatching, which may be done in a future study. And, in this study, the minimization of the total empty travel time was considered as the objective function of the assignment problem. However, other objective functions may be used in different situations. More experiments are needed to study different objective functions.

Appendix: Proof of Property 2

The following four cases exist, any of which triggers either of the procedures:

Case 1: A new vehicle becomes available, and the number of available vehicles is smaller than or equal to the number of available tasks.

Case 2: A new vehicle becomes available, and the number of available vehicles is larger than the number of available tasks.

Case 3: A new task becomes available, and the number of available vehicles is smaller than or equal to the number of available tasks.

Case 4: A new task becomes available, and the number of available vehicles is larger than the number of available tasks.

Case 1: Each unassigned task (k) has the price of p_0, which is so high that every vehicle can get a positive margin when assigned to task k. When a vehicle becomes newly available, there are $(n-1)$ assigned vehicles, $(n-1)$ assigned tasks, and $(m-n+1)$ unassigned and inactivated tasks. The new vehicle arrives in an unassigned but activated (UA) state, and one vehicle remains "UA" until a vehicle becomes finally assigned to one of the $(m-n+1)$ UI tasks. As soon as a vehicle is assigned to one of the $(m-n+1)$ UI tasks, the procedure is terminated, and the results are n assigned vehicles, n assigned tasks, and $m-n$ UI tasks. At every iteration, the price of one assigned task decreases by $v_{ij^*} - w_{ij^*} + \varepsilon$ ($> \varepsilon >$0) from (4). Thus, in the long run, a UI task must be selected as the best task (j^*) by the UA vehicle. Thus, the procedure terminates in finite iterations. Once a vehicle submits a bid to a task, it means that the vehicle is almost happy about the price of the task (from 4) and remains so until the vehicle becomes "UA." It results from the fact that the price of no other task increases during the iteration. For the case of dummy vehicles, because the cost of assignment to every task is the same and the prices of unassigned tasks are p_0, which is high enough to make dummy vehicles have a positive margin, dummy vehicles must be almost happy for being assigned to any of the unassigned tasks.

Case 2: When the number of available vehicles is larger than the number of available tasks, $(n-m)$ vehicles become "unassigned" and cannot find a task that results in a positive margin. Because the prices of dummy tasks are the same and the cost of assigning a vehicle to every dummy task is the same, every unassigned vehicle must be almost happy for being assigned to any of the dummy tasks.

The proof for cases 3 and 4 can be derived as for cases 1 and 2 from the perspective of tasks instead of vehicles. Thus, the conclusion holds.

References

Bartholdi III JJ, Platzman LK (1989) Decentralized control of automated guided vehicles on a single loop. IIE Transactions 21: 76–81

Bertsekas DP (1990) The auction algorithm for assignment and other network flow problems: a tutorial. Interfaces 20(4): 133–149

Bertsekas DP (1981) A new algorithm for the assignment problem. Mathematical Programming 21: 152–171

Bilge Ü, Ulusoy G. (1995) A time window approach to simultaneous scheduling of machines and material handling system in an FMS. Operations Research 43(6): 1058–1070

Bilge, Ü, Tanchoco, JMA (1997) AGV systems with multi-load carriers. Journal of Manufacturing Systems 16(3): 159–173

Co CG., Tanchoco JMA (1991) A review of research on AGVS vehicle management. Engineering Costs and Production Economics 21: 35–42

Egbelu PJ, Tanchoco JMA (1984) Characterization of automatic vehicle dispatching rules. International Journal of Production Research 22: 359–374

Egbelu PJ (1987) Pull versus push strategy for automated guided vehicle load movement in a batch manufacturing system. Journal of Manufacturing Systems 6: 209–221

Hillier FS, Lieberman GJ (1986) Introduction to operations research, 4th edn., Chapter 6. Holden-Day

Kim CW, Tanchoco JMA, Koo PH (1999) AGV dispatching based on workload balancing. International Journal of Production Research 37(17): 4053–4066

Klein CM, Kim J (1996) AGV dispatching. International Journal of Production Research 34: 95–110

Lee J, Tangjarukij M, Zhu Z (1996) Load selecting of automated guided vehicles in flexible manufacturing systems. International Journal of Production Research 34 (12): 3383–3400

Sabuncuoglu I, Hommertzheim DL (1992) Dynamic dispatching algorithm for scheduling machines and automated guided vehicles in a flexible manufacturing system. International Journal of Production Research 30: 1059–1079

Taghaboni-Dutta F (1997) A value-added approach for automated guided vehicle task assignment. Journal of Manufacturing Systems 16 (1): 24–34

Scheduling railway traffic at a construction site*

Peter Brucker, Silvia Heitmann, and Sigrid Knust

Fachbereich Mathematik/Informatik, Universität Osnabrück, Albrechtstraße 28, 49069 Osnabrück, Germany
(e-mail: {peter,sheitman,sigrid}@mathematik.uni-osnabrueck.de)

Received: December 8, 1999 / Accepted: May 2, 2001

Abstract. We consider the problem of rescheduling trains in the case where one track of a railway section consisting of two tracks in opposing directions is closed due to construction activities. After presenting an appropriate model for this situation we derive a polynomial algorithm for the subproblem of finding an optimal schedule with minimal latenesss if the subsequences of trains for both directions outside the construction site are fixed. Based on this algorithm we propose a local search procedure for the general problem of finding good schedules and report test results for some real world instances.

Key words: Scheduling – Railway – Local search

1 Introduction

During the last decade discrete optimization models have been formulated for many problems arising in public rail transport like line and network planning, schedule generation for trains, personal and rolling stock scheduling, etc. For most recent survey articles in this area we refer to Bussieck et al. [3] and Cordeau et al. [6].

One of the challenging tasks railway companies are faced with is the construction of timetables for railway traffic. Until very recently such construction tasks were performed more or less manually. Over the last few years at the German Railways

* Supported by the Deutsche Forschungsgemeinschaft, Project 'Komplexe Maschinen-Schedulingprobleme'.

We would like to thank Andreas Landt for implementing the method proposed in this paper and for testing it on real world data sets (cf. Landt [16]). We would also like to thank the German Railways for providing these data sets. Finally, we would like to acknowledge the fruitful cooperation with TLC and the helpful comments of two anonymous referees.
Correspondence to: P. Brucker

computer systems have been installed to support the timetable construction process using appropriate data bases and interactive graphical systems (cf. Brünger [2]).

The next step would be to create an automatic timetable construction system, or at least routines which support updates of timetables which are necessary in the case of disturbances due to accidents, technical problems, bad weather, construction activities, etc. Schrijver [23] has developed the scheduling system CADANS which constructs periodic timetables, with a period of 60 minutes, for the Dutch Railways.

In this paper we address the updating problem in the case that one track of a railway section consisting of two tracks is closed due to construction activities. Works which are related to our work deal with the problem of scheduling trains on a single-track railway-line with sidings, where trains can cross or overtake each other. This problem is called train-scheduling problem in the literature and was formulated e.g. by Cai and Goh [4] and Higgins et al. [11]. Our problem can be considered as a very special case of the train-scheduling problem: In reach of the construction site only a single track is available and out of reach of the construction site two tracks are available where trains of opposite directions can cross each other. Of course, the dimensions of single- and double-track sections are reversed in our case. In both above mentioned works, the train-scheduling problem was modeled as integer program and on the basis of an unresolved train plan conflicting trains were identified. Cai and Goh [4] proposed a heuristic based on local optimality criteria to solve a conflict, whereas Higgins et al. [11] developed a branch and bound algorithm. Recently, Isaai and Singh [12] presented a lookahead, constraint-based heuristic algorithm for the train-scheduling problem. Other related works on this topic are: Szpigel [24], Peterson and Taylor [20], Peterson et al. [21], Kraft [15], Goh and Mees [8], Jovanovic and Harker [13], Kraay et al. [14], Mees [18], Mills et al. [19], Higgins et al. [9], Higgins et al. [10], Chiu et al. [5], Sahin [22].

In this study, we develop and implement a procedure which finds a schedule which minimizes the maximum lateness. It allows to schedule the trains according to their priorities and can be used interactively. This is important because it is difficult to formulate a function measuring the quality of the constructed schedule.

The procedure is based on an efficient algorithm to solve the following decision problem $DP2$, which is the main result of this paper. Given are two sequences π and $\overline{\pi}$ of the trains (one for each direction) and for each train i a time window $[r_i, d_i]$. We ask whether it is possible to schedule the trains in such a way that

- for each train i the time s_i passing the construction place belongs to $[r_i, d_i]$,
- the order induced by the s_i-values is compatible with π and $\overline{\pi}$,
- given safety distances l_{ij} between successive trains i and j are satisfied, i.e. $s_i + l_{ij} \leq s_j$ if j is scheduled directly after i.

The paper is organized as follows. In the next section we introduce some notations and formulate the train scheduling problem. In Section 3 we derive the algorithm which solves the corresponding problem $DP2$. In Section 3.3 this algorithm is used as a subroutine in local search procedures for finding sequences π and $\overline{\pi}$ such that problem $DP2$ has a solution. In Section 4 these procedures are applied to data provided by the German Railways and the results are compared with the solution provided by the staff of the German Railways. The final section contains some concluding remarks.

2 Formulation of the problem

Consider a railway connection between two stations A and B consisting of two tracks, one for each direction. Assume that parts of one direction are closed due to track-repair activities, as shown in Figure 1. Thus, trains in both directions have to use the same part C of the open track. If C is blocked by a BA-train AB-trains have to stop at a, and if C is blocked by an AB-train BA-trains have to stop at b.

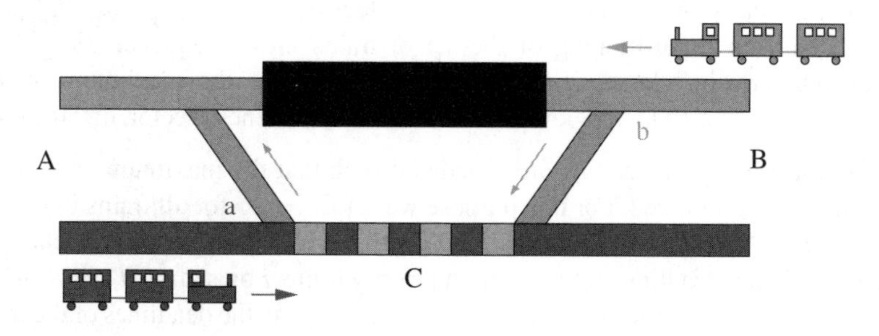

Fig. 1. Two-way track with construction site

We assume that we are given $n := n_1 + n_2$ trains, where n_1 denotes the number of AB-trains and n_2 the number of BA-trains. Associated with each train $i = 1, \ldots, n$ is a time window $[r_i, d_i]$, where the release date r_i denotes the earliest possible arriving time at a/b and the deadline d_i denotes the latest starting time of train i in a/b such that the schedule is acceptable. Furthermore, for all pairs of trains i, j safety distances l_{ij} are given. Let s_i be the time at which an AB-train i passes at a or a BA-train passes at b. Then we consider the following basic decision problem DP: Does a sequence σ in which all trains $i = 1, \ldots, n$ pass C and associated passing times $s_i \in [r_i, d_i]$ exist such that between successive trains in the sequence $\sigma = (\sigma_1, \ldots, \sigma_n)$ the safety distances are satisfied, i.e. $s_{\sigma_k} + l_{\sigma_k, \sigma_{k+1}} \leq s_{\sigma_{k+1}}$ for $k = 1, \ldots, n-1$ holds ?

This decision problem generalizes the problem of finding feasible starting times for a set of jobs with processing times p_i in given time windows $[r_i, d_i]$ on a single machine (choose $l_{ij} = p_i$). Thus, it is strongly NP-complete (for an NP-completeness proof of the single-machine problem see Lenstra et al. [17]). Note that a sequence σ induces subsequences π and $\bar{\pi}$ in which all AB-trains pass a and all BA-trains pass b, respectively.

If we have a procedure which solves problem DP, this procedure can also be used to find a solution $s = (s_i)_{i=1}^n$ which minimizes the maximum lateness $\max_{i=1}^{n} L_i$. The lateness L_i of train i for a given due date d_i is given by the expression $L_i := s_i - d_i$. For a given threshold value L, condition $s_i - d_i \leq L$ holds for $i = 1, \ldots, n$ if and only if condition $s_i \leq d_i + L$ holds for $i = 1, \ldots, n$. Thus, a schedule $s = (s_i)$ with maximum lateness less than or equal to L exists if and only if DP with the time windows $[r_i, d_i + L]$ has a feasible solution. By binary search a solution which minimizes L can be found which is an optimal solution for the general maximum lateness problem.

If trains of different priorities have to pass C, minimizing the maximum lateness may be not an appropriate objective function. In this case one possibility is to replace maximum lateness by maximum weighted lateness $\max_{i=1}^{n} \{w_i(s_i - d_i)\}$, where the weights w_i are positive numbers, depending on the priorities of the trains i. Because $w_i(s_i - d_i) \leq L$ is equivalent to $s_i \leq d_i + \frac{L}{w_i}$, we have to replace the d_i-values by $d_i + \frac{L}{w_i}$ in this case.

Another method to take care of trains with different priorities is to minimize lexicographically the lateness vector. The lateness vector $(L^1_{\max}, \ldots, L^m_{\max})$ is associated with a partitioning of the set of trains into a sequence $P_1, \ldots, P_m$ of disjoint priority classes. For each set P_j we define the maximum lateness $L^j_{\max} := \max_{\nu \in P_j} L_\nu$. To lexicographically minimize the lateness vector, first the trains of highest priority in class P_1 are scheduled such that the maximum lateness for this subset is minimized. For this purpose we set $d_i := \infty$ for all trains in the sets $P_2, \ldots, P_m$. Let L^*_1 be the minimal maximum lateness for the trains in class P_1. Then we fix the deadlines for these high priority trains i by setting $d_i := d_i + L^*_1$, and optimize the lateness of the next priority class P_2, fix the deadlines of the trains in this class, etc.

3 A solution procedure

In this section we will develop a procedure for calculating a schedule $s = (s_i)_{i=1}^n$ which minimizes the maximum lateness. This will be done in three steps.

First we consider a restricted version of the decision problem DP (called $DP1$) where the sequence σ in which all trains pass C is given. We ask for a schedule $s = (s_i)_{i=1}^n$ with $s_i \in [r_i, d_i]$ in which the trains are planned according to σ, and the safety distance constraints are respected. If such a schedule exists, we want to calculate one in which each train starts as early as possible respecting the given sequence σ. We will show that this problem can easily be solved in $O(n)$ time.

In the second step we consider a generalization of $DP1$ (called $DP2$) where not the whole sequence σ is given, but subsequences π and $\overline{\pi}$ for all AB-trains and BA-trains are fixed. We ask whether it is possible to merge both sequences into a joint sequence σ such that the decision problem $DP1$ for the joint sequence has a feasible solution. We will show that also this problem can be solved in polynomial time.

Finally, we solve the general maximum lateness problem by applying local search. The search space consists of all pairs of subsequences $(\pi, \overline{\pi})$. During the search process (using binary search) we calculate for each considered pair the smallest value L such that the corresponding decision problem $DP2$ with deadlines $d_i + L$ has a feasible solution.

3.1 Solving problem $DP1$

W.l.o.g. let $\sigma = (1, 2, \ldots, n)$ be the given sequence in which the trains pass C. Then we calculate s_i-values iteratively by

$$\begin{aligned} s_1 &:= r_1 \\ s_{i+1} &:= \max\{s_i + l_{i,i+1}, r_{i+1}\} \quad i = 1, \ldots, n-1. \end{aligned}$$

Clearly, these values are the smallest s_i-values such that $r_i \leq s_i$ for $i = 1, \ldots, n$ and $s_i + l_{i,i+1} \leq s_{i+1}$ for $i = 1, \ldots, n-1$ hold. Furthermore, $DP1$ has a feasible solution if and only if the calculated values s_i satisfy $s_i \leq d_i$ for $i = 1, \ldots, n$.

3.2 Solving problem $DP2$

W.l.o.g. let $\pi = (1, \ldots, n_1)$ be the given subsequence of all AB-trains and let $\bar{\pi} = (\bar{1}, \ldots, \bar{n}_2)$ be the subsequence of all BA-trains. Furthermore, we are given time windows $[r_i, d_i]$ for $i = 1, \ldots, n_1, \bar{1}, \ldots, \bar{n}_2$, and minimal distances l_{ij}. We have to find a sequence σ of all trains which is compatible with the subsequences π and $\bar{\pi}$ and which solves the corresponding problem $DP1$.

We reduce this problem to a path-finding problem in a directed acyclic graph G, which is defined as follows. The vertices of G are given by triples of the form $(i, \bar{j}, 0)$ and $(i, \bar{j}, 1)$ with $i = 0, 1, \ldots, n_1$ and $j = 0, 1, \ldots, n_2$. Vertex $(i, \bar{j}, \alpha)$ with $\alpha \in \{0, 1\}$ represents all partial sequences of the trains $1, \ldots, i, \bar{1}, \ldots, \bar{j}$ which are compatible with $\pi, \bar{\pi}$ and in which i (if $\alpha = 0$) or j (if $\alpha = 1$) is the last element. If we want to extend such a subsequence by one additional train, this can be done by adding either the train $i + 1$ from subsequence π or the train $\overline{j+1}$ from $\bar{\pi}$. Thus, we have arcs of the form

$$\begin{array}{ll} (i, \bar{j}, 0) \rightarrow (i+1, \bar{j}, 0), & (i, \bar{j}, 1) \rightarrow (i+1, \bar{j}, 0), \\ (i, \bar{j}, 0) \rightarrow (i, \overline{j+1}, 1), & (i, \bar{j}, 1) \rightarrow (i, \overline{j+1}, 1), \end{array}$$

which correspond to these two possible extensions of the partial sequences represented by the vertices $(i, \bar{j}, \alpha)$. Finally, a path from $(1, \bar{0}, 0)$ or $(0, \bar{1}, 1)$ to $(n_1, \bar{n}_2, 0)$ or $(n_1, \bar{n}_2, 1)$ corresponds to a sequence σ of all trains which is compatible with π and $\bar{\pi}$. This is illustrated by the following example.

Example: Graph G for subsequences $\pi = (1, 2, 3)$ and $\bar{\pi} = (\bar{1})$ is shown in Figure 2. All possible compatible sequences σ and the associated paths in G are listed in Table 1.

□

Problem $DP2$ has a solution if and only if in G a path from $(1, \bar{0}, 0)$ or $(0, \bar{1}, 1)$ to $(n_1, \bar{n}_2, 0)$ or $(n_1, \bar{n}_2, 1)$ exists such that problem $DP1$ has a solution for the corresponding sequence σ. Such a path (if there is any) is calculated by a graph search algorithm, which will be described next.

A vertex $v = (i, \bar{j}, 0)$ or $v = (i, \bar{j}, 1)$ is reachable if a path exists from $(1, \bar{0}, 0)$ or from $(0, \bar{1}, 1)$ to v such that problem $DP1$ is solvable for the corresponding partial sequence.

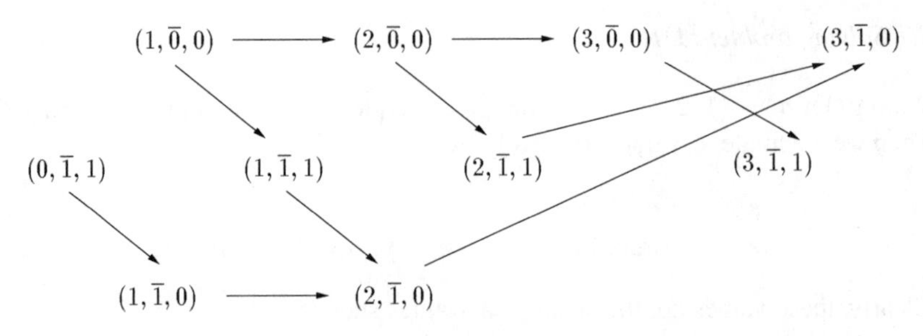

Fig. 2. Graph G for $\pi = (1, 2, 3)$ and $\bar{\pi} = (\bar{1})$

Table 1. Compatible sequences and associated paths

Sequence	Path
$(1,2,3,\bar{1})$	$(1,\bar{0},0) \to (2,\bar{0},0) \to (3,\bar{0},0) \to (3,\bar{1},1)$
$(1,2,\bar{1},3)$	$(1,\bar{0},0) \to (2,\bar{0},0) \to (2,\bar{1},1) \to (3,\bar{1},0)$
$(1,\bar{1},2,3)$	$(1,\bar{0},0) \to (1,\bar{1},1) \to (2,\bar{1},0) \to (3,\bar{1},0)$
$(\bar{1},1,2,3)$	$(0,\bar{1},1) \to (1,\bar{1},0) \to (2,\bar{1},0) \to (3,\bar{1},0)$

The following reaching algorithm calculates all vertices which are reachable. Also the corresponding s_i-values are stored. Problem $DP2$ has a solution if and only if $(n_1, \bar{n}_2, 0)$ or $(n_1, \bar{n}_2, 1)$ is reachable. In this case a sequence σ representing a feasible solution can be found by tracing the $PRED$-array backwards.

Reaching Algorithm

1. $v := (1, \bar{0}, 0); \bar{v} := (0, \bar{1}, 1); s_v := r_1; s_{\bar{v}} := r_{\bar{1}}$;
2. Label v and $\bar{v}$ reachable and investigated;
3. WHILE a non-investigated vertex $v = (i, \bar{j}, \alpha)$ $(\alpha \in \{0, 1\})$ exists such that
   all predecessors of v are investigated DO BEGIN
4. IF $\alpha = 0$ THEN BEGIN
5. IF $u = (i-1, \bar{j}, 0)$ is reachable AND $s_u + l_{i-1,i} \le d_i$ THEN
   $s := \max\{r_i, s_u + l_{i-1,i}\}$ ELSE $s := \infty$;
6. IF $\bar{u} := (i-1, \bar{j}, 1)$ is reachable AND $s_{\bar{u}} + l_{\bar{j}i} \le d_i$ THEN
   $\bar{s} := \max\{r_i, s_{\bar{u}} + l_{\bar{j}i}\}$ ELSE $\bar{s} := \infty$;
7. $s_v := \min\{s, \bar{s}\}$;
8. IF $s_v < \infty$ THEN BEGIN
9. v is reachable;
10. IF $s_v = s$ THEN $PRED(v) := u$ ELSE $PRED(v) := \bar{u}$;
    END
    END
    ELSE BEGIN $(* \, \alpha = 1 \, *)$
11.-16. similar to Steps 5 to 10;
    END;
17. Set v investigated;
18. END

Graph G is acyclic since all arcs have the form $(i, \bar{j}, \alpha) \rightarrow (i+1, \bar{j}, \alpha')$ or $(i, \bar{j}, \alpha) \rightarrow (i, \overline{j+1}, \alpha')$. Since G has $O(r^2)$ arcs with $r := \max\{n_1, n_2\}$, the complexity of the reaching algorithm is also $O(r^2)$.

3.3 Local search

As previously mentioned, for given subsequences π and $\bar{\pi}$ a merged sequence which minimizes the maximum lateness can be calculated by binary search. To find a good pair $(\pi, \bar{\pi})$ of sequences the two local search methods, iterative improvement and tabu search, are applied (for an overview of different local search algorithms see Aarts and Lenstra [1], for tabu search cf. Glover and Laguna [7]). The idea of local search is to navigate through the search space consisting of all pairs $(\pi, \bar{\pi})$ of subsequences with the objective of improving the current pair. The elementary step of this search process is a so-called move where the current pair $(\pi, \bar{\pi})$ is replaced by another pair $(\pi', \bar{\pi}')$. We use moves which interchange two adjacent trains in one of the subsequences π or $\bar{\pi}$. Moves based on shifts are not used because a good schedule is unlikely to be "very" different from the starting schedule since shifting a train far away from its position in the original sequence causes a great change of the corresponding s_i-value. On the other hand, small shifts can be accomplished by a few interchanges. Computational results have confirmed these observations.

While iterative improvement always moves to a better pair in each step and stops if no better pair to move to exists, the tabu search method also allows non-improving moves. In order to avoid cycling, a so-called tabu list is kept in which attributes of the visited solution are stored. All solutions which have an attribute stored in this list are excluded from the current set of possible moves. The techniques used in our implementation of the tabu search procedure were chosen in a standard way.

4 Computational results

The methods described in the previous sections were implemented on a 350 MHz Pentium II processor using the programming language C.

Real world instances provided by the German Railways have been used as test instances. Table 2 shows three of these test instances with different characteristics. The other instances were very similar to the three chosen ones and were therefore omitted after some preliminary tests. For all these instances characteristics of the construction site are listed, such as the period of construction, density of traffic during this period, length of the closed track, passing time, and type of traffic. We derived the time windows $[r_i, d_i]$ from the normal schedule (a scheduler may try different possibilities).

The trains belong to four priority classes (IC, IR, E/N, and GZ, where IC stands for fast long-distance intercity trains, IR for interregio trains, E/N for local traffic and GZ for trains transporting goods). For each instance an initial pair $(\pi^0, \bar{\pi}^0)$ of subsequences is derived from the normal schedule before construction is commenced. Starting with this solution, heuristic solutions $(\pi^*, \bar{\pi}^*)$ are calculated with the local search procedures using the lexicographic method and the method

Table 2. Characteristics of three test instances

Construction work		Data
Kirchhorsten – Stadthagen	Section:	Löhne – Minden – Hannover
	Date:	Sun, 01.11.1998
	Duration:	8 hours
	No. of trains:	62
	Trains per hour:	7.8
	Length of site:	approx. 4 km
	Thoroughfare time:	2–5 minutes
	Use:	Almost exclusively passenger services, both local and long-distance traffic
Lehrte – Immensen	Section:	Lehrte – Fallersleben – Oebisfelde
	Date:	Sat, 28.11. – Mon, 30.11.1998
	Duration:	36 hours
	No. of trains:	132
	Trains per hour:	3.7 (peak 6)
	Length of site:	approx. 8 km
	Thoroughfare time:	3–8 minutes
	Use:	Mainly passenger services, both local and long-distance traffic
Kirchweyhe – HB-Hemelingen	Section:	Osnabrück – Diepholz – Bremen
	Date:	Sat, 24.10. – Sun, 25.10.1998
	Duration:	30 hours
	No. of trains:	200
	Trains per hour:	6.7 (peak 13)
	Length of site:	approx. 10 km
	Thoroughfare time:	4–9 minutes
	Use:	Both goods and passenger services, very intensive use at times

based on maximum weighted lateness, which were described at the end of Section 2. Test results have shown that the lexicographic method (although more time-consuming) provides much better results. Therefore, we only report results based on the lexicographic method.

The main results, derived by extensive test runs, can be described as follows. The tabu search method converges very quickly (approximately 10 iterations). Furthermore, it provides no better results than the iterative improvement method, which is much faster. This unusual fact is in accordance with our observations in Section 3.3 that good schedules are not very different from the starting schedule. In Table 3 the results based on this method are presented and compared with the solutions provided by the staff of the German Railways. To get an idea of the

Table 3. Results of our procedure compared with the practical solutions

Construction work	Result of procedure	Practical solution of the DB
Kirchhorsten – Stadthagen	Lateness vector: [2,4,23,30] Lateness of Type IC: 9 minutes Type IR: 22 minutes Type E/N: 31 minutes Type GZ: 33 minutes Total lateness: 95 minutes Weighted total: 197 minutes No. of late trains: 15 No. of times overtaken: 8 Computing time: 0.15 seconds	Lateness vector: [4,4,23,30] Lateness of Type IC: 14 minutes Type IR: 23 minutes Type E/N: 23 minutes Type GZ: 33 minutes Total lateness: 93 minutes Weighted total: 204 minutes No. of late trains: 16
Lehrte – Immensen	Lateness vector: [0,3,10,41] Lateness of Type IC: 0 minutes Type IR: 14 minutes Type E/N: 63 minutes Type GZ: 60 minutes Total lateness: 137 minutes Weighted total: 228 minutes No. of late trains: 21 No. of times overtaken: 6 Computing time: 1.9 seconds	Lateness vector: [0,3,10,41] Lateness of Type IC: 0 minutes Type IR: 16 minutes Type E/N: 64 minutes Type GZ: 65 minutes Total lateness: 145 minutes Weighted total: 241 minutes No. of late trains: 22
Kirchweyhe – HB-Hemelingen	Lateness vector: [2,4,23,30] Lateness of Type IC: 28 minutes Type IR: 47 minutes Type E/N: 250 minutes Type GZ: 64 minutes Total lateness: 389 minutes Weighted total: 815 minutes No. of late trains: 67 No. of times overtaken: 4 Computing time: 8.7 seconds	Lateness vector: [5,11,17,20] Lateness of Type IC: 25 minutes Type IR: 47 minutes Type E/N: 255 minutes Type GZ: 88 minutes Total lateness: 415 minutes Weighted total: 839 minutes No. of late trains: 71

quality of these solutions the following data are reported: the lateness vector, the sum of lateness-values for each priority class of trains, the (weighted) sum of all lateness-values, and the number of late trains. Furthermore, we state the number of times where trains are overtaken in our sequences $(\pi^*, \bar{\pi}^*)$ compared with the initial sequences $(\pi^0, \bar{\pi}^0)$.

By comparing the practical solutions with the solutions provided by our algorithm, we conclude that our method provides very good solutions. Furthermore, our method is very fast. It can be embedded into an interactive system for finding schedules which are appropriate in a given situation.

5 Concluding remarks

We have presented a method for adapting a train schedule in the case that construction activities cause railway traffic to be blocked in one direction. The method provides good results very quickly and may be embedded into an interactive decision support system.

One method to avoid intolerable lateness of trains due to a construction site is to replace short-distance trains by buses. In such a situation our method may provide some insight into how the elimination of trains affects the schedule.

The method will be an important module in a software package for railway timetabling. It allows to evaluate many alternatives in an efficient way. Due to this the method can save time and tedious work. At the time being the German Railways are working on a general timetabling system. Our method has not yet been embedded in this system.

The proposed procedures can also be applied to the situation where two railway tracks come together to share a common track (cf. Fig. 3). If we consider this situation with a fixed number $k \geq 2$ of tracks, decision problem $DP2$ remains polynomially solvable when the subsequences for all these k tracks are fixed. In the reaching algorithm the vertices $(i, \bar{j}, \alpha)$ only have to be replaced by tuples $(i_1, i_2, \dots, i_k, \alpha)$ with $\alpha \in \{1, \dots, k\}$ and the complexity increases to $O(kr^k)$, where r again denotes the maximal length of the k subsequences.

An interesting variant of problem $DP2$, for which the complexity status remains open, arises if we consider two or more construction sites in series. If we are given the two subsequences outside the construction sites, then for each construction site we have to determine a compatible sequence such that all time windows and

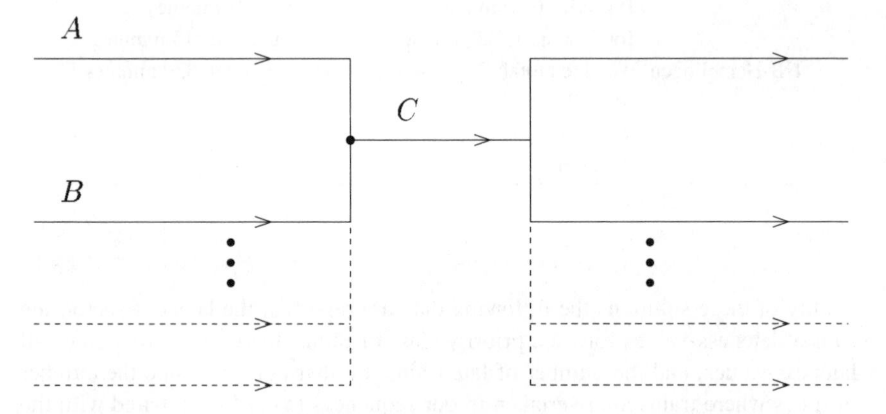

Fig. 3. Sharing a common track

safety distances are respected. The problem with several construction sites in series has the same structure as the single-track scheduling problem considered in the introduction of this paper.

References

1. Aarts E, Lenstra JK (1997) Local search in combinatorial optimization. Wiley, New York
2. Brünger O (1995) Konzeption einer Rechnerunterstützung für die Feinkonstruktion von Eisenbahnfahrplänen. Veröffentlichungen des Verkehrswissenschaftlichen Institutes der RWTH Aachen, Nr. 51
3. Bussieck MR, Winter T, Zimmermann UT (1997) Discrete optimization in public rail transport. Mathematical Programming 79: 415–444
4. Cai X, Goh CJ (1994) A fast heuristic for the train scheduling problem. Computers and Operational Research 21 (5): 499–510
5. Chiu CK, Chou CM, Lee JHM, Leung HF, Leung YW (1996) A constraint-based interactive train rescheduling tool. Proceedings 2nd International Conference Principle and Practice of Constraint Programming-CP'96, Cambridge, MA
6. Cordeau J, Toth P, Vigo D (1998) A survey of optimization models for train routing and scheduling. Transportation Science 32: 380–404
7. Glover F, Laguna M (1997) Tabu search. Kluwer, Amsterdam
8. Goh CJ, Mees AI (1991) Optimal control on a graph with application to train scheduling problems. Mathematica Computing Modeling 15: 49–58
9. Higgins A, Kozan E, Ferreira L (1994) Development of a model to optimise the position of crossing and overtaking loops on a single line railway. Physical Infrastructure Res. Rep., 16-94, Queensland University of Technology, Brisbane
10. Higgins A, Kozan E, Ferreira L (1995) Rescheduling of trains on a single line track: improved heuristic solution techniques. Physical Infrastructure Res. Rep., 1-95, Queensland University of Technology, Brisbane
11. Higgins A, Kozan E, Ferreira L (1996) Optimal scheduling of trains on a single line track. Transportation Research Part B 30 (2): 147–161
12. Isaai MT, Singh MG (2000) An object-oriented, constraint-based heuristic for a class of passenger-train scheduling problems. IEEE Transactions on Systems, Man, and Cybernetics – Part C: Applications and Reviews 30 (1): 12–21
13. Jovanovic D, Harker P (1991) Tactical scheduling of rail operations: the SCAN I system. Transportation Science 25: 46–64
14. Kraay D, Harker P, Chen B (1991) Optimal pacing of trains in freight railroads. Operations Research 39: 82–99
15. Kraft ER (1987) A branch and bound procedure for optimal train dispatching. Journal of Transportation Research Forum 28: 263–276
16. Landt A (1999) Scheduling von Zügen im Baustellenbereich. Diplomarbeit Universität Osnabrück, Fachbereich Mathematik/Informatik
17. Lenstra JK, Rinnooy Kan AHG, Brucker P (1977) Complexity of machine scheduling problems. Annals of Discrete Mathematics 1: 343–362
18. Mees AI (1991) Railway scheduling by network optimization. Mathematical Computing Modeling 15 (1): 33–42
19. Mills RG, Perkins SE, Pudney PJ (1991) Dynamic rescheduling of long haul trains for improved timekeeping and energy. Asia-Pacific Journal of Operational Research 8: 146–165

20. Peterson ER, Taylor J (1982) A structured model for rail line simulation and optimization. Transportation Science 16 (2): 192–206
21. Peterson ER, Taylor J, Martland CD (1986) An introduction to computer assisted train dispatch. Journal of Adv Transn 20: 63–72
22. Sahin I (1999) Railway traffic control and train scheduling based on inter-train conflict management. Transportation Research Part B 33: 511–534
23. Schrijver A (2000) Timetabling for the dutch railways. Proceedings of the 7th International Workshop on Project Management and Scheduling, Osnabrück, http://www.mathematik.uni-osnabrueck.de/research/OR/pms2000
24. Szpigel B (1973) Optimal train scheduling on a single line railway. Operations Research 72: 344–351

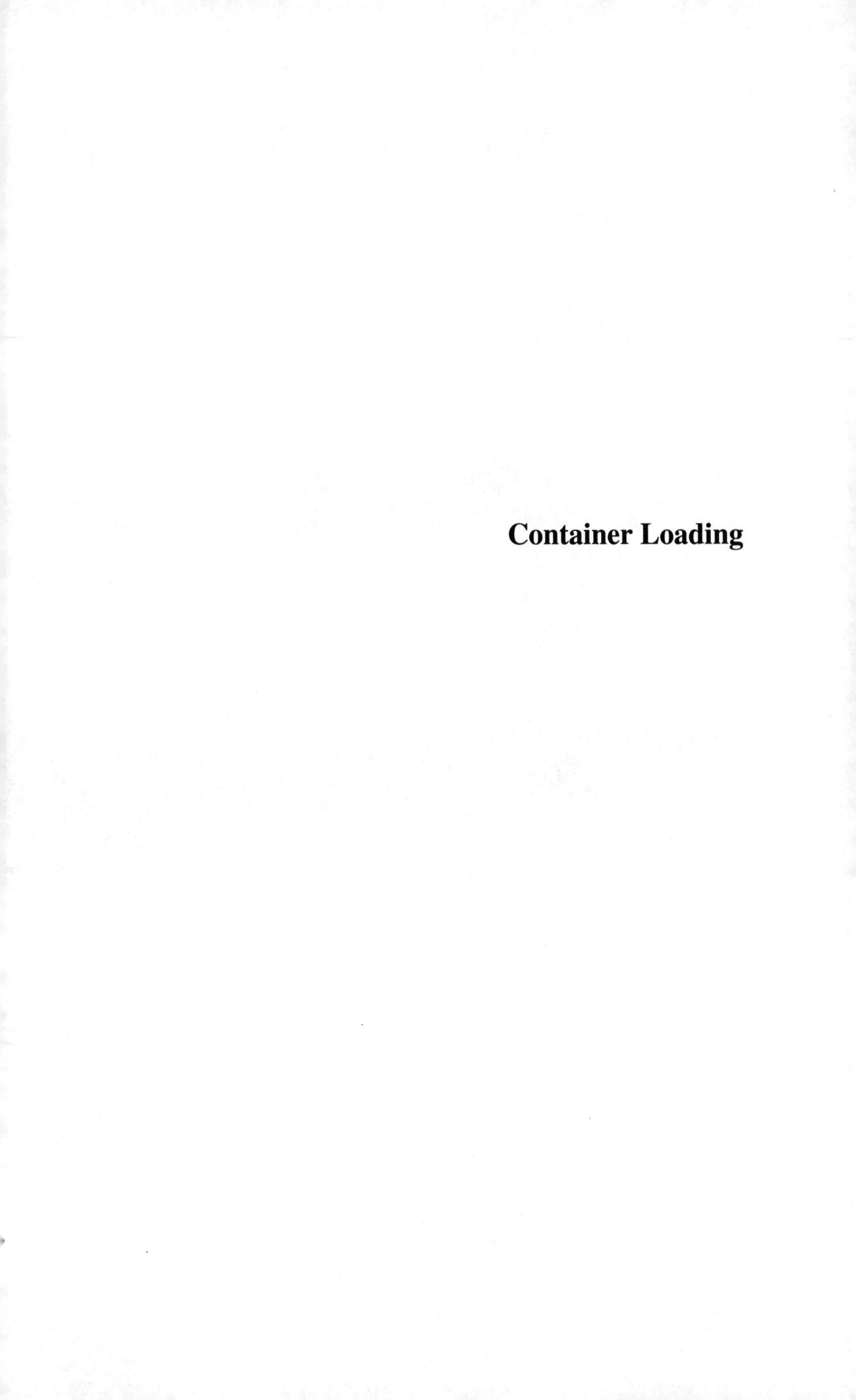

Container Loading

A bottleneck assignment approach to the multiple container loading problem

Michael Eley

Fraunhofer Institut für Techno- und Wirtschaftsmathematik (ITWM),
Gottlieb Daimler Straße 49, 67663 Kaiserslautern, Germany
(e-mail: eley@itwm.fhg.de)

Abstract. The container loading problem addresses the question of how to store several three dimensional, rectangular items (e.g. boxes) in one or more containers in such a way that maximum use is made of the container space. The multiple container problem concentrates on the situation where the consignment to be loaded cannot be accommodated in a single container. To minimize the number of required containers the repeated application of a single container approach is often suggested in the literature. In contrast, in this paper an approach based on a set partitioning formulation of the problem is presented. Within this approach a single container algorithm is used to produce alternative loading patterns. This approach easily allows introducing additional aspects, e.g. separation of boxes or complete shipment of boxes.

Key words: Packing – Container loading – Bottleneck assignment – Additional constraints

1 Introduction

The three dimensional packing problem discussed in this article can be described as follows: A given set of various regular shaped boxes of varying size has to be packed into one or more large containers, also available in different sizes, so as to use the container space as efficient as possible. Further assumptions are:

- Boxes and containers of the same size are grouped into box types or container types, respectively.
- Only orthogonal arrangements are considered, i.e. all boxes are loaded parallel to the edges of the container.
- Boxes can be rotated. Consequently, up to six different orientations are allowed.
- The number of boxes for each box type is restricted (constraint problem).
- All information about boxes and containers are given, i.e. we consider an off-line situation.

Many additional constraints and assumptions can be added to this basic framework in order to take into account some aspects of real-life problems. Examples therefore are discussed below.

Within the multiple container loading problem we distinguish two sub problems. For the three dimensional *knapsack problem* the available container space is not large enough to stow all boxes. Consequently, the task is to determine an optimal selection of the boxes that maximizes volume utilization or the value of the packed boxes. In contrast, for the *bin packing problem* the storage space is sufficient to stow all boxes. The objective here is to minimize the number of containers that are required to ship all boxes. If more than one container type is available and different costs are associated with the container types, then a more appropriate objective would be to minimize total costs. In this case, beside finding packing patterns for the different containers, the optimal selection of containers has also to be determined.

Unlike the single container loading problem, not much attention has been paid in the literature to the multiple container loading problem so far. In fact, it is suggested to adapt single container heuristics to the multiple container case. Therefore, one of the following strategies is used:

1. *Sequential strategy:* By using a single container approach one container is filled after the other (cf. Fig. 1). Whenever no further boxes can be stowed a new container is chosen. Bischoff and Ratcliff showed in [1] that this strategy may lead to poor results, in particular if additional requirements are imposed on the packing arrangements, e.g. the containers should be filled by non straddling walls. Obviously, this strategy has the disadvantage that large or awkwardly formed boxes are often left to the end resulting in poor volume utilization ratios for the containers filled at the end. If different sized containers are available, then the performance of an algorithm also depends on the sequence in which the container types are chosen.
2. *Pre-assignment strategy:* Rather than letting the algorithm choose the most suitable boxes in a greedy manner, in a first step boxes are distributed among different containers. Then a single container heuristic is applied for each such container. Additional mechanisms are required to distribute boxes that could not be stowed in their designated containers to other containers. This pre-assignment can be based on preferences given by the user or like in [2] by trying to distribute large and small boxes evenly among the containers.

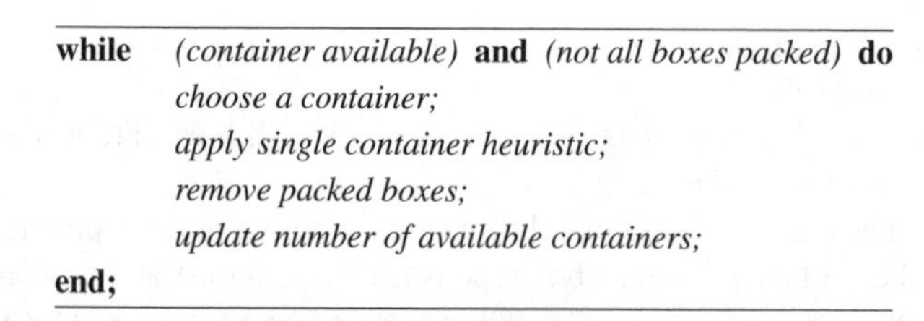
while *(container available)* **and** *(not all boxes packed)* **do**
choose a container;
apply single container heuristic;
remove packed boxes;
update number of available containers;
end;

Fig. 1. Sequential strategy

One crucial point of the pre-assignment strategy is to determine the number of required containers, particularly, if different container types are available.

3. *Simultaneous strategy:* Considering more than one container at a time should result in more efficient packing and avoid in particular poor volume utilization for the containers packed at the end. In [1] and [3] sequential and simultaneous strategies were compared. Surprisingly, the sequential strategy outperformed the simultaneous strategy in both approaches. Again, as for the pre-assignment strategy the number or, in the case of different sized containers, the combination of containers has to be specified in advance for the bin packing problem.

In recent years the following papers have been published on the multiple container loading problem. Ivancic et al. (cf. [4]) focused on the bin packing problem, Mohanty et al. (cf. [5]) on the knapsack problem. Both papers have presented approaches that solve the problem by using a sequential strategy. Bischoff and Ratcliff (cf. [1]) and Eley (cf. [3]) focused on the bin packing problem with only one container type. The same problem type, but with containers of different types and with fixed box orientation, was considered by Martello et al. (cf. [6]). Bortfeldt (cf. [7]) examined both problem types with different container types and extended the sequential strategy by diversifying the search. Terno et al. (cf. [2]) developed an algorithm for the multiple pallet loading problem. They considered only one pallet type and pursued a pre-assignment approach, where large and small boxes are distributed uniformly over all containers. Whenever a solution was found where all boxes are loaded on one pallet, the number of pallets was reduced.

In this paper a new approach is presented based on IP-formulation. In Section 2 a formal problem description is given. The new approach is then described in Section 3. Test results for the knapsack and for the bin packing problem are discussed in Section 4. Two extensions of the problem to cater for additional practical constraints are presented in Section 5. Finally, Section 6 summarizes the results and puts them into a perspective.

2 Problem formulation

A container can be packed with boxes in many different ways. Each such arrangement will be called a packing pattern. It includes information on the container type as well as on the number of boxes of different types that have been packed. Before formally stating the problem, we introduce some notation.

Notation:

$i = 1, ..., I$	index of box types
$k = 1, ..., K$	index of container types
$j = 1, ..., J$	index of packing patterns, where p_k consecutive packing patterns require the same container type k, i.e. $j = 1, ..., p_1$, $p_1 + 1, ..., p_1 + p_2, ..., \alpha(K), ..., \sum_{l=1}^{K} p_l$. The index of the first container of type k is defined by $\alpha(1) := 1$ for $k = 1$ and $\alpha(k) := (\sum_{l=1}^{k-1} p_l) + 1$ for $k \geq 2$.

a_{ij} number of boxes of type i packed in the j-th packing pattern, with matrix $A := (a_{ij})$
x_j decision variable stating how often the j-th packing pattern is used
b_i number of boxes of the i-th box type available
c_j cost of the j-th packing pattern, e.g. volume of the corresponding container type
d_k availability of containers of type k
v_i value of boxes of type i, e.g. its volume

For both problem types we have to solve a set-partitioning problem, i.e. we have to select a combination of columns from matrix A so that the number of available boxes is not violated and that the objective function is optimized. Using this notation, the knapsack problem can then be formulated as follows:

$$\max \sum_{i=1}^{I} \sum_{j=1}^{J} a_{ij} v_i x_j \tag{1}$$

$$s.t. \quad \sum_{j=1}^{J} a_{ij} x_j \leq b_i \quad \forall i = 1, ..., I \tag{2}$$

$$\sum_{j=\alpha(k)}^{\alpha(k+1)-1} x_j \leq d_k \quad \forall k = 1, ..., K \tag{3}$$

$$x_j \geq 0 \quad \text{and integer} \quad \forall j = 1, ..., J \tag{4}$$

The objective function (1) maximizes the sum of the value of the boxes that are contained in the selected packing patterns. Restrictions (2) and (3) guarantee that only available boxes are packed and the number of available boxes is not exceeded, respectively.

The corresponding bin packing problem is stated next:

$$\min \sum_{j=1}^{J} c_j x_j \tag{5}$$

$$s.t. \quad \sum_{j=1}^{J} a_{ij} x_j \geq b_i \quad \forall i = 1, ..., I \tag{6}$$

$$x_j \geq 0 \quad \text{and integer} \quad \forall j = 1, ..., J \tag{7}$$

For a given range of packing patterns the objective function (5) minimizes the total cost of the chosen containers. Constraint (6) ensures that all boxes are packed.

3 The bottleneck approach

The problem formulation for the bin packing problem presented in the last section is similar to the one dimensional cutting stock problem formulated by Gilmore and Gomory (cf. [8]). Since the matrix A that contains all possible patterns might have a large number of columns, they proposed a column generation approach. In the one dimensional case, the resulting sub problem can be modeled as an one dimensional knapsack problem that can, although it is np-hard, be solved efficiently. But here the sub problem is already a single container loading problem. Since no exact solution approach is known that can solve realistic problem instances in reasonable time, a straightforward way is to use a heuristic that generates an appropriate number of columns for the matrix A and to solve the IP-problem with this limited number of packing patterns. Obviously, this cannot guarantee that the optimal solution is found. But if the number of generated packing patterns is sufficiently large we might get good results. Furthermore, this two-step approach has the advantage that the single container heuristic already allows considering additional practical requirements, e.g. orientation constraints for the boxes, load stability of the packing patterns or an even weight distribution within each container.

In [4] and [5] similar IP-formulations were presented for the bin packing and the knapsack problem, respectively, like the ones in the last section. But in both papers the problem is solved by using a sequential strategy. In contrast, the new approach is based on the problem formulations (1) to (4) and (5) to (7) which are solved by using the commercial solver CPLEX 7.5.

For the sake of brevity, we will not provide a full description of the single container algorithm, but only outline its structure. The interested reader is referred to [3]. The single container heuristic consists of two components:

- A greedy heuristic that generates a solution for a given sequence of boxes. One box after another is stowed at the most suitable position in a container, i.e. where the remaining space that cannot be filled with additional boxes is minimal.
- A tree search procedure that allows considering different loading sequences of boxes.

Each node in the tree represents a partially filled container. Branching is carried out by stowing one additional box by using the greedy heuristic. The potential of each node leading to a good final solution is estimated by applying the greedy heuristic that fills the remaining space with remaining boxes where the loading sequence is determined by the boxes' volume. Consequently, in each node a complete solution for the single container loading problem is generated. In order to limit the search, only the p best solutions are expanded. All solutions generated during the tree search are inserted into matrix A.

For the knapsack problem, in some cases it might be advantageous to choose a combination of packing patterns that contains more than the maximum number of available boxes of a particular type. Thus, not all packing patterns have to be realized completely, but some boxes may be left out. Therefore, the constraint (2) was replaced by the two following constraints, (8) and (9). The additional decision variable Δ_i states, how many boxes of type i are packed more than its availability

b_i. Consequently, we can omit dominated patterns from matrix A, i.e. columns $\bar{j}$ of matrix A for which another column $\tilde{j}$ exist with $a_{i\tilde{j}} \geq a_{i\bar{j}}$ $\forall i \in 1, ...mI$.

$$\sum_{j=1}^{J} a_{ij} x_j \leq b_i + \Delta_i \quad \forall i = 1, ..., I \tag{8}$$

$$\Delta_i \geq 0 \quad \text{and integer} \quad \forall i = 1, ..., I \tag{9}$$

Of course, these additional boxes should not contribute to the objective function value. Therefore the objective function (1) was also modified as follows:

$$\max \sum_{i=1}^{I} \sum_{j=1}^{J} a_{ij} v_i x_j - \sum_{i=1}^{I} \Delta_i v_i \tag{10}$$

But leaving out some boxes of a solution for a single container might cause some problems when this reduced solution is realized. This is in particular the case when additional restrictions, e.g. stability constraints, have to be considered. For example, leaving out a box that is stowed on the container floor with a larger box placed on its top might cause that the remaining arrangement is not stable any more. Indeed, during the test runs it was always possible to find a new solution for the reduced number of boxes.

4 Test results

For benchmarking, test cases from the literature were taken to evaluate the performance of the proposed algorithm. In [4] the following 64 test cases for the bin packing problem have been provided:

- B1C1 - B1C47: 47 test cases with only one container type
- B2C1 - B2C4: 4 test cases with two container types
- B3C1 - B3C13: 13 test cases with three container types

Each test case consists of two to five box types and has between 47 and 180 boxes. In all test cases the cost of a container was set to its volume.

In [5] 16 additional test cases for the knapsack problem have been generated:

- K2C1 - K2C5: 5 test cases with 2 different container types
- K3C1 - K3C11: 11 test cases with 3 different container types

As in [3], the parameter p, that indicates how many of the partial solutions were expanded in the tree search, was set to seven. All test cases were solved on a Pentium II PC with a 266 MHz clock and 192 MB memory. The required cpu-time for a single problem was less than 30 seconds. To solve the set partitioning problem CPLEX has never taken more than five seconds.

To evaluate the performance of the bottleneck algorithm, bounds on the optimal solution are needed. If only one container type is available, then the number of required containers can be calculated for the bin packing problem by rounding up the quotient of the sum of the volume of all boxes and the container volume V.

$$\left\lceil \left(\sum_{i=1}^{I} v_i / V \right) \right\rceil \tag{11}$$

For the case of multiple container types a bound can be computed by solving the following problem.

$$\min \sum_{l=1}^{L} V_l z_l \tag{12}$$

$$s.t. \quad \sum_{i=1}^{I} v_i y_{il} \leq V_l z_l \quad \forall l = 1, ..., L \tag{13}$$

$$\sum_{l=1}^{L} y_{il} = b_i \quad \forall i = 1, ..., I \tag{14}$$

$$z_l \in \{0, 1\} \quad \forall l = 1, ..., L \tag{15}$$

$$y_{il} \geq 0 \quad \text{and integer} \quad \forall i = 1, ..., I \quad \forall l = 1, ..., L \tag{16}$$

Notation:

$l = 1, ..., L$ index of containers, where q_k consecutive containers belong to the same container type k, i.e. $L := \sum_{k=1}^{K} q_k$. The values q_k can be calculated by applying a simple bin packing heuristic, e.g. first-fit (cf. [10]), for each single container type.

V_l volume of container l

z_l decision variable stating if the l-th container is used

y_{il} decision variable stating how many boxes of type i are assigned to the l-th container

b_i number of the i-th box type available

w_i volume of box type i

The objective function (12) minimizes the required container volume. Constraints (13) and (14) ensure that the sum of the volume of the boxes assigned to each single container does not exceed the capacity of the containers and that exactly b_i boxes are packed for every box type i, respectively.

If we assume that the value of box type i, v_i, is not necessarily equal to its volume w_i, a bound for the knapsack problem can be obtained by solving the following problem.

$$\max \sum_{i=1}^{I} \sum_{l=1}^{L} v_i y_{il} \tag{17}$$

$$s.t. \quad \sum_{i=1}^{I} w_i y_{il} \leq V_l \quad \forall l = 1, ..., L \tag{18}$$

Table 1. Number of containers required to pack all boxes for the bin packing problem with only one container type

Test case	Ivancic et al. Containers	Bortfeldt Containers	Bottleneck Containers	Bottleneck #packing patterns
B1C1	26	25	25[a]	8
B1C2	11	10	10	24
B1C3	20	20	20	89
B1C4	27	28	26[a]	42
B1C5	65	51	51	28
B1C6	10	10	10[a]	18
B1C7	16	16	16[a]	15
B1C8	5	4	4[a]	24
B1C9	19	19	19[a]	20
B1C10	55	55	55[a]	5
B1C11	18	18	17	21
B1C12	55	53	53[a]	12
B1C13	27	25	25	26
B1C14	28	28	27[a]	25
B1C15	11	11	11[a]	70
B1C16	34	26	26[a]	28
B1C17	8	7	7[a]	122
B1C18	3	2	2[a]	218
B1C19	3	3	3[a]	234
B1C20	5	5	5[a]	101
B1C21	24	21	20	120
B1C22	10	9	8[a]	370
B1C23	21	20	20	135
B1C24	6	6	6	298
B1C25	6	5	5	263
B1C26	3	3	3[a]	432
B1C27	5	5	5	233
B1C28	10	10	10	82
B1C29	18	17	17	109
B1C30	24	22	22	62
B1C31	13	13	13	152
B1C32	5	4	4[a]	145
B1C33	5	5	5	83
B1C34	9	8	8	70
B1C35	3	2	2[a]	96
B1C36	18	14	14[a]	9
B1C37	26	23	23[a]	41
B1C38	50	45	45	26
B1C39	16	15	15	86
B1C40	9	9	8	324
B1C41	16	15	15	113
B1C42	4	4	4[a]	202
B1C43	3	3	3[a]	213
B1C44	4	3	4	127
B1C45	3	3	3	777
B1C46	2	2	2[a]	359
B1C47	4	3	3[a]	600
Sum	763	705	699	

[a]Optimal solution

$$\sum_{l=1}^{L} y_{il} \leq b_i \quad \forall i = 1, ..., I \tag{19}$$

$$y_{il} \geq 0 \quad \text{and integer} \quad \forall i = 1, ..., I \quad \forall l = 1, ..., L \tag{20}$$

The objective function (17) maximizes the sum of the value of the packed boxes. Constraints (18) and (19) ensure that the volume of the containers is not exceeded, and that no more than the available boxes of each box type are packed.

Table 1 shows the test results for the 47 bin packing problems from [4] where only one container type is available. The results of the bottleneck approach are compared with the approaches presented in [4] and [7]. The last column gives the number of considered packing patterns. The optimality of the solution was proven by (11) or by solving the problem (12) to (16), respectively.

In total, the bottleneck approach needed fewer containers to solve all problems than any other approach. Compared to the second best approach of Bortfeldt (cf. [7]), for six test problems a better solution was found and only for the test case *B1C44* Bortfeldt's approach obtained a better result. Additionally, for the three test cases *B1C14*, *B1C22* and *B1C4* new solutions were found, that could be proven to be optimal.

Table 2 shows the results for the bin packing problem with different container types. The bounds displayed in the last column were calculated by solving the problem (12) to (16).

Again, the bottleneck approach obtained the highest average volume utilization over all 17 test cases and only for the test case *B3C9* the approach of Bortfeldt achieved a better solution. For test cases *B3C1* and *B3C4*, optimal solutions were found. It is interesting to mention that for the test case *B3C2*, the solutions of the bottleneck approach and the approach of Bortfeldt achieved the same volume utilization although both approaches generated different combinations of container types.

Finally, Table 3 shows the results for the knapsack problem. Here the objective was not to maximize the sum of the volume of the loaded boxes, but the sum of the boxes' values. So each box type has a value that does not necessarily correspond to its volume. The values for the bounds in the second column were calculated by solving the IP-problem (17) to (20).

Also, for the knapsack problem an improvement of the solutions can be observed. Although the average objective function value over all problems was only slightly higher than for Bortfeldt's approach, a better solution was found for six test cases. In contrast, the approach of Bortfeldt outperformed the bottleneck approach six times.

5 Considering additional constraints

Bischoff and Ratcliff (cf. [9]) claimed that most of the approaches for container loading problems discussed in the literature are only applicable to a narrow part of the spectrum of situations encountered in practice since additional requirements

Table 2. Number of containers required to pack all boxes and average volume utilization for the bin packing problem with different container types

Test	Ivancic et al.		Bortfeldt		Bottleneck			Bound
case	Containers[a]	Avg.	Containers[a]	Avg.	Containers[a]	NPP	Avg.	
B2C1	26 / 0	71.8	25 / 0	74.7	25 / 0	30	74.7	99.73
B2C2	2 / 5	87.3	2 / 5	87.3	1 / 7	406	88.5	98.9
B2C3	1 / 8	74.3	2 / 0	92.2	2 / 0	137	92.2	99.5
B2C4	1 / 14	84.1	0 / 15	89.0	0 / 15	421	89.0	97.3
B3C1	7 / 13 / 6	97.6	1 / 19 / 11	95.1	2 / 13 / 17	125	99.9	99.9
B3C2	4 / 6 / 1	97.6	4 / 1 / 2	99.7	7 / 4 / 0	74	99.7	99.7
B3C3	10 / 1 / 7	85.8	16 / 0 / 2	86.8	2 / 0 / 14	41	87.4	99.1
B3C4	3 / 0 / 26	95.8	7 / 0 / 23	97.9	3 / 0 / 25	79	99.4	99.4
B3C5	7 / 6 / 1	92.2	8 / 5 / 0	96.6	6 / 9 / 0	192	96.6	99.6
B3C6	1 / 0 / 2	90.6	1 / 0 / 2	90.6	1 / 0 / 2	549	90.6	95.9
B3C7	3 / 3 / 11	81.2	0 / 4 / 10	85.9	9 / 2 / 5	814	88.4	98.0
B3C8	5 / 1 / 0	75.0	2 / 1 / 1	90.2	3 / 0 / 1	810	93.5	98.2
B3C9	9 / 1 / 5	85.3	14 / 1 / 1	87.8	13 / 1 / 2	323	86.3	99.4
B3C10	0 / 2 / 4	88.7	2 / 1 / 1	94,0	2 / 1 / 1	298	94.0	100.0
B3C11	3 / 2 / 11	76.3	3 / 3 / 10	79.1	2 / 3 / 11	128	79.2	99.3
B3C12	4 / 0 / 0	82.7	2 / 1 / 0	91.6	2 / 1 / 0	685	91.6	97.8
B3C13	1 / 0 / 2	77.1	0 / 0 / 3	84.7	0 / 0 / 3	1327	84.7	98.0
Average		84.9		89.6			90.3	

[a]Number of used containers for every container type; NPP: number of considered packing patterns

are often not taken into account. They present a comprehensive overview of possible extensions. Some of them affect the packing patterns directly and have to be considered during the single container heuristic. Others have a bearing only on the distribution of boxes among the different containers. Examples for the first case are box orientation, packing arrangement, load stability or the weight distribution within a single container. Separation of boxes or the complete shipment of groups of boxes are instances for the latter case. In particular the last two aspects have not been examined yet. The remainder of this paper outlines how these additional considerations can easily be incorporated into the bottleneck approach.

5.1 Separation of boxes

In many practical applications it is often desirable that boxes of two types must not be stowed in the same container. For example foodstuff and perfumery articles should be transported separately.

In this section, we will assume that boxes of the two box types $\hat{\imath}$ and $\bar{\imath}$ should be separated. Since information on separation of boxes is not included in the single container heuristic, a straightforward way to solve the problem is to pursue a three-step approach:

Table 3. Sum of value of packed boxes for the knapsack problem

Test	Bound	Mohanty et al.		Bortfeldt		Bottleneck		
Case		Absolute	PB	Absolute	PB	NPP	Absolute	PB
K2C1	11112.0	8640.0	77.8	8640.0	77.8	28	8640.0	77.8
K2C2	20203.2	16668.0	82.5	17409.0	86.2	364	17004.0	84.2
K2C3	15360.0	15360.0	100.0	15360.0	100.0	313	15360.0	100.0
K2C4	36556.8	36556.8	100.0	36556.8	100.0	102	36556.8	100.0
K2C5	42922.8	37558.8	87.5	39727.2	92.6	491	39382.2	91.8
K3C1	86016.0	83494.4	97.1	85120.0	99.0	208	85376.0	99.3
K3C2	53500.0	53262.5	99.6	53262.5	99.6	252	53262.5	99.6
K3C3	653750.0	495500.0	75.8	581250.0	88.9	92	583750.0	89.3
K3C4	143424.0	138240.0	96.4	139584.0	97.3	296	141216.0	98.5
K3C5	77986.8	65741.6	84.3	68645.6	88.0	940	69121.2	88.6
K3C6	139356.0	119772.0	85.9	128952.0	92.5	1081	133632.0	95.9
K3C7	68353.2	49995.0	73.1	53202.8	77.8	362	52873.6	77.4
K3C8	24964.0	23529.0	94.3	24235.2	97.1	448	23673.0	94.8
K3C9	71552.0	56492.8	79.0	65316.8	91.3	215	68723.2	96.0
K3C10	666829.6	556458.0	83.4	595770.0	89.3	491	591535.0	88.7
K3C11	2720640.00	2333440.0	85.8	2333440.0	85.8	46	2307840.0	84.8
Average			87.6		91.4			91.8

PB: In % of bound; NPP: number of considered packing patterns

STEP 1: Solve single container loading problem for box types $\{1, ..., I\} \setminus \{\hat{\imath}\}$

STEP 2: Solve single container loading problem for box types $\{1, ..., I\} \setminus \{\bar{\imath}\}$

STEP 3: Solve set partitioning problem by using the solutions generated in steps 1 and 2

In order to evaluate the performance of this approach the following scenario was examined. An additional separation constraint stating that boxes of the first two types must not be stowed in the same container was imposed.

Since benchmarks from the literature are missing, the results are compared with the solutions obtained with bottleneck approach presented in Section 3. Additionally, a bound was calculated by extending the problem (12) to (16) for the bin packing problem and (17) to (20) for the knapsack problem, respectively, by the following constraints.

$$y_{il} \leq M\gamma_{il} \quad \forall l = 1, ..., L \quad i \in \{\hat{\imath}, \bar{\imath}\} \tag{21}$$

$$y_{il} \geq \gamma_{il} \quad \forall l = 1, ..., L \quad i \in \{\hat{\imath}, \bar{\imath}\} \tag{22}$$

$$\gamma_{\hat{\imath}l} + \gamma_{\bar{\imath}l} \leq 1 \quad \forall l = 1, ..., L \tag{23}$$

$$\gamma_{\hat{\imath}l}, \gamma_{\bar{\imath}l} \in \{0, 1\} \quad \forall l = 1, ..., L \tag{24}$$

Here $\gamma_{\hat{i}l}$ and $\gamma_{\bar{i}l}$ represent decision variables that are equal to 1 if $y_{\hat{i}l} > 0$ or $y_{\bar{i}l} > 0$, respectively, and M is a sufficient large number, i.e.

$$M \geq \max_{l=1,\ldots,L} (y_{\hat{i}l}, y_{\bar{i}l}). \tag{25}$$

Tables 4, 5 and 6 show the test results for the bin packing and the knapsack problem, respectively. The values in the last columns of these three tables show the deterioration of the solutions if the additional separation constraint is imposed. To make a comparison of the results for the bin packing problem with multiple container types possible, not the number of required containers but the total container volume is shown in Table 5.

As expected, imposing an additional separation constraint resulted in poor solutions. But this effect is not very pronounced. For the test cases in Table 4 for example, the results are still better than those obtained by the approach of Ivancic et al., that does not consider separation constraints at all (cf. Table 1). Similarly, for the knapsack problems the bottleneck approach outperformed the approach of Mohanty et al. in ten out of 16 test cases. For the vast majority of test cases the gap between the calculated solution and the bound was less than 20 percent. Of course, it strikes that for some test problems, e.g. test cases *B2C3*, *K2C1* and *K3C11*, results are much worse compared to the situation where no additional separation constraint was considered. But it is worth mentioning that for these instances only two different box types were available. Imposing a separation constraint results in solving two independent homogeneous single container loading problems where it is not possible to improve volume utilization by finding good combinations of different box types.

5.2 Complete shipment

If a machine has been packed into different boxes it makes no sense if not all boxes are loaded completely. Hence, it is often necessary to ensure that if any part of a sub-set is packed, then all other boxes belonging to it are also included in the shipment. Obviously, this additional requirement is only relevant for the knapsack problem since for the bin packing problem all boxes have to be loaded anyway. Possible variants of complete shipment restrictions are:

- A single box type has to be loaded completely.
- A subset of boxes has to be loaded or left out completely.
- The number of boxes of a specific type must be multiples of a given lot size (e.g. 5, 10, 15 if the lot size is 5).
- A combination of boxes of different types, a so called complete shipment group, must be multiples of a given lot size. For example, a single disassembled machine may consist of 2 boxes of type 1 and 4 boxes of type 2. If one decides to pack u machines, then $2u$ boxes of type 1 and $4u$ boxes of type 2 have to be loaded completely.

In order to evaluate the bottleneck algorithm the following scenario was examined. A lot size for the first two box types was imposed. For the first box type this

Table 4. Number of containers required to pack all boxes for the bin packing problem with an additional separation constraint

Test case	Bound	With separation	Without separation	Difference
B1C1	19	27	25	2
B1C2	7[a]	11	10	1
B1C3	19	20	20	0
B1C4	26	28	26	2
B1C5	46[a]	51	51	0
B1C6	10	10	10	0
B1C7	16	16	16	0
B1C8	4	4	4	0
B1C9	18	19	19	0
B1C10	47	55	55	0
B1C11	16	18	17	1
B1C12	45[a]	56	53	3
B1C13	22	25	25	0
B1C14	28	28	27	1
B1C15	11	12	11	1
B1C16	21	26	26	0
B1C17	7	9	7	2
B1C18	2	2	2	0
B1C19	3	3	3	0
B1C20	4	5	5	0
B1C21	17[a]	21	20	1
B1C22	8	8	8	0
B1C23	17[a]	21	20	1
B1C24	5	6	6	0
B1C25	4	5	5	0
B1C26	3	3	3	0
B1C27	4	5	5	0
B1C28	9	10	10	0
B1C29	15[a]	17	17	0
B1C30	18[a]	23	22	1
B1C31	11	13	13	0
B1C32	4	4	4	0
B1C33	4	5	5	0
B1C34	7	9	8	1
B1C35	3	3	2	1
B1C36	11	18	14	4
B1C37	12	23	23	0
B1C38	26	45	45	0
B1C39	12	15	15	0
B1C40	7	9	8	1
B1C41	14	16	15	1
B1C42	4	5	4	1
B1C43	3	4	3	1
B1C44	3	4	4	0
B1C45	2	3	3	0
B1C46	2	2	2	0
B1C47	3	3	3	0
Sum	599	725	699	26

[a]Values were obtained by solving the linear relaxation

Table 5. Sum of container volumes necessary to pack all boxes for the bin packing problem with an additional separation constraint

Test case	Bound	With separation Absolute	In (%) of bound	Without separation	Decrease (%)
B2C1	18104.0	25920.0	143.2	24000.0	8.0
B2C2	49500.0	56250.0	113.6	54750.0	2.7
B2C3	25275.0	34995.0	138.5	26250.0	33.3
B2C4	49392.0	54087.0	109.5	54000.0	0.2
B3C1	116160.0[a]	119040.0	102.5	116224.0	2.4
B3C2	70500.0	70500.0	100.0	70500.0	0.0
B3C3	2952960.0	3152640.0	106.8	3141120.0	0.4
B3C4	932500.0[a]	973000.0	104.3	938000.0	3.7
B3C5	117360.0	120960.0	103.1	120960.0	0.0
B3C6	9792.0	22031.0	225.0	21132.0	4.3
B3C7	89059.0[a]	101520.0	114.0	100800.0	0.7
B3C8	108000.0	117560.0	108.9	113200.0	3.9
B3C9	128979.0[a]	150186.0	116.4	150186.0	0.0
B3C10	25200.0	27000.0	107.1	26800.0	0.7
B3C11	98400.0[a]	124320.0	126.3	124320.0	0.0
B3C12	363430.0	430080.0	118.3	388092.0	10.8
B3C13	167200.0	198044.0	118.4	193500.0	2.3
Average					4.3

[a]Values obtained by solving the linear relaxation

lot size was set to 7 percent of the number of boxes of that type available. The value for the second box type was set to 12 percent.

To consider complete shipment requirements the following additional constraints have to be added to the problem (10), (8), (3), (4) and (9):

$$\sum_{j=1}^{J} a_{ij}x_j - \Delta_i = \lambda_{ri}m_r \quad \forall r = 1, ..., R \quad \forall i \in I_r \tag{26}$$

$$m_r \geq 0 \quad \text{and integer} \quad \forall r = 1, ..., R \tag{27}$$

Notation:

$r = 1, ..., R$	index of complete shipment groups, $I_r \subseteq \{1, ..., I\}$
λ_{ri}	lot size of box type i in group r
m_r	decision variable stating how many lots of complete shipment group r are packed

Table 7 shows the results for the test cases from [5]. Values for the bounds in column 2 were calculated by adding the following constraints to problem (17) till (20).

$$\sum_{j=1}^{J} a_{ij}x_j = \lambda_{ri}m_r \quad \forall r = 1, ..., R \quad \forall i \in I_r \tag{28}$$

Table 6. Sum of value of packed boxes for the knapsack problem with an additional separation constraint

Test case	Bound	With separation Absolute	With separation In(%) of bound	Without separation	Increase (%)
K2C1	10560.0	5120.0	48.5	8640.0	68.8
K2C2	19941.0	16707.0	83.8	17004.0	1.8
K2C3	15360.0	15360.0	100.0	15360.0	0.0
K2C4	36556.8	34022.4	93.1	36556.8	7.4
K2C5	42124.5	39382.2	93.5	39382.2	0.0
K3C1	85555.2	85376.0	99.8	85376.0	0.0
K3C2	53262.5	53262.5	100.0	53262.5	0.0
K3C3	599000.0	536250.0	89.5	583750.0	8.9
K3C4	143232.0	139968.0	97.7	141216.0	0.9
K3C5	77783.6	69121.2	88.9	69121.2	0.0
K3C6	139176.0	128088.0	92.0	133632.0	4.3
K3C7	65149.2	52873.6	81.2	52873.6	0.0
K3C8	24782.4	22730.4	91.7	23673.0	4.1
K3C9	70416.0	66995.2	95.1	68723.2	2.6
K3C10	663954.0	568482.0	85.6	591535.0	4.1
K3C11	2589440.0	1354752.0	52.3	2307840.0	70.4
Average					10.8

$$m_r \geq 0 \quad \text{and integer} \quad \forall r = 1, ..., R \tag{29}$$

Although the average deterioration of the solutions compared to the situation where the additional constraint is not considered is with 19.4% larger than in the last section, the total value packed was more than 80% of the upper bound value for most test cases. For eleven test cases even more than 90% were obtained. Like in the last section, test cases *K2C1* and *K3C11* obtained the worst results.

6 Outlook

In this paper a new approach for solving multiple container problems was presented. A greedy heuristic from the literature was deployed to generate solutions for the single container loading problem. These solutions were then joined together in a second stage to obtain a solution for the multiple container problem.

Beside the fact that the proposed approach outperformed benchmark algorithms from the literature, it has the advantage that additional constraints found in many real world applications can be implemented easily. Two scenarios catering for additional separation constraints and for complete shipment restrictions were examined. A self-evident extension of these additional constraints would be to consider them not as hard restrictions but as soft ones that allow recognizing certain deviations. This violation fshould of course then be penalized in the objective function.

Table 7. Sum of value of packed boxes for the knapsack problem with an additional complete shipment constraint

Test case	Bound	With separation		Without separation	Increase (%)
		Absolute	In(%) of bound		
K2C1	10656.0	4864.0	45.6	8640.0	77.6
K2C2	19500.0	16584.0	85.0	17004.0	2.5
K2C3	14003.2	14003.2	100.0	15360.0	9.7
K2C4	27148.8	27148.8	100.0	36556.8	22.5
K2C5	38344.5	34987.6	91.2	39382.2	12.6
K3C1	79820.8[a]	78643.2	98.5	85376.0	8.6
K3C2	48900.0	48900.0	100.0	53262.5	8.9
K3C3	559000.0	557500.0	99.7	583750.0	4.7
K3C4	124704.0	118080.0	94.7	141216.0	19.6
K3C5	77783.6	68705.2	88.3	69121.2	0.6
K3C6	132516.0	126990.0	95.8	133632.0	5.2
K3C7	64381.2	52873.6	82.1	52873.6	0.0
K3C8	24537.6	22809.6	93.0	23673.0	3.8
K3C9	69689.6	65612.8	94.2	68723.2	4.7
K3C10	577004.0	571002.0	99.0	591535.0	3.6
K3C11	2381926.4	1020825.6	42.9	2307840.0	126.1.8
Average					19.4

[a]Value obtained by solving the linear relaxation

References

1. Bischoff E E, Ratcliff M S W (1995) Loading multiple pallets. Journal of the Operational Research Society 46: 1322–1336
2. Terno J, Scheithauer G, Sommerweiß U, Riehme, J (2000) An efficient approach for the multi-pallet loading problem. European Journal of Operational Research 123: 372–381
3. Eley M (2001) Solving container loading problems by block arrangement. European Journal of Operational Research 141: 393–409
4. Ivancic N J, Mathur K, Mohanty B B (1989) An integer-programming based heuristic approach to the three-dimensional packing problem. Journal of Manufacturing and Operations Management 2: 268–298
5. Mohanty B B, Mathur K, Ivancic N J (1994) Value considerations in three-dimensional packing – A heuristic procedure using the fractional Knapsack problem. European Journal of Operational Research 74: 143–151
6. Martello S, Pisinger D, Vigo D (1997) The three-dimensional bin packing problem. Technical Report OR-97–6, DIES Operations Research Group, University of Bologna, Bologna. Operations Research (to appear)
7. Bortfeldt A (2000) Eine Heuristik für Multiple Containerladeprobleme. OR Spektrum 22: 239–262
8. Gilmore P C, Gomory R E (1961) A linear programming approach to the cutting stock problem. Operations Research 9: 849–859
9. Bischoff E E, Ratcliff M S W (1995) Issues in the development of approaches to container loading. Omega – International Journal of Management Science 23: 377–390
10. Martello S, Toth P (1990) Knapsack problems – algorithms and computer implementations. Knapsack problems – algorithms and computer implementations Wiley, New York